AF538618

COMPUTATIONAL PHYSICS

COMPUTATIONAL PHYSICS

Compiled & Edited By

Dr. J.D. Dubey

Dr. Prajwalit Shikha

KNOWLEDGE BOOK DISTRIBUTORS

NEW DELHI (INDIA)

ISBN : 978–93–80350–18–9

Published By :
KNOWLEDGE BOOK DISTRIBUTORS
Flat No. 311, Ist Floor,
Sector 16-B, Pocket-3-C,
Dwarka, New Delhi-110 078

Edition : 2011

Printed in India

Preface

In modern science, Computational physics is the study and implementation of numerical algorithms to solve problems in physics, for which a quantitative theory already exists. It is often regarded as a subdiscipline of theoretical physics, but some consider it an intermediate branch between theoretical and experimental physics. Physicists often have a very precise mathematical theory, describing how a system would behave.

Unfortunately, it is often the case that solving the theory equations, ab initio, in order to produce a useful prediction is not practical. This is especially true with quantum mechanics, where only a handful of simple models have complete analytic solutions. In cases, where the systems only have numerical solutions, computational methods are used. Computation now represents an essential component of modern research in accelerator physics, astrophysics, fluid mechanics, lattice field theory and lattice gauge theory, especially lattice quantum chromodynamics, plasma physics and solid state physics.

Computational solid state physics, uses density functional theory to calculate properties of solids, a method similar to that used by chemists to study molecules. Many more other general numerical problems fall loosely under the domain of computational physics, although they could easily be considered, to be, pure mathematics or part of any number of applied areas. Computational Physics also encompasses the tuning of the software and hardware structure to solve the problems.

This semi-textbook covers all dimensions of the subject, particularly aimed at the students, who intend to master in this field of Mathematics. Enlightening comments from concerned circles are solicited.

Book writing-whether original or compilation is a complicated job requiring academic labour.

The present work is purely a research compilation of literature from authoritative publications/readings/excerpts/notes/reviews/researches/literature, authored by eminent scholars/distinguished writers.

During research and compilation, the author haa also taken assistance/references/literature from various Websites through Internet.

The author, therefore humbly, acknowledge the contribution of all those eminent writers/scholars alongwith their respective publishers and Websites from whose learned writings/displays references/literature has been taken while preparing this book.

—*J.D. Dubey*

Contents

1

Introduction

The science of physics assumes that physical phenomena may be explained and understood as a result of the functioning of physically real systems structured in certain ways and constituted of elements possessing certain properties. The exact nature of the real system in itself is usually unknown, so to gain an understanding of it a model (whose exact nature is known) is constructed.

Role of Models in Physics

An abstract model is a system (normally conceptual but possibly physical) of parts and whole which are held to interact in certain ways, described so as to be clear to any interested and competent observer, and to interact in such ways consistently over time (thereby allowing prediction as well as explanation). The model incorporates analogues of those aspects of the real system which are relevant to the phenomena to be explained.

Explanation is by analogy: the conceptual analogue of the physical phenomenon to be explained is understood as occurring as a result of the properties of the model, and the phenomenon itself is then explained as a result of assumed properties of the real system which are analogous to the properties of the model.

Successful prediction based upon such understanding encourages us to believe that the model is true (although there is the possibility that future experience will reveal that we were mistaken in so believing).

This does not prove that the real system in fact has the structure and functioning attributed to it in the model. Generally we can never in any case know what the structure and functioning of the real system in itself is except by analogy with some model. To the degree in which the properties of the model are mirrored by the properties of the real system (its "adequacy") we are justified in believing that the model gives us knowledge of the constitution of the real system.

This is all fine so long as we don't have two different, equally adequate, models. Especially if the two models are not just "different" but "apparently irreconcilable". Unfortunately this may happen. A well-known instance of two such apparently irreconcilable models is the theory of light as consisting of waves and the theory of light as consisting of particles ("photons"). The wave model of light and the particle model of light each adequately explain certain observed phenomena. Yet cognitive dissonance may be felt if both models are accepted as true because conceptually waves and particles are very different things. Thus arises the problem of the real nature of light: does it consist of waves or of particles? Or both?

Computational Physics: Perceptions

The research described in this study is part of the field of *condensed matter physics*, that branch of physics concerned with explaining how combinations of extremely large numbers of particles such as atoms or electrons behave in certain interesting ways under certain conditions. It is a science which seeks for laws describing the overall properties of systems in which many particles interact. This work is also in the field of *statistical physics*, which attempts to predict the properties of complex systems containing many interacting components undergoing random thermal motions.

Condensed matter physics (like other fields of physics) may be subdivided into three kinds: theoretical, experimental and computational.

Experimental Physics

This involves observers interacting directly with those aspects of physical reality which are under investigation. The practice of modern experimental physics is to set up one or a series of experiments designed to decide among one of (usually) many possible outcomes. The point of an experiment is that the outcome is not known in advance; rather it is Nature which decides the outcome and we observe.

Theoretical Physics

Theoretical physics consists in the formulation and testing of theories about the structure and functioning of systems believed or supposed to be present in physical reality. Such theories often make use of conceptual models. Part of the model may consist in the assertion that certain aspects of the model may be described mathematically and exhibit certain mathematical relations, thereby allowing properties of the model to be deduced on the basis of results established by mathematicians. If these deduced properties are confirmed by experiment then we have reason to trust the model.

Computational Physics

This is a branch of physics distinct from the other two because (a) unlike experimental physics it does not query Nature directly and (b) unlike theoretical physics its results are not arrived at primarily by theorising but rather by computation.

Most results arrived at in computational physics are not statements that could ever be logically derived from a description of the model. E.g., values of the so-called critical temperature, critical exponents, etc., in a spin model of magnetic material. So the computational experiment typically does not just reveal what could have been deduced (were it not for complexity and mass of data) by theorising.

There are, however, some exceptions to this, notably, the analytical solution of the 2-dimensional Ising model on a square lattice by Lars Onsager in 1944, which provides, e.g. the exact value of the critical temperature. For other models no analytic solution has been found except by the imposition of assumptions which simplify the mathematical description so that it becomes amenable to currently existing mathematical techniques.

Given that the construction and querying of computational models is the main activity of a computational physicist, what distinguishes him from a simple computer programmer? Given that a certain computer model may have a certain plausibility and elegance, what saves a computational physicist from becoming captured by his model, so to speak?

Computational physics is not a science of the properties of selected computer models, however interesting they may be to study. Such a science, though it might be a part of the science of computational physics, would be in itself a branch of computer science, not a branch of physics.

Computational physics, like the other branches of physics, aims at knowledge of the physical world. It arrives at this knowledge by showing that a model has the same properties (within limits of uncertainty in measurement), as revealed by computation, as the system modelled, as revealed by direct querying of Nature (the province of the experimental physicist).

Insofar as a model incorporates theoretical hypotheses, and facilitates predictions, this process provides a test of the theory. Thus there is a synergistic relationship among the three branches of physics, the experimental, the theoretical and the computational.

Finally we may consider the claim that computational physics is really just the handmade of theoretical physics, not a distinct branch of physical science, on the grounds that the computational physicist simply obtains results which are implicit in the theoretical model (upon which the computer model is based), results which the theoretical physicist cannot himself obtain because of the complexity of the model and the large amount of computation needed.

But most results arrived at by the computational physicist are not, even in principle, capable of being derived by logical deduction from a description of the model — at least not in the case of Monte Carlo experiments. These are experiments (and the principal kind of experiment conducted by computational physicists) which depend upon a stream of many random numbers. E.g., we have a spin system which is a model of magnetic material, in which the spins are either up or down and the system evolves over time according to certain rules involving "chance".

Each spin is considered for possible flipping about as often as any other spin, and it flips or it doesn't flip according to a criterion which involves the generation of a random number.

Since the random numbers, although generated by a deterministic algorithm, behave as true random numbers (if the algorithm is properly designed), they cannot be predicted, so whether a given spin at a given time will flip or not flip is completely unpredictable.

The system is allowed to evolve over time according to this dynamics, and when the system attains equilibrium certain measurements are made, e.g., the analogues in the model of magnetisation, specific heat, etc. By considering a range of temperatures (or conducting other such computational experiments) we find out how these properties depend on temperature or other factors.

But the measurements, and so the properties of the system, are not, even in principle, deducible from the theoretical description of the system and the rules for its evolution. Therefore the computational physicist can arrive at results which the theoretical physicist cannot arrive at. Thus computational physics cannot be characterised simply as theoretical physics aided by computing power, and so is a branch of physics distinct both from the experimental and the theoretical.

Modelling Magnetic Material

Magnetism (or electromagnetism) is one of the fundamental forces of Nature. A field of magnetic force is produced by the motion of an electrically charged particle, and so an electric current (which consists of moving electrons) produces a magnetic field.

In 1925 G. E. Uhlenbeck and S. Goudsmit hypothesised that the electron has a "spin", and so behaves like a small bar magnet, and that in an external magnetic field the direction of the electron's magnetic field is either parallel or anti-parallel to that of the external field.

> In the same year Lenz suggested to his student that if an interaction was introduced between spins so that parallel spins in a crystalline lattice attracted one another, and anti-parallel spins repelled one another, then at sufficiently low temperatures the spins would all be aligned, and the model might provide an atomic description of ferromagnetism.

Thus arose the "Ising model" in which "spins", located on the sites of a regular lattice, have one of two values, +1 and –1, and spins with spin values S_i and S_j on adjacent sites interact with an energy $-J.S_i.S_j$ (where J is a positive real number), so that spins with similar values interact with an energy $-J$, and those with dissimilar values interact with the (higher) energy J (the whole system tending towards a state of lower energy, unless energy is added from outside, e.g. by means of a "heat bath").

The magnetisation per spin of a system of N spins is defined as $\Sigma_i(S_i / N)$ (which thus lies between –1 and +1) and the total energy (the "Hamiltonian") is defined as the sum of the interaction energies, i.e.,

$$\Sigma_{ij}(-J.S_i.S_j) = -J.\Sigma_{ij}\left(S_i.S_j\right)$$

where the sum is over all pairs of nearest neighbour spins.

If the spins are allowed to have more than two states we have the "Potts model", named after the Australian mathematician R. B. Potts who published a study of this model in 1952.

Magnetic materials exhibit interesting temperature-dependent behaviour. In particular, above a certain temperature (characteristic of the particular material) any magnetisation present disappears. When the temperature is reduced below that temperature regions of stable magnetisation spontaneously appear (even in the absence of an external magnetic field). This temperature is called the "critical" or "Curie" temperature. Various discontinuities (called "phase transitions") in measurable properties are found to occur at the critical temperature.

> The criterion with which to test the model is whether it gives rise to the characteristic pattern of singularities associated with ferromagnetism, and more particularly whether a Curie temperature, T_c, can be defined with the property that there exists a non-zero spontaneous magnetisation for $T < T_c$.

Ising studied the simplest possible model consisting simply of a linear chain of spins, and showed that for this 1-dimensional case there is no (non-zero) critical temperature (i.e., the spins become aligned only at $T = 0$).

> Ten years elapsed before Peierls showed that the two-dimensional model does have a non-zero spontaneous magnetisation, and can therefore be regarded as a valid model of a ferromagnet.

In 1944 the physicist Lars Onsager, studying the 2-dimensional Ising model on a square lattice, was able to demonstrate by analytical means the existence of a phase transition in the model, a result considered to be a landmark in the physics of critical phenomena.

In the mid-20th century it became possible to use high-speed electronic computers to set up models of magnetic materials, to study the corresponding behaviour of those models, and to compare the computational results with observations of real systems. In recent years there have been many computational studies of the behaviour of magnetic materials at the critical temperature using Ising and Potts spin models.

These studies often use so-called "Monte Carlo" techniques, which are methods relying on a stream of random numbers to drive a stochastic process, in this case the generation of a succession of many states of the spin model (as is the case in this research). The course of this process is governed by the choice of a "dynamics algorithm", a concept which will be explicated.

We now turn to a detailed discussion of the ways in which the Ising and Potts models are implemented for the purpose of simulating the behaviour of ferromagnetic material.

First the concept of a lattice geometry is discussed, then precise definitions of the Ising and the Potts spin models are given. We then consider how temperature enters the model, and how the model is made to evolve over time (by means of one or another algorithm controlling the dynamics). Further related issues are discussed, such as the fact that we use finite models in an attempt to gain knowledge of the behaviour of the ideal infinite system.

Lattice Geometries

In a system of spins the spins must have some location, or at least there must be some way to relate them to each other and to decide which of them interact (usually only "nearest neighbour" spins interact). This may be done by postulating a regular geometric "lattice" structure consisting of lines joining vertices. A simple example of a lattice geometry is the "square" lattice:

```
  |    |    |    |    |
-O--O--O--O--O-
  |    |    |    |    |
-O--O--O--O--O-
  |    |    |    |    |
-O--O--O--O--O-
  |    |    |    |    |
```

The analogue of this in three dimensions is the "cubic" lattice, and 4- and higher-dimensional analogues are known as "hypercubic" lattices.

Other geometries which have been used in spin models are the "honeycomb" lattice:

```
   -O—O      O—O-
   /    \    /    \
-O       O—O       O-
   \    /    \    /
    O—O      O—O
   /    \    /    \
-O       O—O       O-
```

the "triangular" lattice:

```
  -O—O—O—O—O-
  / \ / \ / \ / \ / \
-O—O—O—O—O—O-
  \ / \ / \ / \ / \ /
  -O—O—O—O—O-
  / \ / \ / \ / \ / \
-O—O—O—O—O—O-
```

and the 3-dimensional lattice known as the "diamond lattice", which is the structure of the carbon atoms in a diamond crystal.

A vertex in the lattice which may be occupied by a spin is known as a "site". Thus spins occupy the sites of a lattice, but not every site must be occupied by a spin. A line joining two sites is called a "lattice bond".

One of the characteristic properties of a lattice geometry is the number of sites directly connected to a site. This is usually the same for all sites. If it is then that number is known as the *coordination number* of the lattice.

The coordination numbers of various possible lattice geometries are given below:

Dimensionality	*Name*	*Coordination number*
2	honeycomb	3
2	square	4
2	triangular	6
3	diamond	4
3	cubic	6
3	tetrahedral	12
4	hypercubic	8

More complex lattice structures are, of course, possible, but in computational physics they are seldom studied because they have not been found to occur in real physical systems.

In most studies a spin system includes a lattice in which all sites are occupied by spins and in which all lattice bonds are interpreted as energetic bonds (i.e., associated with that bond is an energy which contributes to the total energy of the system). Such a lattice (or spin system) is called a "pure" or "non-diluted" lattice.

In some recent studies (and in this research) not all lattice sites must be occupied by spins and not all lattice bonds are energetic bonds. In a *site-diluted* lattice not all lattice sites are occupied by spins. In a *bond-diluted* lattice not all lattice bonds are energetic bonds.

In the case of a bond-diluted lattice we may distinguish between lattice bonds which are *open,* meaning that the spins connected interact energetically, and those which are *closed,* meaning that the spins connected by such lattice bonds do not interact energetically, so in a bond-diluted lattice there is a mixture of open bonds and closed bonds.

It would also be possible to consider both kinds of dilution simultaneously, whereby not all lattice sites are occupied and not all spins on neighbouring sites are connected by an open bond. So far there seem to have been no studies published of this form of dilution.

Ising and Potts Spin Models

A *spin system* consists of:

- A "lattice", being a non-empty set, whose elements are called "lattice sites", and a binary, irreflexive, symmetric relation (of "bonds") among those sites.
- A non-empty subset of lattice sites, said to be "occupied"; each such lattice site is occupied by a single "spin".
- A non-empty subset of the set of lattice bonds connecting occupied sites; such bonds are said to be "energetic" bonds (a.k.a. "open" bonds).
- A set of real numbers, interpreted as "spin values"; each spin possesses a single spin value (which may change).

- A definition of the interaction energy between any two spins in terms of their spin values.
- A definition of the total energy of the spin system (the "Hamiltonian") in terms of these interaction energies (and thus ultimately in terms of the spin values of the spins).

A spin system in which all sites are occupied by spins, and all bonds are open, is said to be "pure". A spin system with some sites not occupied by spins is said to be "site-diluted". A spin system with some spins directly connected by closed lattice bonds is said to be "bond-diluted".

Two lattice sites which are directly connected by a lattice bond are said to be "nearest neighbour" lattice sites. Two spins on sites directly connected by an open lattice bond are said to be "nearest neighbour" spins. In a bond-diluted lattice it is possible for two spins to occupy nearest neighbour lattice sites even though those spins are not nearest-neighbour spins (in the case that they are connected by a closed lattice bond), e.g., spins a and b in the following lattice (where open bonds are marked by — and |, and closed bonds by ....):

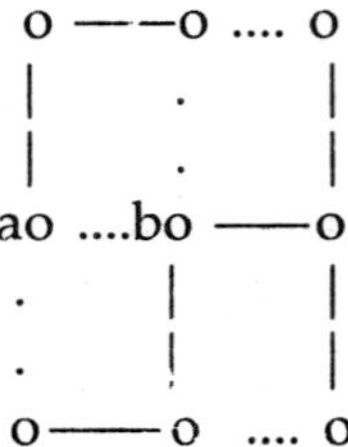

A spin system is said to be a "nearest neighbour" spin system if the interaction energy between any two spins which are not nearest neighbour spins is zero. All spin systems studied in this research are nearest neighbour spin systems.

A "state" of a spin system is a mapping of each spin to a spin value.

The standard Ising spin model is a spin model (on some lattice), such that:

- The set of spin values is { +1, -1 }.
- The value of the interaction energy associated with any pair of nearest neighbour spins S_i and S_j is $-J.S_i.S_j$, where J is a non-zero real number.

(c) The Hamiltonian (assuming the absence of any external magnetic field) is the sum of the interaction energies over all pairs of nearest neighbour spins (only), i.e., $-J.\Sigma_{ij}(S_i.S_j)$.

For a model of a ferromagnetic material $J > 0$, and for a model of an anti-ferromagnetic material $J < 0$. This research concerns only models of ferromagnetic material.

The standard q-state Potts spin model is a spin model in which there can be q different spin values, with $q \geq 2$. The spin values are normally denoted by the numerals 1, 2, ..., q.

There are two variants of the q-state Potts model, depending on how the interaction energy is defined.

The q-state Potts spin model, Version A, is a spin model, as defined above, such that the interaction energy associated with any pair of spin values S_i and S_j is $-J.\delta(S_i, S_j)$, where $\delta()$ is the Kronecker delta function: $\delta(x,y) = 1$ if and only if $x = y$ (otherwise 0).

The q-state Potts spin model, Version B, is a spin model such that the interaction energy associated with any pair of spin values S_i and S_j is

$$-J.\left[2.\delta\left(S_i, S_j\right) - 1\right]$$

In both versions, as in the Ising model, the Hamiltonian is the sum of the interaction energies over all nearest neighbour spins (assuming the absence of any external magnetic field).

The critical properties (other than the critical temperature) of these two versions of the q-state Potts model are the same.

Dynamics and the Principle of Detailed Balance

All physical systems undergo change. The definition of a spin model allows for change in the values of the spins, and we now consider this aspect of the model more closely.

Given a particular spin system (e.g., an Ising spin system on a square lattice of size 10x10) we may speak of the *phase space* of that system, which is the set of all possible assignments of spin values (+1 or -1) to the spins. Clearly in this example there are 2^{100} (c. 10^{30}) states in the phase space.

It might be thought that we could measure a quantity such as magnetisation by random sampling of the phase space, i.e., by sampling many states at random, calculating the magnetisation and averaging them. But there are two objections to this:

- In a physical system (at a certain temperature) some states are far more (or less) likely than others. At higher temperature states which are disordered are far more likely than states which are ordered (i.e., which have regions of aligned spins). A random sampling of states (or more exactly, a sampling of states by randomly assigning spin values to the spins) will strongly favour disordered states, with magnetisation close to zero (where magnetisation is defined as usual as $M = (\Sigma_i S_i)/N$, where N is the number of spins and each spin S_i has spin value +1 or –1) .
- In any case, so far there is no analogue of temperature in our definition of a spin model. The magnetisation of a physical system clearly depends on the temperature, and random sampling of the phase space takes no account of the possibility of magnetisation being dependent on temperature.

An important concept in thermodynamics is that of *thermal equilibrium*. A system at thermal equilibrium exhibits no tendency to change its bulk properties — its extensive properties such as mass or its intensive properties such as magnetisation and temperature.

To obtain a realistic measurement of a quantity such as magnetisation we need to sample the phase space in such a way that the probability of selecting a state is the same as (or very close to) the probability of that state occurring when the system is in equilibrium. Then the values of the quantities for the states sampled contribute to the average in just the degree that those values occur in the real system.

Now (according to thermodynamic theory) the probability $P(\mathbf{X})$ of a system at thermodynamic equilibrium being in state X is $e^{(-\beta E_X)}/Z$ where $\beta = 1/(k_B T)$, with T the temperature and k_B the Boltzmann constant, 1.380658 x 10^{-23} J/K, E_x is the energy of state X and Z is the *partition function*, $Z = \Sigma_Y e^{(-\beta E_y)}$, where the summation is over all possible states Y (with energy denoted by E_Y) of the system. In the case of a spin model E_x is the Hamiltonian.

But how to sample the phase space so that each state occurs with the same probability as its equilibrium probability?

The analogue in the model of the temporal evolution of a real magnetic system is a sequence of states (with some kind of connection between successive states). Such a sequence of states is called a *trajectory* through the phase space.

If a spin system starts in some initial state and is allowed to evolve according to some algorithm which generates a successor state from any given state (a *dynamics algorithm*) then we may measure the values of quantities such as magnetisation, absolute magnetisation, the second moment of the magnetisation, etc., as the system evolves.

When the spin system has evolved to the point where the quantities of interest (magnetisation, etc.) are stable, the system is in equilibrium. We can then measure the equilibrium properties of the system by taking an average of quantities measured at each state over a sequence of states sufficient in number to give us a reasonable measurement of mean and standard deviation. This is called a *time average* of a quantity.

Combining a dynamics algorithm (for producing a sequence of states) with a spin model allows us to obtain trajectories through the phase space. This dynamic evolution may occur either as a result of a succession of single spins changing their value ("single spin flip" or "local" dynamics) or as a result of groups of connected spins changing their values together ("cluster flip dynamics").

However, *a priori*, we have no assurance that the dynamics algorithm used will actually produce (eventually) a stable equilibrium condition. How can we find a dynamics algorithm which is guaranteed to produce an equilibrium condition (thereby allowing us to measure the equilibrium properties of the system)?

Suppose a spin system evolves as described above, and suppose that we have a dynamics algorithm such that for any given state A there is a definite probability that it will be succeeded in the trajectory by state B (for any possible state B), and let $P(A \rightarrow \mathrm{B})$ denote that probability (called a *transition probability*).

Let $P(A,t)$ denote the probability that the spin system is in state A at time t. Then the rate of change of $P(A,t)$ with respect to t is given by

$$\Sigma_B\left[P(B,t).P(B \rightarrow A)\right] - \Sigma_B\left[P(A,t).P(A \rightarrow B)\right]$$

i.e., the sum, over all states B, of the probability that the system is in state $B(\neq A)$ at time t and moves to state A minus the sum, over all states $B(\neq A)$, of the probability that the system is in state A at time t and moves to state B.

At equilibrium the rate of change of $P(A,t)$ is zero (if it were not then some macroscopic property of the system would be changing, and so the system would not be in equilibrium). Thus we can speak of $P(A)$, the equilibrium probability of state A (i.e., the probability, once the system has reached equilibrium, that it is in state A).

Thus for any time t at equilibrium, for any state A

$$\Sigma_B[P(B,t).P(B \to A)] = \Sigma_B[P(A,t).P(A \to B)]$$

where the summation is over all states $B \neq A$. Thus at equilibrium for any state A

$$\Sigma_B[P(B).P(B \to A)] = \Sigma_B[P(A).P(A \to B)]$$

where the summation is over all states $B \neq A$. Then

$$\Sigma_B[P(B).P(B \to A)] + P(A).P(A \to A) = \Sigma_B[P(A).P(A \to B)] + P(A).P(A \to A)$$

so we may drop the restriction in the summation that $B \neq A$ and thus at equilibrium for all states A

$$\Sigma_B[P(B).P(B \to A)] = \Sigma_B[P(A).P(A \to B)]$$

where the summation is over all states B.

This suggests what is called the principle of *microscopic reversibility,* a.k.a. the principle of *detailed balance,* which asserts that at thermodynamic equilibrium the likelihood of a system being in any state A and changing to any other state B is equal to the likelihood of its being in state B and changing to state A. This principle may be stated as:

$$P(A).P(A \to B) = P(B).P(B \to A)$$

for all states A and B.

In a landmark paper Metropolis *et al.* (1953) showed that if a dynamics algorithm, which is such that $P(X \to Y)$ is well-defined for any states X and Y, satisfies the principle of detailed balance then in the long run (in any trajectory) the probability of any state X occurring will be the same as its equilibrium probability $e^{(-\beta E_X)}/Z$.

Single Spin Flip Dynamics Algorithms

Dynamics algorithms used in spin model studies may be divided into *single spin flip* algorithms and *cluster flip* algorithms. Among the former are the so-called Metropolis algorithm and the Glauber (or "heat bath") algorithm. Among the latter are the Swendsen-Wang algorithm and the Wolff algorithm.

In a single spin flip algorithm a spin is selected (in such a way that in the long run each spin is selected about as often as any other spin) and a decision is then made as to whether it should take on a new spin value.

We then calculate the change in the energy of the system that would result if the spin changed its spin value to some other, and from this we calculate a so-called *transition probability*. This is such that the larger is the energy increase which would result from the spin flip the smaller is the probability that the spin will be flipped (the transition probability). We then choose a random number between 0 and 1. If this is less than the transition probability then the spin adopts the new spin value, otherwise the spin remains unchanged.

The change in the energy (from before the spin flip to after) can be calculated from the static properties of the spin system. The dynamics of the system is governed by the manner in which the transition probability is calculated, given this change in energy. This is done in different ways by these two dynamics algorithms, and the calculation depends upon a parameter T, which is the analogue of temperature.

Suppose a spin is in state S_i and it is under consideration for change to state S_j and let the transition probability be denoted by $P(S_i \rightarrow S_j)$. According to the principle of detailed balance, we must have:

$$P(S_i).P(S_i \rightarrow S_j) = P(S_j).P(S_j \rightarrow S_i)$$

that is,

$$P(S_i \rightarrow S_j)/P(S_j \rightarrow S_i) = P(S_j)/P(S_i)$$

The probability $P(X)$ of a system at thermodynamic equilibrium being in state X is $e^{(-\beta E_X)}/Z$ where $\beta = 1(k_B T)$ and Z is the partition function. For the standard Ising model, E_X is $-J.\Sigma_{ij}(S_i.S_j)$ for a state X with spin values S_i = +1 or –1 for all spins S_i.

Thus for any two states X and Y,

$$P(X)/P(Y) = \left[e^{(-\beta E_X)}/Z\right]/\left[e^{(-\beta E_Y)}/Z\right] = e^{(-\beta E_X)}/e^{(-\beta E_Y)} = e^{[-\beta(E_X - E_Y)]} = e^{-\beta\Delta E}$$

where $\Delta E = E_X - E_Y$ is the change in energy of the system when it moves to state X from state Y. Thus a single spin flip dynamics algorithm satisfies detailed balance if and only if

$$P(S_i \rightarrow S_j)/P(S_j \rightarrow S_i) = e^{-\beta\Delta E_{ji}}$$

where S_i (resp. S_j) is the state of the system with the spin under consideration having spin value S_i (resp. S_j), and ΔE_{ji} is the energy change resulting from the spin's changing from value S_i to S_j.

In order to calculate ΔE_{ji} it is not necessary to calculate the total energies of the "before" and "after" states. If ΔE_{ji} is the energy difference produced by changing the spin value of a single spin (as in the Metropolis and the Glauber algorithms) then ΔE_{ji} can be calculated by considering the sum of the interaction energies of that spin with its nearest neighbours before and after the contemplated change.

The transition probabilities (i.e., the likelihood that a spin will change its value) for the Metropolis algorithm and for the Glauber algorithm are defined as follows:

Metropolis algorithm: $P(S_i \rightarrow S_j) = 1$ if $\Delta E_{ji} \leq 0$

$= e^{-\Delta Eji(k_BT)}$ otherwise.

Glauber algorithm: $P(S_i \rightarrow S_j) = 1 / [\, 1 + e^{-\Delta Eji(k_BT)}\,]$

It is not difficult to prove that in both cases $P(S_i \rightarrow S_j)/P(S_j \rightarrow S_i) = e^{-\beta\Delta Eji}$, so both the Metropolis and the Glauber algorithms satisfy the condition of detailed balance.

In practice, when considering whether to flip a spin, one compares the value of the transition probability $P(S_i \rightarrow S_j)$, as defined above, with a random number between 0 and 1, and flips the spin if and only if it is less than the transition probability. The transition probability is calculated from the energy difference, ΔE_{ji}. This value is arrived at on the basis of part (e) of the definition of the spin model, i.e., the definition of the interaction energy between any two spins in terms of their spin values. This interaction energy is defined in terms of the value of a spin and the value of its nearest neighbours (assuming a nearest neighbour spin model).

Since (in an Ising or a Potts spin model) the number of possible spin values is finite, the number of possible values of the interaction energies, $-J.S_i.S_j$, is also finite. Since the number of nearest neighbours is finite, the number of possible values for ΔE_{ji} is also finite, and so there are only a finite number of possible transition probability values. Thus a table of transition probabilities can be constructed for any given spin model.

2

Computer Algebra

A Computer Algebra system is a type of software package that is used in manipulation of mathematical formulae. The primary goal of a Computer Algebra system is to automate tedious and sometimes difficult algebraic manipulation tasks. The principal difference between a Computer Algebra system and a traditional calculator is the ability to deal with equations symbolically rather than numerically. The specific uses and capabilities of these systems vary greatly from one system to another, yet the purpose remains the same: manipulation of symbolic equations. Computer Algebra systems often include facilities for graphing equations and provide a programming language for the user to define his/her own procedures.

Computer Algebra systems have not only changed how mathematics is taught at many universities, but have provided a flexible tool for mathematicians worldwide. Examples of popular systems include Maple, Mathematica, and MathCAD. Computer Algebra systems can be used to simplify rational functions, factor polynomials, find the solutions to a system of equation, and various other manipulations. In Calculus, they can be used to find the limit of, symbolically integrate, and differentiate arbitrary equations.

Attempting to expand the equation

$$(x-100)^{1000}$$

using the binomial theorem by hand would be a daunting task, nearly impossible to do without error. However, with the aid of Maple, this equation can be expanded in less than two seconds. Differentiating the result term-by-term can then be performed in milliseconds. The usefulness of such a system is obvious: not only does it act as a time saving device, but problems which simply were not reasonable to perform by hand can be performed in seconds.

Leibniz and Newton developed calculus in terms of algorithmic processes. Computer Algebra systems can now take these methods and remove the human from the process. However, in studying Calculus and even simple algebraic operations, it would seem that computers would be extraordinarily inept at performing such tasks. After all, most of us consider there to be a great deal of problem solving involved in the mathematics taught in grade school and beyond. How is it that a computer, a mindless composition of binary digits, is able to perform such complex tasks? It would seem that the computer would be unsuitable for such tasks, but the success of popular Algebra software packages show that this is not the case. On the contrary, Computer Algebra systems often know how to perform more operations on equations than the user!

Rather than discuss the many ways that Computer Algebra systems have altered the education and use of Calculus, we were most intrigued by how these systems actually worked. Our approach was to begin by researching the theories and issues involved in creating a Computer Algebra system. Coinciding with our research, we began writing our own Computer Algebra system in C++. The rest of this section is dedicated to a summary of our research and the specifics of the implementation we chose.

Symbolic Computing

Algebraic or Symbolic computing is very different from the sort of numerical computing you have learned so far. Algebraic computing systems have been available for a long time, almost as long as the older conventional computer languages such as FORTRAN or ALGOL. Systems such as REDUCE or MACSYMA were available on mainframes in the 1960's as general purpose packages. There were also some specialist packages available for various branches of physics and mathematics, such as the programme FORM which is widely used in theoretical particle physics. With the development of PCs packages such as muMATH became available and more recently MAPLE and MATHEMATICA.

The newer packages not only perform algebraic manipulations but have powerful built in numerical analysis and plotting capabilities. Most systems can be used as *interpreters*, with the commands typed in one by one with an instant (?) response, but many also include the possibility to define and run your own macros or procedures, so that they have many of the attributes of a conventional compiler as well.

As many of these packages were designed with different specialised criteria in mind, some packages are better for particular classes of problems than others.

Basic Principles

In this section we consider some of the basic commands of Mathematica and how they can be used. We will make no attempt at completeness. Any work done using computer algebra will require a copy of the Mathematica handbook.

Note firstly the general point that the system is *Case Sensitive*: pi and Pi are different (the latter representing π). Different sorts of brackets are used for different purposes

()	for grouping	$(a+b)(a-b)$
[]	for functions	$f[x]$ or $\sin[x]$
{}	for lists	see examples later
[[]]	for indexing	$a[[n]] \equiv a_n$

In the examples to be given below prompts by the computer are shown in *italics*, things typed by the user in typewriter font and responses from the computer in normal type.

Variables are manipulated as in algebra unless they have been given a particular value:

In[1]:= $y=(1+x)^2$

Out[1]= $(1+x)^2$

In[2]:= $x=2$

Out[2]= 2

In[3]:= y

Out[3]= 9

As well as being given a permanent value variables can be given a temporary value

---

In[4]:= Clear[x]

In[5]:= y = (1 + x)^2

Out[5]= $(1+x)^2$

In[6]:= y/. x -> 2

Out[6]= 9

In[7]:= y

Out[7]= $(1+x)^2$

Note the command Clear[x] which makes sure that x does not have a value.

Basic algebra can be done by using the functions Expand[] and Factor[] as follows

---

In[8]:= Expand[y]

Out[8]= $1+2x+x^2$

In[9]:= Factor[%]

Out[9]= $(1+x)^2$

Note the use of % as shorthand for the result of the last operation. We could also have written Factor[Out[8]] here.

There are several other commands available which do specific manipulations on such expressions, in particular:

- Simplify[] tries to find the simplest form of an expression.
- Together[] put all terms over a common denominator.
- Collect[expr,x] groups together powers of x in the expression.
- Coefficient[expr,x^4] picks out the coefficient of x^4 in the expression.

Differentiation and integration are carried out using the *D*[*f*,*x*] and Integrate[*f*,*x*] operators.

In[10]:= D[y,x]

Out[10]= $2(1+x)$

In[11]:= Integrate[%,x]

Out[11]= $2x+x^2$

In[12]:= Factor[%]

Out[12]= $x(2+x)$

In[13]:= Integrate[y,x]

Out[13]= $x+x^2+\frac{x^3}{3}$

---

or, using special functions,

In[14]:= D[Sin[x],x]

Out[14]= Cos[x]

In[15]:= Integrate[%,x]

Out[15]= Sin[x]

In[16]:= x = Sin[z]

Out[16]= Sin[z]

In[17]:= y

Out[17]= $(1+\mathrm{Sin}[z])^2$

In[18]:= D[y,z]

Out[18]= $2\mathrm{Cos}[z](1+\mathrm{Sin}[z])$

In[19]:= Integrate[y,z]

Out[19]= $\frac{3z}{2}-2\mathrm{Cos}[z]-\frac{\mathrm{Sin}[2z]}{4}$

Higher derivatives are written as

- *D*[*f*, *x*, *y*, *z*,...] multiple derivative $\frac{\partial}{\partial x}\frac{\partial}{\partial y}\frac{\partial}{\partial z}\cdots$.
- *D*[*f*, {*x*,*n*}] repeated derivative $\frac{\partial^n}{\partial x^n}$.

whereas definite integrals have the format

- Integrate[*f*, {*x*,*a*,*b*}] definite integral $\int_a^b dx$.
- Integrate[*f*, {*x*,0,Infinity}] definite integral $\int_a^\infty dx$.

Whereas differentiation always has a well-defined answer, sometimes *Mathematica* is unable to find an analytical result. In such cases the function *N*[] can be used to obtain a numerical result.

In[20]:=	Integrate[Sin[Sin[*x*]], {*x*,0,1}]
Out[20]=	Integrate [Sin[Sin[*x*]], {*x*,0,1}]
In[21]:=	N[%]
Out[21]=	0.430606
In[22]:=	N[%%,40]
Out[22]=	0.4306061031206906049123773552484666

where the last command evaluates the 2nd last (%%) result to 40 significant figures. This function can also be used in other situations such as

In[23]:=	N[Pi, 100]
Out[23]=	3.14159265358979323846264338327950288419716939937510582097494459230781640628620899862803482534211706 8

Mathematica has a number of built in constants (all beginning with capital letters) such as

In[24]:=	E^(I Pi)
Out[24]=	–1

Solving Equations

Before considering how to solve simple equations using Mathematica we have to note one other detail of notation, == is the symbol for Boolean equality. So, for example

In[25]:=	*x*=4
Out[25]=	4
In[26]:=	*x*==4
Out[26]=	True
In[27]:=	*x*==5
Out[27]=	False
In[28]:=	*x*
Out[28]=	4

from which we see that the result of $x{==}y$ is either "True" or "False". Contrast this with the symbol which is used to assign a value to a symbol.

Now we introduce the Solve function for solving equations. The syntax can be seen from the following examples:

In[29]:=	Solve[4x + 8 == 0, x]
Out[29]=	$\{\{x->-2\}\}$
In[30]:=	Solve[x^2 + 2x - 7 == 0, x]
Out[30]=	$\left\{\left\{x->\frac{-2-4\text{Sqrt}[2]}{2},x->\frac{-2+4\text{Sqrt}[2]}{2}\right\}\right\}$
In[31]:=	N[%]
Out[31]=	$\{x->-3.82843,x->1.82843\}\}$
In[32]:=	Solve[a x^2 + b x + c == 0, x]
Out[32]=	$\left\{\left\{x->\frac{-b+\text{Sqrt}\left[b^2-4ac\right]}{2a},x->\frac{-b-\text{Sqrt}\left[b^2-4ac\right]}{2a}\right\}\right\}$

where both numerical and *symbolic* solutions may be obtained. We can also find more complicated problems, such as

In[33]:=	Solve[{x^2 + y^2 == 1, x + 3 y == 0}, {x, y}]
Out[33]=	$\left\{\left\{x->\frac{-3}{\text{Sqrt}[10]},y->\frac{1}{\text{Sqrt}[10]}\right\},\left\{x->\frac{3}{\text{Sqrt}[10]},y->\left(\frac{1}{\text{Sqrt}[10]}\right)\right\}\right\}$
In[34]:=	x/.%
Out[34]=	$\left\{x->\frac{-3}{\text{Sqrt}[10]},x->\frac{3}{\text{Sqrt}[10]}\right\}$
In[35]:=	y/.%%
Out[35]=	$\left\{y->\frac{1}{Sqrt[10]},y->-\left(\frac{1}{\text{Sqrt}[10]}\right)\right\}$

in which we should note the syntax for solving systems of equations by grouping things in braces, {}, and for extracting parts of the solution. Note also how the different, distinct solutions are grouped together.

Unfortunately there are equations which Mathematica cannot solve so simply, but which with a little help it could do so. One very useful command in this context is Eliminate, used as follows

In[36]:=	Eliminate[{*a x* + *y* == 0, 2 *x* + (1 - *a*) *y* == 1}, *y*]
Out[36]=	$-(ax)+a^2x==1-2x$
In[37]:=	Solve[%, *x*]
Out[37]=	$\left\{\left\{x->\frac{1}{2-a+a^2}\right\}\right\}$

Often Mathematica is unable to solve a problem in a single step but can do so if the user *guides* it through the steps. This is particularly the case when an equation may have unphysical solutions which can be ignored. The user can recognise the physically important solutions and concentrate on those.

Differential Equations

The function DSolve is used to solve differential equations. Note the following examples

In[38]:=	DSolve[y'[x] == a y[x], y[x], x]
Out[38]=	$\{\{y[x]->E^{ax}C[1]\}\}$
In[39]:=	DSolve[{*y*'[*x*] == *a y*[*x*], *y*[0] == 1}, *y*[*x*], *x*]
Out[39]=	$\{\{y[x]->E^{ax}\}\}$

where we can see several important features. The notation *y*'[*x*] has the obvious meaning of 1st derivative of the function *y*[*x*] with respect to *x*. The 2nd and 3rd parameters of the DSolve function are the unknown function and the variable of integration respectively. Note also the appearance of the arbitrary constant C[1] in the 1st example and the inclusion of a boundary condition *y*[0] == 1 in the 2nd.

Mathematica can handle more general systems of differential equations, but solutions are usually only found if the equations are linear, unless the user is able to fill in some intermediate steps.

There are also functions NSolve and NDSolve which provide numerical solutions when the analytical solutions are not available from Solve or DSolve.

Other Features

We do not have enough time here to list all the powerful facilities within Mathematica, but we present a short list of some of the more useful ones.

- Functions to expand, sum, and take the limit of series and functions: Series, Sum, Limit.
- Plotting and Fitting numerical data and solutions of (e.g.) differential equations.
- Facilities to manipulate vectors and matrices in different coordinate systems.
- Statistical functions.

- A programming language using which it is possible to write your own symbolic programmes and hence to expand Mathematica's facilities.
- A library of prewritten extensions for various specialist purposes as well as some specially written code available from other sources.

Computer Algebra or Symbolic Computing is particularly useful when a problem involves a large amount of very tedious work where it is easy to miss or lose important terms. One such is the solution of non-linear differential equations using a series expansion. We therefore illustrate such a solution where we are trying to solve the equation

$$\frac{dy}{dx} = ay^2. \qquad ...2.1$$

In[40]:= y = a0 + a1 x + a2 x^2 + a3 x^3 + a4 x^4

Out[40]= $a0 + a1x + a2x^2 + a3x^3 + a4x^4$

In[41]:= D[y, x] - a (y)^2

Out[41]= $a1 + 2a2x + 3a3x^2 + 4a4x^3$

$-a\left(a0 + a1x + a2 + x^2 + a3x^3 + a4x^4\right)^2$

In[42]:= equ = Expand[%]

Out[42]= $-a\left(a\,a0^2\right) + a1 - 2a\,a0\,a1\,x + 2\,a2x - a\,a1^2\,x^2 - 2a\,a0\,a2\,x^2$

$+3a3x^2 - 2a\,a1\,a2x^3 - 2a\,a0\,a3\,x^3 + 4\,a4\,x^3$

$-a\,a2^2\,x^4 - 2a\,a1\,a^3x^4 - 2a\,a0\,a4\,x^4 - 2a\,a2\,a3\,x^5$

$-2a\,a1\,a4\,x^5 - a\,a3^2\,x^6 - 2a\,a2\,a4\,x^6 - 2a\,a3\,a4\,x^7$

$-a\,a4^2\,x^8$

In[43]:= equ/.x –> 0

Out[43]= $-\left(a\,a0\right)^2 + a1$

In[44]:= Solve[%==0, a1]

Out[44]= $\{\{a1 -> a\,a0^2\}\}$

In[45]:= a1 = a1/.%[[1]]

Out[45]= $a\,a0^2$

In[46]:= equ

Out[46]= $-2a^2a0^3x+2a2x-a^3a0^4x^2-2a\,a0\,a2\,x^2+3a3\,x^2$
$-2a^2a0^2a2x^3-2a\,a0\,a3\,x^3+4a4\,x^3-a\,a2^2x^4$
$-2a^2a0^2a3\,x^4-2a\,a0\,a4\,x^4-2a\,a2\,a3\,x^5-2a^2a0^2a4\,x^5$
$-a\,a3^2x^6-2a\,a2\,a4\,x^6-2a\,a3\,a4\,x^7-a\,a4^2x^8$

In[47]:= Coefficient[equ, x]

Out[47]= $-2a^2a0^3+2a^2$

In[48]:= Solve[%==0, a2]

Out[48]= $\{\{a2->a^2a0^3\}\}$

In[49]:= a2 = a2/.%[[1]]

Out[49]= a^2a0^3

In[50]:= equ

Out[50]= $-3a^3a0^4x^2+3a3\,x^2-2a^4a0^5x^3-2a\,a0\,a3\,x^3$
$+4a4\,x^3-a^5a0^6x^4-2a^2a0^2a3\,x^4-2a\,a0\,a4\,x^4$
$-2a^3a0^3a3\,x^5-2a^2a0^2a4\,x^5-a\,a3^2x^6-2a^3a0^3a4\,x^6$
$-2a\,a3\,a4\,x^7-a\,a4^2x^8$

In[51]:= Coefficient[equ, x^2]

Out[51]= $-3a^3a0^4+3a^3$

In[52]:= Solve[%==0, a3]

Out[52]= $\{\{a3->a^3a0^4\}\}$

In[53]:= a3 = a3/.%[[1]]

Out[53]= a^3a0^4

In[54]:= equ

Out[54]= $-4a^4a0^5x^3+4a4\,x^3-3a^5a0^6x^4-2a\,a0\,a4\,x^4$
$-2a^6a0^7x^5-2a^2a0^2a4\,x^5-a^7a0^8x^6-2a^3a0^3a4\,x^6$
$-2a^4a0^4a4\,x^7-a\,a4^2x^8$

In[55]:= Coefficient[equ, *x*^3]

Out[55]= $-4a^4a0^5+4a4$

In[56]:= Solve[%==0, *a*4]

Out[56]= $\{\{a4->a^4a0^5\}\}$

In[57]:= a4 = a4/.%[[1]]

Out[57]= a^4a0^5

In[58]:= *y*

Out[58]= $a0+a\,a0^2x+a^2a0^3x^2+a^3a0^4x^3+a^4a0^5x^4$

In[59]:= Factor[%]

Out[59]= $a0\left(1+a\,a0\,x+a^2a0^2x^2+a^3a0^3x^3+a^4a0^4x^4\right)$

This looks like the first few terms of the series for

$$y(x)=\frac{a_0}{1-a_0ax} \qquad ...2.2$$

which is indeed the solution of the differential equation (2.1. So we see that in this sort of *bookkeeping* exercise the use of computer algebra can be of considerable assistance).

Thomas-Fermi Approximation

Consider a system of free electrons or of electrons in a semi-conductor within the effective mass approximation. Since the density of states for free electrons, $\rho(E)\propto E^{1/2}$, the density of electrons, *n*, below a chemical potential μ is $n\propto\mu^{3/2}$. If an electrical potential energy, *V* is added while keeping the electrochemical potential energy (or Fermi level), $E_F=\mu+V$, constant, then the electron density changes to $n\propto(E_FV)^{3/2}$.

In the Thomas-Fermi approximation this charge density is assumed to be a local quantity. Poisson's equation for the electrical potential then takes the form

$$\frac{d^2V}{dx^2}\propto(E_F-V)^{3/2}, \qquad ...2.3$$

which can be simplified into the form

$$\frac{d^2\phi}{dx^2}=A\phi^{3/2}. \qquad ...2.4$$

This is a second order differential equation, so the general solution must contain 2 unknowns. One solution is known, namely

$$\phi=\frac{400}{A^2(x-x_0)^4}, \qquad ...2.5$$

which contains the single unknown x_0.

The object of the exercise here is to find a more general solution, or at least something which may be used as such (i.e. a series expansion).

Some Ideas

One possible approach would be to use the Frobenius method by assuming

$$\phi(x) = x^s \sum_{n=0}^{\infty} a_n x^n, \qquad \text{...2.6}$$

but this presents 2 major difficulties.

- It doesn't adequately take account of the known solution, (2.5).
- The 3/2 power cannot be easily expanded.

For this reason I suggest an alternative approach: make the substitution $\phi(x) = y^2(x)$ so that (2.4) then takes the form

$$\frac{d^2y^2}{dx^2} = Ay^3, \qquad \text{...2.7}$$

and then use the Frobenius expansion in $z = x - x_0$ to solve the equation. Thus

$$y(z) = z^s \sum_{n=0}^{\infty} b_n z^n. \qquad \text{...2.8}$$

By considering the limit of small and the known solution (2.5) we see immediately that $s = -2$ and that $b_0 = 20/A$.

You should try to find some more coefficients in the expansion (2.8) using *Mathematica*. The technique is similar to that described in the notes. You should bear in mind that we have already found one arbitrary coefficient, x_0, so that your solution should contain at least one other such arbitrary coefficient. Some of the coefficients may be zero; you should try to find a few non-trivial ones.

Simulation Programme and Its Validation

Simulation Programme

The simulation programme (ISM.EXE), which is written in C, is designed to construct in computer memory an Ising or a q-state Potts spin model of a specified type and to run that model (with a specified dynamics algorithm and lattice geometry, using periodic boundary conditions) so as to simulate a system of magnetic spins evolving over time. In accord with the aims of computational physics is the spin model is studied to attain knowledge of real physical systems which are believed to have an analogous structure and functioning.

Good simulation software relies on a good source of random numbers. It is impractical to use a truly random source based on a physical process (e.g., radioactive decay) so a source based upon a deterministic computational source is used. Beginning with some "seed" integer a sequence of positive integers is generated; these can be converted to numbers between 0 and 1. The generator should have the following properties:

- Reproducibility (a given seed number always produces the same sequence of random numbers, so simulations can be repeated exactly if desired).
- Efficiency (very little computer memory and very little computation time are required to produce the next number in the sequence, since a simulation usually requires the generation of many millions of random numbers).
- Long period (any sequence of random numbers should continue for a long time without repeating itself).
- Independence and unpredictability (if the seed number is not known then there should be no way to predict the continuation of a sequence of random numbers based on knowledge of the particular numbers generated up to that point).
- Uniform distribution (the random numbers, whether integers or numbers between 0 and 1, should be uniformly distributed over the range of possible numbers).

The presence of properties (iv) and (v) can be tested by the application of statistical tests, which look for regularities or anomalies in the random number sequence.

The random number generator used in this simulation programme generates random numbers at a speed of 3.39 million per second (on a 330 Mhz PC), and has a period of more than 10^{18}. "The Random Number Generator", where the results of a chi-square test are given.

Running the simulation programme with no command line arguments gives:

```
Syntax:  ISM.EXE  input_file_1.IN  [input_file_2.IN  [...]]

Output  data  file  has  same  name  with  extension  .OUT.

For  the  format  of  the  input  data  use:  ISM.EXE  /F
```

The parameters for a simulation are specified in an input file, and several simulations may be run in succession by including several input file names on the command line. The simulation programme has been written to handle a variety of models, lattice geometries, types of dilution, etc. The programme provides a summary of the input parameters if it is run with the command line: ISM /F, as follows:

```
———  ISM V1.93 INPUT PARAMETERS (SPIN MODEL DYNAMICS)  ———

Parameter name              Permitted values

lattice type:               SQU | HON | TRI | CUB | DIA | HC4

lattice size:               4 - 4096 (2d), 4 - 256 (3d), 4 - 64 (4d)

model:                      Is[ing] | q-[state Potts]

q-value:                    2 - 15 //  q-state Potts model only

dynamics:                   ME[tropolis] | GL[auber] | SW[endsen-Wang] |
                            WO[lff]

temperature:                0.001 - 50.000 // or "critical"
```

```
temperatures:                  // Separate multiple temperatures with commas
initial magnetisation:         -1.0 - +1.0 // >= -1/(q-1) for q-state Potts
site concentration:            0.001 - 1.000
bond concentration:            0.001 - 1.000
number of configurations:      1 - 100,000  //  1 for a non-diluted lattice
number of spin assignments:    1 - 100,000
number of repetitions:         1 - 500   // Using different random numbers
number of timeslices:          2 - 10,001
step length:                   1 - 10,000 // timesteps between timeslices
percentage range for mean:     1 - 100
Use "Y[es]" or "N[o]" for the following (1st three required):
precompute nn sites:           absolute magnetisation:      timeslice values:
second moment:                 autocorrelation:             map file:
internal energy:               Binder cumulant:             log values:
adjust zero initial magnetisation:
```

Or in other words:

Lattice type:	Square, honeycomb, triangular, cubic, diamond, hypercubic
Lattice size:	Maximum lattice size depends on dimensionality and dynamics algorithm (smaller with cluster algorithms).
Model:	Ising or q-state Potts (for q in the range 2 through 15).
Dynamics:	Metropolis, Glauber, Swendsen-Wang or Wolff.

Once the type of lattice is specified, the initial state of the system is defined as follows:

Temperature(s):	0.0 – 20.0. The temperature can also be specified as "critical", in which case the critical temperature for the type of spin system is used, if known. Multiple temperatures are accepted.
Initial magnetisation:	–1.0 through +1.0 (Initial magnetisation must be ≥ –1/(q–1) for the q-state Potts model.)
Site concentration:	0.001 – 1.0 (the concentration of occupied sites).
Bond concentration:	0.001 – 1.0 (the concentration of open bonds among pairs of spins located on nearest neighbour sites).

For a pure lattice both concentrations are specified as 1.0 (the default value). It is possible to simulate a system which is simultaneously both site- and bond-diluted.

Normally the programme performs many simulation runs. Each such run is termed a "sample". It is desirable to distribute these samples over a range of possible initial configurations.

In the case of a site-diluted (resp. bond-diluted) lattice, with a certain concentration of sites (resp. bonds), the occupied sites (resp. open bonds) may be varied, giving different *configurations*. Given a particular configuration of occupied sites and open bonds the manner of assigning spin values to the spins (on occupied sites) may be varied, giving different *spin assignments*. Given a particular configuration and a particular assignment of spin values, a run may be repeated with different random numbers, giving different *repetitions*. The programme allows:

Number of configurations:	1 – 100,000 (but only 1 for a non-diluted lattice)
Number of spin assignments:	1 – 100,000
Number of repetitions:	1 – 500

The simulation programme allows four forms of dynamics: Metropolis, Glauber, Swendsen-Wang and Wolff. Each form of dynamics is a method for updating the spins in the spin model. In the first two methods spins are updated individually, and in the second two methods they are updated in clusters. In all four methods, however, the equilibration process may be divided into "timesteps" and "timepoints".

A timestep is one "sweep" through the lattice, in which each spin is considered (for possible change) once and once only. The number of timesteps is, in other terminology, the number of "Monte Carlo steps per spin". A "timepoint" is a state of the model after n timesteps $(n \geq 0)$.

Timepoints are analogous to points in time, and the change in the state of the spin model is analogous to the temporal evolution of the system being modelled.

A "timeslice" is a timestep at which measurements are carried out. Three quantities are measured:

- the magnetisation per spin (optionally, the absolute value thereof),
- the actual internal energy of the spin system
- the autocorrelation.

The first timeslice is the initial timepoint. Measurements are then taken after every n timesteps, for $n \geq 1$, for as many timesteps as were requested.

The results that the simulation programme prints are averages at each timeslice over a (possibly large) number of samples (where a "sample" is one run of the requested number of timesteps starting from some spin assignment and, in the case of a dilute lattice, from some configuration of occupied sites).

Two of the input parameters used by the simulation programme are "number of timeslices" and "step length". The step length is the number n of timesteps between timeslices. The number of timeslices, including the initial state, can range from 11 to 10,001, and the step length can range from 1 to 10,000.

The simulation programme also takes as input parameters whether or not to calculate the following four quantities: the internal energy per spin, the standard deviation of the magnetisation, the second moment of the magnetisation, the autocorrelation and the Binder cumulant. If internal energy is measured then results for specific heat per spin are given also. In the remainder of this section we will assume that all quantities are to be measured.

Magnetisation per spin is a quantity dependent on the values of the spins in a particular state of the spin model. It is always a value between +1 and -1. It is derived from the values of the spins. The derivation of this value from the actual number of spins in various states depends on the model (Ising or q-state Potts).

At each timeslice during a given run the magnetisation (per spin) of the spin system is monitored, and the value is added to a running total (which, of course, begins at zero). A running total is also kept of the square of the magnetisation (per spin), and the fourth power thereof, at each timeslice.

A running total is kept of the autocorrelation value, which is also calculated differently in the two kinds of spin model.

For each timeslice a running total is kept of the actual internal energy of the spin system (not the internal energy per spin) and the square of this quantity. The internal energy, like the magnetisation, depends on the values of the spins (and also on the spins in the immediate neighbourhood of a spin) and is calculated differently for the Ising and the q-state Potts models.

After all runs are completed we have, for each timeslice, six numbers, as follows, where n >= 0 denotes the timeslice number:

m(n)	sum of the magnetisation (per spin) values
$m^2(n)$	sum of the squares of the magnetisation values
$m^4(n)$	sum of the fourth powers of the magnetisation values
a(n)	sum of the autocorrelation values
e(n)	sum of the actual internal energy values
$e^2(n)$	sum of the squares of the internal energy values

Analysis, resulting in the values printed in the report (indicated below by italics), is then performed as follows, where N is the number of samples:

For each timeslice n:

- The *mean magnetisation (per spin)* = M = m(n) / N.
- The *mean square magnetisation* = $M^{(2)}$ = $m^2(n)$ / N
- The *standard deviation of the magnetisation* is then (assuming N > 1) given by

$$\sqrt{\left[\left(M^{(2)} - M^2\right)/(N-1)\right]}$$

- The *second moment of the magnetisation* is, in the case of the Ising model, the same as the mean square magnetisation. In the case of the Potts model it is calculated in a more complex manner.

- The *Binder cumulant* = 1 - $[m^4(n)/N]/[3.(M^{(2)})^2]$.
- The *mean autocorrelation* = a(n)/N.
- The *mean actual internal energy* = U = e(n)/N.
- The *internal energy per spin* is the mean actual internal energy divided by the number of spins in the spin system.
- The *mean square of the actual internal energy* = $U^{(2)} = e^2(n) / N$.
- The *actual specific heat* = $(U^{(2)} - U^2)/T^2$, where T denotes temperature.
- The *specific heat per spin* is this divided by the number of spins in the spin system.

Percentage range for mean: 1 - 100

When a quantity such as the magnetisation or the Binder cumulant is measured the value given is an average over a range of timepoints occurring after the system has reached equilibrium (i.e., a time average). E.g., if (using Wolff dynamics) it is known by observeration that measurements of quantities in a particular system stabilise after 30 timesteps, and measurements are made over 100 timesteps, then one might specify *percentage range for mean* as 40, meaning that the final value for the quantities measured is an average over timepoints 60 through 100.

Some input parameters have Yes or No values, in particular: Absolute magnetisation, Timeslice values and Precompute nn sites.

If *absolute magnetisation* is Yes then the absolute value of the magnetisation is measured. This is necessary when Swendsen-Wang or Wolff dynamics are used, since in these cases the magnetisation often flips between positive and negative within a few timesteps.

If *timeslice values* is Yes then the values of the quantities measured are reported for each timeslice (otherwise only the final time averages are reported).

If *precompute nn sites* is set to No then the locations of nearest neighbour sites are repeatedly calculated for many spin updates, which requires a significant amount of computer time. However if *precompute nn sites* is set toYes then the locations of the nearest neighbours of all sites are precomputed, stored in a large array and retrieved as required. This increases significantly the amount of memory required and is helpful only if the additional memory used does not cause memory to be swapped out to disk (which requires much more time than repeated computation of nearest neighbour locations).

Before using the simulation programme to obtain data as the basis of new research it is necessary to ensure that the programme works correctly, that is, that it is a valid implementation of the Ising and the q-state Potts spin models. This has been done here by reproducing some known results, details of which are given in the remainder of this chapter.

It contains information about several input files for the simulation programme which are given on the accompanying computer disk. These specify some of the many simulations carried out during the course of this work. The output obtained from these simulations is displayed in various graphs in later sections, and may be reproduced by running the programme with the supplied input files (although due to random variations and sometimes small numbers of samples such simulations will not reproduce the reported results exactly).

Comparison of Results for Internal Energy and Specific Heat with Those from High Temperature Series Expansion

Simulations were performed for the pure Ising square model for T = 1.0 to 5.0 in steps of 0.5 (plus the critical temperature, T ≅ 2.269). The internal energy per spin and the specific heat per spin were measured.

Lattice sizes of 32×32, 64×64 and 100×100 were used, with the number of samples ranging from 1000 to 8000. Figure 2.1 shows a plot of internal energy per spin against temperature (the internal energy per spin approaches zero asymptoticly) for lattice size 64×64. (The values of the internal energy for the other lattice sizes are almost the same.) Figure 2.2 shows a plot of specific heat per spin against temperature for the three lattice sizes.

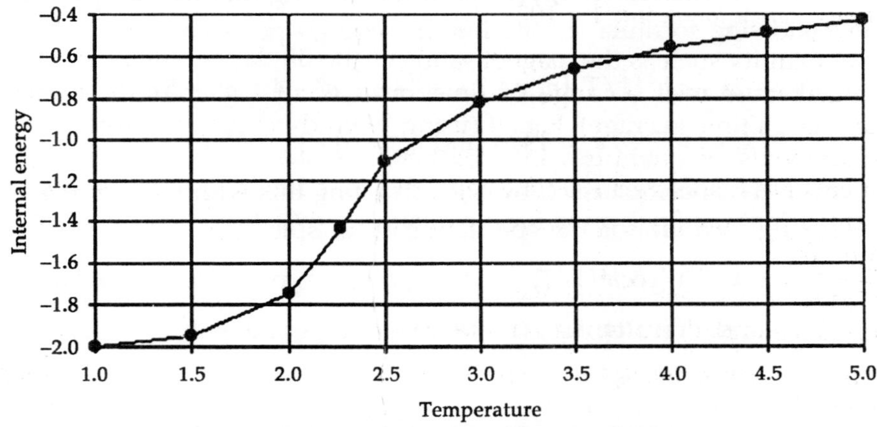

Figure 2.1: *Lsing square lattice, 64×64, Glauber dynamics*

Error bars were not calculated. The point of the simulations in this section were not so much to establish results concerning internal energy and specific heat as such but rather to ascertain whether the simulation software gives results in accord with high temperature series expansions (which, of course, have no error estimates associated with them).

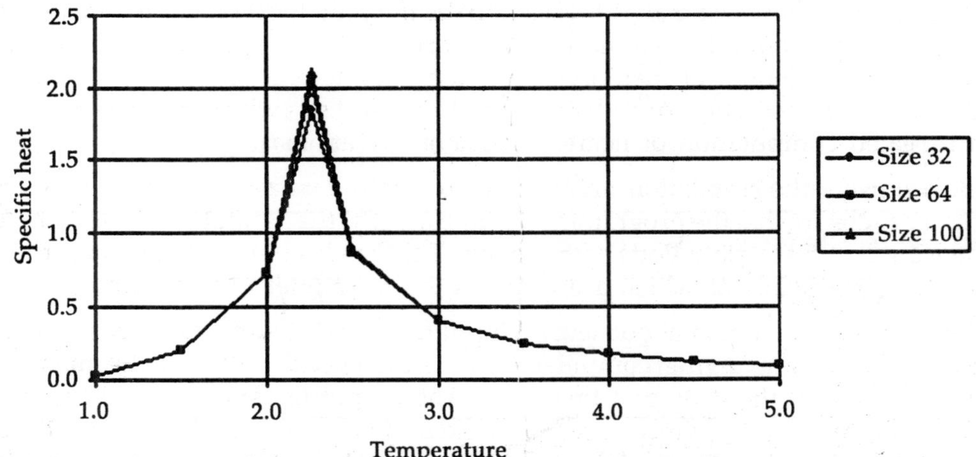

Figure 2.2: *Lsing square lattice, Glauber dynamics*

The method of high temperature series expansions leads to a formula for the free energy for temperatures higher than the critical temperature. From this can be derived formulae for the internal energy and the specific heat.

Yeomans gives the following expression for the free energy of a spin system with N spins at temperature T :

$$F = -Nk_BT\left(\ln 2 + v^2 + 3/2v^4 + 7/3v^6 + 19/4v^8 + 61/5v^{10} + ...\right)$$

where k_B is Boltzman's constant, $v = \tanh(\beta J)$ and $\beta = 1(k_BT)$. It is usual to put $J = k_B = 1$.

We may obtain the internal energy U by differentiating this with respect to b. So doing, we obtain the following formula for the internal energy per spin:

$$UN = -2k\,\mathrm{sec}h^2(1/T)\left(a_2.v + 2.a_4.v^3 + 3.a_6.v^5 + 4.a_8.v^7 + 5.a_{10}.v^9 + ...\right)$$

where $a_2 = 1$, $a_4 = 3/2$, $a_6 = 7/3$, $a_8 = 19/4$ and $a_{10} = 61/5$.

We may obtain the specific heat C by differentiating this with respect to T. So doing, we obtain the following formula for the specific heat per spin:

$$C/N = (2k/T^2)\ [\mathrm{sec}h^4(1/T)\ (\ 1 + 3.b_3.v^2 + 5.b_5.v^4 + 7.b_7.v^6 + 9.b_9.v^8 + ...)$$
$$-\ 2\ \mathrm{sec}h^2(v)\ \tan h(v)\ (v + b_3.v^3 + b_5.v^5 + b_7.v^7 + b_9.v^9 + ...)\ ]$$

where $b_3 = 3$, $b_5 = 7$, $b_7 = 19$ and $b_9 = 61$.

Values for the internal energy and the specific heat for T = 3.0, 3.5, 4.0, 4.5 and 5.0 were obtained from these formulae for comparison with the simulation data.

The internal energy per spin results are (with the contributions of the powers of v):

temp	v	a2*v	2*a4*v^3	3*a6*v^5	4*a8*v^7	5*a10*v^9	U/N
3.00	0.321513	0.321513	0.099705	0.024049	0.006747	0.002239	-0.814593
3.50	0.278185	0.278185	0.064584	0.011662	0.002450	0.000609	-0.659649
4.00	0.244919	0.244919	0.044074	0.006169	0.001004	0.000193	-0.557165
4.50	0.218635	0.218635	0.031353	0.003497	0.000454	0.000070	-0.483733
5.00	0.197375	0.197375	0.023067	0.002097	0.000222	0.000028	-0.428220

Thus the results for the two sources for U/N are:

Temperature	3.0	3.5	4.0	4.5	5.0
Simulation	-0.817	-0.660	-0.557	-0.484	-0.428
Series exp.	-0.815	-0.660	-0.557	-0.484	-0.428

The specific heat per spin results are (with the even-power terms, which increase the value, and the odd-power terms, which decrease it, given separately):

temp	3*b3v^2	5*b5*v^4	7*b7*v^6	9*b9*v^8	C/N
3.00	0.930334	0.373991	0.146906	0.062684	0.392425
3.50	0.696485	0.209607	0.061639	0.019690	0.246868
4.00	0.539866	0.125938	0.028707	0.007108	0.171184
4.50	0.430212	0.079974	0.014527	0.002866	0.126495
5.00	0.350613	0.053118	0.007863	0.001264	0.097712

temp	b1*v	b3*v^3	b5*v^5	b7*v^7	b9*v^9	C/N
3.00	0.321513	0.099705	0.024049	0.006747	0.002239	0.392425
3.50	0.278185	0.064584	0.011662	0.002450	0.000609	0.246868
4.00	0.244919	0.044074	0.006169	0.001004	0.000193	0.171184
4.50	0.218635	0.031353	0.003497	0.000454	0.000070	0.126495
5.00	0.197375	0.023067	0.002097	0.000222	0.000028	0.097712

Thus the results for the two sources for C/N are:

Temperature	3.0	3.5	4.0	4.5	5.0
Simulation	0.405	0.245	0.171	0.125	0.097
Series exp.	0.392	0.247	0.171	0.126	0.098

The measurements of internal energy per spin for the temperatures 3.0, 3.5, 4.0, 4.5 and 5.0 were found to agree almost exactly with the results obtained from high temperature series expansion.

The measurements of specific heat for the same five temperatures were also found to agree almost exactly except for T = 3.0 (where simulation gave 0.405 and series expansion gave 0.392), which can be explained as a result of taking an insufficient number of terms in the series expansion for that temperature.

This comparison shows that the simulation programme gives results which are in accord with those from high temperature series expansion, confirming the correctness of the software.

Comparison of Simulation Results from Different Dynamics Algorithms

To test the implementation of the Swendsen-Wang and Wolff algorithms the simulations were repeated using these algorithms with lattice size 64x64.

These simulation results are given in Figures 2.3 and 2.4 (which also show the results for the Glauber algorithm for lattice size 64).

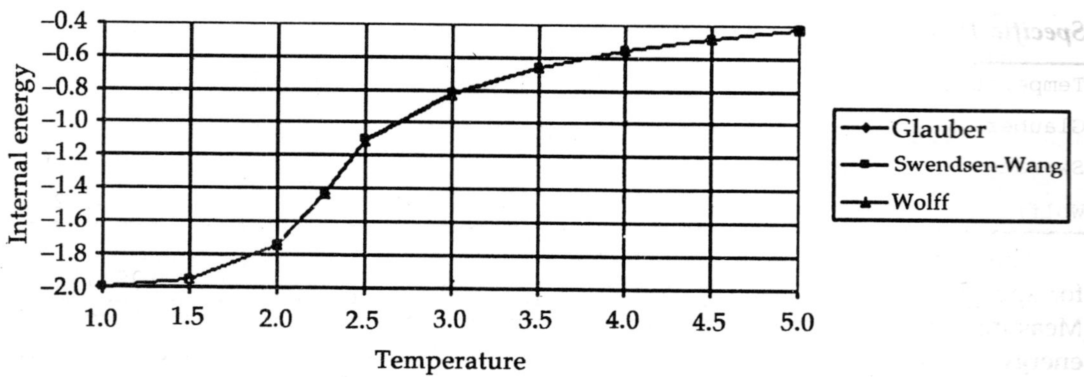

Figure 2.3: Ising square lattice, 64×64

Error bars are not shown because the errors were too small. For the Swendsen-Wang algorithm the largest error is 0.0013 at the critical temperature (2.269...) and for the Wolff algorithm it is 0.0015 (also at T_c).

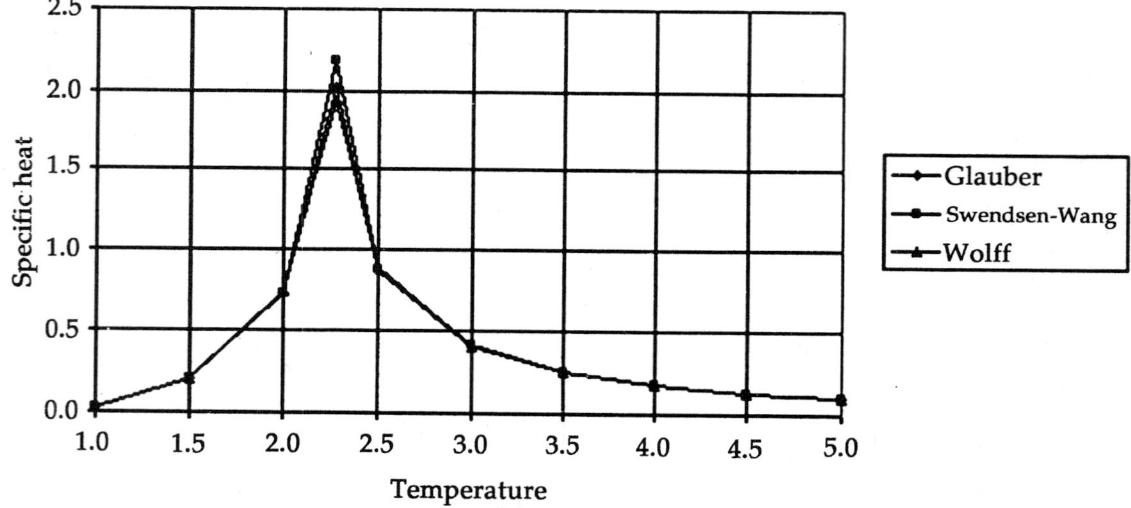

Figure 2.4: Ising square lattice, 64×64

Errors were not estimated in the Swendsen-Wang case. For the Wolff algorithm errors in the measurement of specific heat ranged from a minimum of 0.001 (for T = 1.0) to a maximum of 0.10 (for the critical temperature, T = 2.269...).

The data in is summarised below:

Internal Energy

Temperature	1.00	1.50	2.00	2.27	2.50	3.00	3.50	4.00	4.50	5.00
Glauber	-1.997	-1.951	-1.746	-1.431	-1.106	-0.817	-0.661	-0.557	-0.484	-0.428
Sw-Wang	-1.997	-1.951	-1.746	-1.424	-1.106	-0.817	-0.659	-0.557	-0.483	-0.429
Wolff	-1.997	-1.951	-1.745	-1.434	-1.113	-0.819	-0.661	-0.558	-0.484	-0.429

Specific Heat

Temperature	1.00	1.50	2.00	2.27	2.50	3.00	3.50	4.00	4.50	5.00
Glauber	0.023	0.197	0.730	2.031	0.860	0.405	0.249	0.171	0.125	0.097
Sw-Wang	0.025	0.197	0.730	2.190	0.881	0.408	0.254	0.176	0.126	0.101
Wolff	0.023	0.197	0.738	1.937	0.885	0.402	0.253	0.169	0.126	0.098

The measurements of internal energy for the three dynamics algorithms are close. Those for specific heat are also close except for the critical temperature ($T \cong 2.269$) and $T = 2.50$. Measurements of specific heat normally vary over a wider range than those for internal energy. These results indicate that the Swendsen-Wang and Wolff algorithms have been implemented correctly, assuming that the Glauber algorithm has been.

3

Scientific Programming in C

C is a flexible, extremely powerful, high-level programming language which was initially designed for writing operating systems and system applications. In fact, all UNIX operating systems, as well as most UNIX applications (*e.g.*, text editors, window managers, etc.) are written in C. However, C is also an excellent vehicle for scientific programming, since, almost by definition, a good scientific programming language must be powerful, flexible, and high-level. Having said this, many of the features of C which send computer scientists into raptures are not particularly relevant to the needs of the scientific programmer. Hence, in the following, we shall only describe that subset of the C language which is really necessary to write scientific programmes. It may be objected that our cut-down version of C bears a suspicious resemblance to FORTRAN. However, this resemblance is hardly surprising. After all, FORTRAN is a high-level programming language which was specifically designed with scientific computing in mind.

C++ is an extension of the C language whose main aim is to facilitate object-orientated programming. The object-orientated features of C++ are superfluous to our needs in this course. However, C++ incorporates some new, non-object-orientated features which are extremely useful to the scientific programmer. We shall briefly discuss these features towards the end of this section. Finally, we shall describe some prewritten C++ classes which allow us to incorporate complex arithmetic (which is not part of the C language), variable size arrays, and graphics into our programmes.

Variables

Variable names in C can consist of letters and numbers in any order, except that the first character must be a letter. Names are *case sensitive*, so upper- and lower-case letters are not interchangeable. The underscore character (_) can also be included in variable names, and is treated as a letter. There is no restriction on the length of names in C. Of course, variable names are not allowed to clash with keywords that play a special role in the C language, such as int, double, if, return, void, etc. The following are examples of valid variable names in C:

```
x    c14    area    electron_mass    TEMPERATURE
```

The C language supports a great variety of different data types. However, the two data types which occur most often in scientific programmes are *integer*, denoted int, and *floating-point*, denoted double. (Note that variables of the most basic floating-point data type float are not generally stored to sufficient precision by the computer to be of much use in scientific programming.) The data type (int or double) of every variable in a C programme must be declared *before* that variable can appear in an executable statement.

Integer constants in C are denoted, in the regular fashion, by strings of arabic numbers: *e.g.*,

```
0    57    4567    128933
```

Floating-point constants can be written in either regular or scientific notation: *e.g.*,

```
0.01    70.456    3e+5    .5067e-16
```

Strings are mainly used in scientific programmes for data input and output purposes. A string consists of any number of consecutive characters (including blanks) enclosed in double quotation marks: *e.g.*,

```
"red"    "Austin TX, 78723"    "512-926-1477"
```

Line-feeds can be incorporated into strings via the *escape sequence* \n: *e.g.*,

```
"Line 1\nLine 2\nLine 3"
```

The above string would be displayed on a computer terminal as

```
Line 1
Line 2
Line 3
```

A *declaration* associates a group of variables with a specific data type. All variables must be declared before they can appear in executable statements. A declaration consists of a data type followed by one or more variable names, ending in a semicolon. For instance,

```
int       a, b, c;
double    acc, epsilon, t;
```

In the above, a, b, and c are declared to be integer variables, whereas acc, epsilon, and t are declared to be floating-point variables.

A type declaration can also be used to assign initial values to variables. Some examples of how to do this are given below:

```
int       a = 3, b = 5;
double    factor = 1.2E-5;
```

Here, the integer variables a and b are assigned the initial values 3 and 5, respectively, whereas the floating-point variable factor is assigned the initial value 1.2×10^{-5}.

Note that there is no restriction on the length of a type declaration: such a declaration can even be split over many lines, so long as its end is signaled by a semicolon. However, all declaration statements in a programme (or programme segment) must occur *prior* to the first executable statement.

Expressions and Statements

An *expression* represents a single data item—usually a number. The expression may consist of a single entity, such as a constant or variable, or it may consist of some combination of such entities, interconnected by one or more *operators*. Expressions can also represent logical conditions which are either true or false. However, in C, the conditions true and false are represented by the integer values 1 and 0, respectively. Several simple expressions are given below:

```
a + b

x = y

t = u + v

x <= y

++j
```

The first expression, which employs the *addition operator* (+), represents the sum of the values assigned to variables a and b. The second expression involves the *assignment operator* (=), and causes the value represented by y to be assigned to x. In the third expression, the value of the expression (u + v) is assigned to t. The fourth expression takes the value 1 (true) if the value of x is less than or equal to the value of y. Otherwise, the expression takes the value 0 (false). Here, <= is a *relational operator* that compares the values of x and y. The final example causes the value of j to be increased by 1. Thus, the expression is equivalent to

```
j = j + 1
```

The *increment* (by unity) operator ++ is called a *unary* operator, because it only possesses one operand.

A *statement* causes the computer to carry out some definite action. There are three different classes of statements in C: *expression statements, compound statements,* and *control statements.*

An expression statement consists of an expression followed by a semicolon. The execution of such a statement causes the associated expression to be evaluated. For example:

```
a = 6;

c = a + b;

++j;
```

The first two expression statements both cause the value of the expression on the right of the equal sign to be assigned to the variable on the left. The third expression statement causes the value of j to be incremented by 1. Again, there is no restriction on the length of an expression statement: such a statement can even be split over many lines, so long as its end is signaled by a semicolon.

A compound statement consists of several individual statements enclosed within a pair of braces { }. The individual statements may themselves be expression statements, compound statements, or control statements. Unlike expression statements, compound statements do *not* end with semicolons. A typical compound statement is shown below:

```
{
   pi = 3.141593;
   circumference = 2. * pi * radius;
   area = pi * radius * radius;
}
```

This particular compound statement consists of three expression statements, but acts like a single entity in the programme in which it appears.

A *symbolic constant* is a name that substitutes for a sequence of characters. The characters may represent either a number or a string. When a programme is compiled, each occurrence of a symbolic constant is replaced by its corresponding character sequence. Symbolic constants are usually defined at the beginning of a programme, by writing

```
#define   NAME   text
```

where NAME represents a symbolic name, typically written in upper-case letters, and text represents the sequence of characters that is associated with that name. Note that text does *not* end with a semicolon, since a symbolic constant definition is not a true C statement. In fact, during compilation, the resolution of symbolic names is performed (by the C *preprocessor*) before the start of true compilation. For instance, suppose that a C programme contains the following symbolic constant definition:

```
#define   PI   3.141593
```

Suppose, further, that the programme contains the statement

```
area = PI * radius * radius;
```

During the compilation process, the preprocessor replaces each occurrence of the symbolic constant PI by its corresponding text. Hence, the above statement becomes

```
area = 3.141593 * radius * radius;
```

Symbolic constants are particularly useful in scientific programmes for representing constants of nature, such as the mass of an electron, the speed of light, etc. Since these quantities are fixed, there is little point in assigning variables in which to store them.

Operators

As we have seen, general expressions are formed by joining together constants and variables via various operators. Operators in C fall into five main classes: *arithmetic operators*, *unary operators, relational and logical operators, assignment operators,* and the *conditional operator.* Let us, now, examine each of these classes in detail.

There are *four* main arithmetic operators in C. These are:

```
addition           +
subtraction        -
multiplication     *
division           /
```

Unbelievably, there is no built-in exponentiation operator in C (C was written by computer scientists)! Instead, there is a *library function* (pow) which carries out this operation.

It is poor programming practice to mix types in arithmetic expressions. In other words, the two operands operated on by the addition, subtraction, multiplication, or division operators should both be either of type int or type double. The value of an expression can be converted to a different data type by prepending the name of the desired data type, enclosed in parenthesis. This type of construction is known as a *cast.* Thus, to convert an integer variable j into a floating-point variable with the same value, we would write

```
(double) j
```

Finally, to avoid mixing data types when dividing a floating-point variable x by an integer variable i, we would write

```
x / (double) i
```

Of course, the result of this operation would be of type double.

The operators within C are grouped hierarchically according to their *precedence* (*i.e.,* their order of evaluation). Amongst the arithmetic operators, * and / have precedence over + and -. In other words, when evaluating expressions, C performs multiplication and division operations prior to addition and subtraction operations. Of course, the rules of precedence can always be bypassed by judicious use of parentheses. Thus, the expression

```
a - b / c + d
```

is equivalent to the unambiguous expression

```
a - (b / c) + d
```

since division takes precedence over addition and subtraction.

The distinguishing feature of unary operators is that they only act on single operands. The most common unary operator is the *unary minus,* which occurs when a numerical constant, variable, or expression is preceded by a minus sign. Note that the unary minus is distinctly different from the arithmetic operator (-) which denotes subtraction, since the latter operator acts on two separate operands. The two other common unary operators are the *increment operator,* ++, and the *decrement operator,* —. The increment operator causes its operand to be increased by 1, whereas the decrement operator causes its operand to be decreased by 1. For example, -i is equivalent to i = i - 1. A *cast* is also considered to be a unary operator. Note that unary operators have precedence over arithmetic operators. Hence, - x

+ y is equivalent to the unambiguous expression (-x) + y, since the unary minus operator has precedence over the addition operator.

Note that there is a subtle distinction between the expressions a++ and ++a. In the former case, the value of the variable a is returned *before* it is incremented. In the latter case, the value of a is returned *after* incrementation. Thus,

```
b = a++;
```

is equivalent to

```
b = a;
a = a + 1;
```

whereas

```
b = ++a;
```

is equivalent to

```
a = a + 1;
b = a;
```

There is a similar distinction between the expressions a— and —a.

There are four relational operators in C. These are:

less than	<
less than or equal to	<=
greater than	>
greater than or equal to	>=

The precedence of these operators is lower than that of arithmetic operators.

Closely associated with the relational operators are the two *equality operators*:

equal to	==
not equal to	!=

The precedence of the equality operators is below that of the relational operators.

The relational and equality operators are used to form logical expressions, which represent conditions that are either true or false. The resulting expressions are of type int, since true is represented by the integer value 1 and false by the integer value 0. For example, the expression i < j is true (value 1) if the value of i is less than the value of j, and false (value 0) otherwise. Likewise, the expression j == 3 is true if the value of j is equal to 3, and false otherwise.

C also possess two *logical operators*. These are:

&&	and
||	or

The logical operators act on operands which are themselves logical expressions. The net effect is to combine the individual logical expressions into more complex expressions that are either true or false. The result of a *logical and* operation is only true if both operands are true, whereas the result of a *logical or* operation is only false if both operands are false. For instance, the expression (i >= 5) && (j == 3) is true if the value of i is greater than or equal to 5 *and* the value of j is equal to 3, otherwise it is false. The precedence of the *logical and* operator is higher than that of the *logical or* operator, but lower than that of the equality operators.

C also includes the unary operator that negates the value of a logical expression: i.e., it causes an expression that is originally true to become false, and *vice versa*. This operator is referred to as the *logical negation* or *logical not* operator. For instance, the expression (k == 4) is true if the value of k is not equal to 4, and false otherwise.

Note that it is poor programming practice to rely too heavily on operator precedence, since such reliance tends to makes C programmes very hard for other people to follow. For instance, instead of writing

```
i + j == 3 && i * l >= 5
```

and relying on the fact that arithmetic operators have precedence over relational and equality operators, which, in turn, have precedence over logical operators, it is better to write

```
((i + j) == 3) && (i * l >= 5)
```

whose meaning is fairly unambiguous, even to people who cannot remember the order of precedence of the various operators in C.

The most common assignment operator in C is =. For instance, the expression

```
f = 3.4
```

causes the floating-point value 3.4 to be assigned to the variable f. Note that the assignment operator = and the equality operator == perform completely different functions in C, and should not be confused. Multiple assignments are permissible in C. For example,

```
i = j = k = 4
```

causes the integer value 4 to be assigned to i, j, and k, simultaneously. Note, again, that it is poor programming practice to mix data types in assignment expressions. Thus, the data types of the constants or variables on either side of the = sign should always match.

C contains four additional assignment operators: +=, -=, *=, and /=. The expression

```
i += 6
```

is equivalent to i = i + 6. Likewise, the expression

```
i -= 6
```

is equivalent to i = i - 6. The expression

```
i *= 6
```

is equivalent to i = i * 6. Finally, the expression

```
i /= 6
```

is equivalent to i = i / 6.

Simple conditional operations can be carried out with the *conditional operator* (? :). An expression that makes use of the conditional operator is called a *conditional expression*. Such an expression takes the general form

```
expression 1   ?   expression 2   :   expression 3
```

If expression 1 is true (*i.e.*, if its value is non-zero) then expression 2 is evaluated and becomes the value of the conditional expression. On the other hand, if expression 1 is false (*i.e.*, if its value is zero) then expression 3 is evaluated and becomes the value of the conditional expression. For instance, the expression

```
(j < 5) ? 12 : -6
```

takes the value 12 if the value of j is less than 5, and the value -6 otherwise. The assignment statement

```
k = (i < 0) ? n : m
```

causes the value of n to be assigned to the variable k if the value of i is less than zero, and the value of m to be assigned to k otherwise. The precedence of the conditional operator is just above that of the assignment operators.

As we have already mentioned, scientific programmes tend to be extremely resource intensive. Scientific programmers should, therefore, always be on the lookout for methods of speeding up the execution of their codes. It is important to realise that multiplication (*) and division (/) operations consume *considerably more* CPU time that addition (+), subtraction (-), comparison, or assignment operations. Thus, a simple rule of thumb for writing efficient code is to try to avoid redundant multiplication and division operations. This is particularly important for sections of code which are executed repeatedly: *e.g.*, code which lies within control loops. The classic illustration of this point is the evaluation of a polynomial. The most straightforward method of evaluating (say) a fourth-order polynomial would be to write something like:

```
p = c_0 + c_1 * x + c_2 * x * x + c_3 * x * x * x + c_4 * x *
x * x * x
```

Note that the above expression employs *ten* expensive multiplication operations. However, this number can be reduced to *four* via a simple algebraic rearrangement:

```
p = c_0 + x * (c_1 + x * (c_2 + x * (c_3 + x * c_4)))
```

Clearly, the latter expression is far more computationally efficient than the former.

Library Functions

The C language is accompanied by a number of standard *library functions* which carry out various useful tasks. In particular, all input and output operations (*e.g.*, writing to the terminal) and all math operations (*e.g.*, evaluation of sines and cosines) are implemented by library functions.

In order to use a library function, it is necessary to call the appropriate *header file* at the beginning of the programme. The header file informs the programme of the name, type, and number and type of arguments, of all of the functions contained in the library in question. A header file is called via the preprocessor statement

```
#include    <filename>
```

where filename represents the name of the header file.

A library function is accessed by simply writing the function name, followed by a list of *arguments*, which represent the information being passed to the function. The arguments must be enclosed in parentheses, and separated by commas: they can be constants, variables, or more complex expressions. Note that the parentheses must be present even when there are no arguments.

The C *math library* has the header file math.h, and contains the following useful functions:

Function	***Type***	***Purpose***
acos(d)	double	Return arc cosine of d (in range 0 to pi)
asin(d)	double	Return arc sine of d (in range -pi/2 to pi/2)
atan(d)	double	Return arc tangent of d (in range -pi/2 to pi/2)
atan2(d1, d2)	double	Return arc tangent of d1/d2 (in range -pi to pi)
cbrt(d)	double	Return cube root of d
cos(d)	double	Return cosine of d
cosh(d)	double	Return hyperbolic cosine of d
exp(d)	double	Return exponential of d
fabs(d)	double	Return absolute value of d
hypot(d1, d2)	double	Return sqrt(d1 * d1 + d2 * d2)
log(d)	double	Return natural logarithm of d
log10(d)	double	Return logarithm (base 10) of d
pow(d1, d2)	double	Return d1 raised to the power d2
sin(d)	double	Return sine of d
sinh(d)	double	Return hyperbolic sine of d
sqrt(d)	double	Return square root of d
tan(d)	double	Return tangent of d
tanh(d)	double	Return hyperbolic tangent of d

Here, Type refers to the data type of the quantity that is returned by the function. Moreover, d, d1, etc. indicate arguments of type double.

A programme that makes use of the C math library would contain the statement

```
#include    <math.h>
```

close to its start. In the body of the programme, a statement like

```
x = cos(y);
```

would cause the variable x to be assigned a value which is the cosine of the value of the variable y (both x and y should be of type double).

Note that math library functions tend to be *extremely expensive* in terms of CPU time, and should, therefore, only be employed when absolutely necessary. The classic illustration of this point is the use of the pow() function. This function assumes that, in general, it will be called with a *fractional* power, and, therefore, implements a full-blown (and very expensive) series expansion. Clearly, it is not computationally efficient to use this function to square or cube a quantity. In other words, if a quantity needs to be raised to a small, positive integer power then this should be implemented *directly*, instead of using the pow() function: i.e., we should write x * x rather than pow(x, 2), and x * x * x rather than pow(x, 3), etc. (Of course, a properly designed exponentiation function would realise that it is more efficient to evaluate small positive integer powers by the direct method. Unfortunately, the pow() function was written by computer scientists!)

The C math library comes with a useful set of predefined mathematical constants:

Name	Description
M_PI	Pi, the ratio of a circle's circumference to its diameter.
M_PI_2	Pi divided by two.
M_PI_4	Pi divided by four.
M_1_PI	The reciprocal of pi (1/pi).
M_SQRT2	The square root of two.
M_SQRT1_2	The reciprocal of the square root of two (also the square root of 1/2).
M_E	The base of natural logarithms.

The other library functions commonly used in C programmes will be introduced, as appropriate, during the remainder of this discussion.

Data Input and Output

Data input and output operations in C are carried out by the standard input/output library (header file: stdio.h) via the functions scanf, printf, fscanf, and fprintf, which read and write data from/to the terminal, and from/to a data file, respectively. The additional functions fopen and fclose open and close, respectively, connections between a C programme and a data file. In the following, these functions are described in detail.

The scanf function reads data from standard input (usually, the terminal). A call to this function takes the general form

```
scanf(control_string,   arg1,   arg2,   arg3,   ...)
```

where control_string refers to a character string containing certain required formatting information, and arg1, arg2, etc., are arguments that represent the individual input data items.

The control string consists of individual groups of characters, with one character group for each data input item. In its simplest form, each character group consists of a per cent sign (%), followed by a set of *conversion characters* which indicate the type of the corresponding data item. The two most useful sets of conversion characters are as follows:

Character	***Type***
d	int
lf	double

The arguments are a set of variables whose types match the corresponding character groups in the control string. For reasons which will become apparent later on, *each variable name must be preceded by an ampersand* (&). Below is a typical application of the scanf function:

```
#include   <stdio.h>

. . .

int    k;

double   x,  y;

. . .

scanf("%d  %lf  %lf",  &k,  &x,  &y);

. . .
```

In this example, the scanf function reads an integer value and two floating-point values, from standard input, into the integer variable k and the two floating-point variables x and y, respectively.

The scanf function returns an integer equal to the number of data values successfully read from standard input, which can be fewer than expected, or even zero, in the event of a matching failure. The special value EOF (which on most systems corresponds to –1) is returned if an end-of-file is reached before any attempted conversion occurs. The following code snippet gives an example of how the scanf function can be checked for error-free input:

```
#include   <stdio.h>

. . .

int   check_input;

double   x,  y,  z;

. . .

check_input  =  scanf("%lf  %lf  %lf",  &x,  &y,  &z);

if  (check_input  <  3)

  {

    printf("Error  during  data  input\n");

    . . .

  }

. . .
```

The printf function writes data to standard output (usually, the terminal). A call to this function takes the general form

```
printf(control_string,   arg1,   arg2,   arg3,   ...)
```

where control_string refers to a character string containing formatting information, and arg1, arg2, etc., are arguments that represent the individual output data items.

The control string consists of individual groups of characters, with one character group for each output data item. In its simplest form, each character group consists of a per cent sign (%), followed by a *conversion character* which controls the format of the corresponding data item. The most useful conversion characters are as follows:

Character	**Meaning**
d	Display data item as signed decimal integer
f	Display data item as floating-point number without exponent
e	Display data item as floating-point number with exponent

The arguments are a set of variables whose types match the corresponding character groups in the control string (*i.e.*, type int for d format, and type double for f or e format). In contrast to the scanf function, the arguments are *not* preceded by ampersands. Below is a typical application of the scanf function:

```
#include   <stdio.h>

. . .

int   k  =  3;

double   x  =  5.4,  y  =  -9.81;

. . .

printf("%d  %f  %f\n",  k,  x,  y);

. . .
```

In this example, the programme outputs the values of the integer variable k and the floating-point variables x and y to the terminal. Executing the programme produces the following output:

```
3  5.400000  -9.810000

%
```

Note that the purpose of the escape sequence \n in the control string is to generate a line-feed after the three data items have been written to the terminal.

Of course, the printf function can also be used to write a simple text string to the terminal: *e.g.*,

```
printf(text_string)
```

Ordinary text can also be incorporated into the control string described above.

An example illustrating somewhat more advanced use of the printf function is given below:

```
#include    <stdio.h>

. . .

int   k  =  3;

double   x  =  5.4,  y  =  -9.81;

. . .

printf("k  =  %3d   x  +  y  =  %9.4f   x*y  =  %11.3e\n",  k,  x  +  y,  x*y);

. . .
```

Executing the programme produces the following output:

```
k  =    3   x  +  y  =    -4.4100   x*y  =   -5.297e+01
%
```

Note that the final two arguments of the printf function are arithmetic expressions. Note, also, the incorporation of explanatory text into the control string.

The character sequence %3d in the control string indicates that the associated data item should be output as a signed decimal integer occupying a field whose width is *at least* 3 characters. More generally, the character sequence %nd indicates that the associated data item should be output as a signed decimal integer occupying a field whose width is *at least* n characters. If the number of characters in the data item is less than n characters, then the data item is preceded by enough leading blanks to fill the specified field. On the other hand, if the data item exceeds the specified field width then additional space is allocated to the data item, such that the entire data item is displayed.

The character sequence %9.4f in the control string indicates that the associated data item should be output as a floating-point number, in non-scientific format, which occupies a field of at least 9 characters, and has 4 figures after the decimal point. More generally, the character sequence %n.mf indicates that the associated data item should be output as a floating-point number, in non-scientific format, which occupies a field of at least n characters, and has m figures after the decimal point.

Finally, the character sequence %11.3e in the control string indicates that the associated data item should be output as a floating-point number, in scientific format, which occupies a field of at least 11 characters, and has 3 figures after the decimal point. More generally, the character sequence %n.me indicates that the associated data item should be output as a floating-point number, in scientific format, which occupies a field of at least n characters, and has m figures after the decimal point.

The printf function returns an integer equal to the number of printed characters, or a negative value if there was an output error.

When working with a *data file*, the first step is to establish a *buffer area*, where information is temporarily stored whilst being transferred between the programme and the file. This buffer area allows information to be read or written to the data file in question more rapidly than would otherwise be possible. A buffer area is established by writing

```
FILE *stream;
```

where FILE (upper-case letters required) is a special structure type that establishes a buffer area, and stream is the identifier of the created buffer area. Note that a buffer area is often referred to as an input/output *stream*. The meaning of the asterisk (*) that precedes the identifier of the stream, in the above statement, will become clear later on. It is, of course, possible to establish multiple input/output streams (provided that their identifiers are distinct).

A data file must be opened and attached to a specific input/output stream before it can be created or processed. This operation is performed by the function fopen. A typical call to fopen takes the form

```
stream = fopen(file_name, file_type);
```

where stream is the identifier of the input/output stream to which the file is to be attached, and file_name and file_type are character strings that represent the name of the data file and the manner in which the data file will be utilised, respectively. The file_type string must be one of the strings listed below:

file_type	**Meaning**
"r"	Open existing file for reading only
"w"	Open new file for writing only (Any existing file will be overwritten)
"a"	Open existing file in append mode. (Output will be appended to the file)

The fopen function returns the integer value NULL (which on most systems corresponds to zero) in the event of an error.

A data file must also be closed at the end of the programme. This operation is performed by the function fclose. The syntax for a call to fclose is simply

```
fclose(stream);
```

where stream is the name of the input/output stream which is to be deattached from a data file. The fclose function returns the integer value 0 upon successful completion, otherwise it returns the special value EOF.

Data can be read from an open data file using the fscanf function, whose syntax is

```
fscanf(stream, control_string, arg1, arg2, arg3, ...)
```

Here, stream is the identifier of the input/output stream to which the file is attached, and the remaining arguments have exactly the same format and meaning as the corresponding arguments for the scanf function. The return values of fscanf are similar to those of the scanf function.

Likewise, data can be written to an open data file using the fprintf function, whose syntax is

```
fprintf(stream, control_string, arg1, arg2, arg3, ...)
```

Here, stream is the identifier of the input/output stream to which the file is attached, and the remaining arguments have exactly the same format and meaning as the corresponding arguments for the printf function. The return values of fprintf are similar to those of the printf function.

An example of a C programme which outputs data to the file "data.out" is given below:

```
#include   <stdio.h>

. . .

int   k  =  3;

double   x  =  5.4,  y  =  -9.81;

FILE  *output;

. . .

output  =  fopen("data.out",  "w");

if  (output  ==  NULL)

  {

    printf("Error  opening  file  data.out\n");

    . . .

  }

. . .

fprintf(output,  "k  =  %3d   x  +  y  =  %9.4f   x*y  =  %11.3e\n",  k,  x  +  y,
x*y);

. . .

fclose(output);

. . .
```

On execution, the above programme will write the line

```
k  =     3   x  +  y  =     -4.4100   x*y  =   -5.297e+01
```

to the data file "data.out".

Structure of a C Programme

The syntax of a complete C programme is given below:

```
. . .

int  main()

{

  . . .

  return  0;

}

. . .
```

The meaning of the statements int main() and return will become clear later on. Preprocessor statements (*e.g.*, #define and #include statements) are conventionally placed *before* the int main() statement. All executable statements must be placed *between* the int main() and return statements. Function definitions are conventionally placed *after* the return statement.

A simple C programme (quadratic.c) that calculates the real roots of a quadratic equation using the well-known quadratic formula is listed below.

```
/* quadratic.c */
/*
   Programme to evaluate real roots of quadratic equation
        2
   a x    + b x + c = 0
   using quadratic formula
                                              2
   x = ( -b +/- sqrt(b  - 4 a c) ) / (2 a)
*/
#include <stdio.h>
#include <math.h>
int main()
{
   double a, b, c, d, x1, x2;
   /* Read input data */
   printf("\na = ");
   scanf("%lf", &a);
   printf("b = ");
   scanf("%lf", &b);
   printf("c = ");
   scanf("%lf", &c);
   /* Perform calculation */
   d = sqrt(b * b - 4. * a * c);
   x1 = (-b + d) / (2. * a);
   x2 = (-b - d) / (2. * a);
   /* Display output */
   printf("\nx1 = %12.3e    x2 = %12.3e\n", x1, x2);
   return 0;
}
```

Note the use of *comments* (which are placed between the delimiters /* and */) to first explain the function of the programme and then identify the programme's major sections. Note, also, the use of *indentation* to highlight the executable statements. When executed, the above programme produces the following output:

```
a = 2

b = 4

c = 1

x1 =      -2.929e-01     x2 =      -1.707e+00

%
```

Of course, the 2, 4, and 1 were entered by the user in response to the programme's prompts.

It is important to realise that there is more to writing a complete computer programme than simply arranging the individual declarations and statements in the right order.

Attention should also be given to making the programme and its output as *readable* as possible, so that the programme's function is *immediately obvious* to other people. This can be achieved by judicious use of indentation and whitespace, as well as the inclusion of comments, and the generation of clearly labelled output.

It is hoped that this approach will be exemplified by the example programmes used in this course.

Control Statements

The C language includes a wide variety of powerful and flexible control statements. The most useful of these are described in the following.

The if-else statement is used to carry out a logical test and then take one of two possible actions, depending on whether the outcome of the test is true or false. The else portion of the statement is optional. Thus, the simplest possible if-else statement takes the form:

```
if (expression)  statement
```

The expression must be placed in parenthesis, as shown. In this form, the statement will only be executed if the expression has a non-zero value (*i.e.*, if expression if true). If the expression has a value of zero (*i.e.*, if expression is false) then the statement will be ignored. The statement can be either simple or compound.

The programme quadratic.c, is incapable of dealing correctly with cases where the roots are complex (*i.e.*, $b^2 < 4ac$), or cases where $a = 0$.

It is good programming practice to test for situations which fall outside the domain of validity of a programme, and produce some sort of error message when these occur.

An amended version of quadratic.c which uses if-else statements to reject invalid input data is listed below.

```
/* quadratic1.c */
/*
   Programme to evaluate real roots of quadratic equation
         2
   a x    + b x + c = 0
   using quadratic formula
                                       2
   x = ( -b +/- sqrt(b - 4 a c) ) / (2 a)
   Programme rejects cases where roots are complex
   or where a = 0.
*/
#include <stdio.h>
#include <math.h>
#include <stdlib.h>
int main()
{
   double a, b, c, d, e, x1, x2;
   /* Read input data */
   printf("\na = ");
   scanf("%lf", &a);
   printf("b = ");
   scanf("%lf", &b);
   printf("c = ");
   scanf("%lf", &c);
   /* Test for complex roots */
   e = b * b - 4. * a * c;
   if (e < 0.)
     {
      printf("\nError: roots are complex\n");
      exit(1);
     }
```

```
    /* Test for a = 0. */
    if (a == 0.)
      {
        printf("\nError: a = 0.\n");
        exit(1);
      }
    /* Perform calculation */
    d = sqrt(e);
    x1 = (-b + d) / (2. * a);
    x2 = (-b - d) / (2. * a);
    /* Display output */
    printf("\nx1 = %12.3e    x2 = %12.3e\n", x1, x2);
    return 0;
}
```

Note the use of indentation to highlight statements which are conditionally executed (*i.e.*, statements within an if-else statement). The standard library function call exit(1) (header file: stdlib.h) causes the programme to abort with an error status. Execution of the above programme for the case of complex roots yields the following output:

```
a = 4
b = 2
c = 6
Error: roots are complex
%
```

The general form of an if-else statement, which includes the else clause, is

```
if (expression)   statement 1   else   statement 2
```

If the expression has a non-zero value (*i.e.*, if expression is true) then statement1 is executed. Otherwise, statement2 is executed. The programme listed below is an extended version of the previous programme quadratic.c which is capable of dealing with complex roots.

```
/* quadratic2.c */
/*
    Programme to evaluate all roots of quadratic equation
          2
    a x   + b x + c = 0
```

```
   using quadratic formula
                                          2
   x = ( -b +/- sqrt(b - 4 a c) ) / (2 a)
   Programme rejects cases where a = 0.
*/
#include <stdio.h>
#include <math.h>
#include <stdlib.h>
int main()
{
   double a, b, c, d, e, x1, x2;
   /* Read input data */
   printf("\na = ");
   scanf("%lf", &a);
   printf("b = ");
   scanf("%lf", &b);
   printf("c = ");
   scanf("%lf", &c);
   /* Test for a = 0. */
   if (a == 0.)
    {
     printf("\nError: a = 0.\n");
     exit(1);
    }
   /* Perform calculation */
   e = b * b - 4. * a * c;
   if (e > 0.) // Test for real roots
    {
     /* Case of real roots */
     d = sqrt(e);
     x1 = (-b + d) / (2. * a);
     x2 = (-b - d) / (2. * a);
     printf("\nx1 = %12.3e    x2 = %12.3e\n", x1, x2);
    }
```

```
    else
      {
        /* Case of complex roots */
        d = sqrt(-e);
        x1 = -b / (2. * a);
        x2 = d / (2. * a);
        printf("\nx1 = (%12.3e, %12.3e)    x2 = (%12.3e, %12.3e)\n",
                x1, x2, x1, -x2);
      }
    return 0;
}
```

Note the use of an if-else statement to deal with the two alternative cases of real and complex roots. Note also that the C compiler ignores all characters on a line which occur after the // construct. Hence, this construct can be used to comment individual lines in a programme. The output from the above programme for the case of complex roots looks like:

```
a = 9
b = 2
c = 2
x1 = (   -1.111e-01,      4.581e-01)   x2 = (   -1.111e-01,    -4.581e-01)
%
```

The while statement is used to carry out looping operations, in which a group of statements is executed repeatedly until some condition is satisfied. The general form of a while statement is

```
while  (expression)  statement
```

The statement is executed repeatedly, as long as the expression is non-zero (*i.e.*, as long as expression is true). The statement can be either simple or compound. Of course, the statement must include some feature that eventually alters the value of the expression, thus providing an escape mechanism from the loop.

The programme listed below (iteration.c) uses a while statement to solve an algebraic equation via iteration, as explained in the initial comments.

```
/* iteration.c */
/*
    Programme to solve algebraic equation
     5        2
    x  + a x   - b = 0
```

```
    by iteration. Easily shown that equation must have at least
    one real root. Coefficients a and b are supplied by user.
    Iteration scheme is as follows:
                              2   0.2
    x      =   ( b - a x    )
     n+1                     n
   where x_n is nth iteration. User must supply initial guess for x.
    Iteration continues until relative change in x per iteration is
   less than eps (user supplied) or until number of iterations exceeds
    NITER. Programme aborts if (b - a x*x) becomes negative.
*/
#include <stdio.h>
#include <math.h>
#include <stdlib.h>
/* Set max. allowable no. of iterations */
#define NITER 30
int main()
{
   double a, b, eps, x, x0, dx = 1., d;
   int count = 0;
   /* Read input data */
   printf("\na = ");
   scanf("%lf", &a);
   printf("b = ");
   scanf("%lf", &b);
   printf("eps = ");
   scanf("%lf", &eps);
   /* Read initial guess for x */
   printf("\nInitial guess for x = ");
   scanf("%lf", &x);
   x0 = x;
   while (dx > eps)   // Start iteration loop: test for convergence
     {
```

```
      /* Check for too many iterations */
      ++count;
      if (count > NITER)
        {
         printf("\nError: no convergence\n");
         exit(1);
        }
      /* Reject complex roots */
      d = b - a * x * x;
      if (d < 0.)
        {
         printf("Error: complex roots - try another initial guess\n");
         exit(1);
        }
      /* Perform iteration */
      x = pow(d, 0.2);
      dx = fabs( (x - x0) / x );
      x0 = x;
      /* Output data on iteration */
     printf("Iter = %3d    x = %8.4f    dx = %12.3e\n", count, x, dx);
    }
  return 0;
}
```

The typical output from the above programme looks like:

```
a = 3
b = 10
eps = 1.e-6
Initial guess for x = 1
Iter =    1    x =    1.4758    dx =    3.224e-01
Iter =    2    x =    1.2823    dx =    1.509e-01
Iter =    3    x =    1.3834    dx =    7.314e-02
Iter =    4    x =    1.3361    dx =    3.541e-02
Iter =    5    x =    1.3595    dx =    1.720e-02
```

```
Iter  =   6   x  =  1.3483   dx  =  8.350e-03
Iter  =   7   x  =  1.3537   dx  =  4.056e-03
Iter  =   8   x  =  1.3511   dx  =  1.969e-03
Iter  =   9   x  =  1.3524   dx  =  9.564e-04
Iter  =  10   x  =  1.3518   dx  =  4.644e-04
Iter  =  11   x  =  1.3521   dx  =  2.255e-04
Iter  =  12   x  =  1.3519   dx  =  1.095e-04
Iter  =  13   x  =  1.3520   dx  =  5.318e-05
Iter  =  14   x  =  1.3519   dx  =  2.583e-05
Iter  =  15   x  =  1.3520   dx  =  1.254e-05
Iter  =  16   x  =  1.3520   dx  =  6.091e-06
Iter  =  17   x  =  1.3520   dx  =  2.958e-06
Iter  =  18   x  =  1.3520   dx  =  1.436e-06
Iter  =  19   x  =  1.3520   dx  =  6.975e-07
%
```

When a loop is constructed using a while statement, the test for the continuation of the loop is carried out at the *beginning* of each pass. Sometimes, however, it is desirable to have a loop where the test for continuation takes place at the *end* of each pass. This can be accomplished by means of a do-while statement. The general form of a do-while statement is

```
do   statement   while   (expression);
```

The statement is executed repeatedly, as long as the expression is true. Note, however, that the statement is always executed at least once, since the test for repetition does not take place until the end of the first pass through the loop. The statement can be either simple or compound, and should, of course, include some feature that eventually alters the value of the expression.

The programme listed below is a marginally improved version of the previous programme (iteration.c) which uses a do-while loop to test for convergence at the end (as opposed to the beginning) of each iteration loop.

```
/* iteration1.c */
/*
   Programme to solve algebraic equation
    5        2
   x  + a x   - b = 0
   by iteration. Easily shown that equation must have at least
   one real root. Coefficients a and b are supplied by user.
```

```
    Iteration scheme is as follows:
                                     2   0.2

    x       =    ( b  -  a  x    )

      n+1                        n

   where x_n is nth iteration. User must supply initial guess for x.
    Iteration continues until relative change in x per iteration is
   less than eps (user supplied) or until number of iterations exceeds
    NITER. Programme aborts if (b - a x*x) becomes negative.
*/
#include <stdio.h>
#include <math.h>
#include <stdlib.h>
/* Set max. allowable no. of iterations */
#define NITER 30
int main()
{
   double a, b, eps, x, x0, dx, d;
   int count = 0;

   /* Read input data */
   printf("\na = ");
   scanf("%lf", &a);
   printf("b = ");
   scanf("%lf", &b);
   printf("eps = ");
   scanf("%lf", &eps);
   /* Read initial guess for x */
   printf("\nInitial guess for x = ");
   scanf("%lf", &x);
   x0 = x;
   do    // Start iteration loop
     {
       /* Check for too many iterations */
```

```
        ++count;
        if (count > NITER)
          {
           printf("\nError: no convergence\n");
           exit(1);
          }
        /* Reject complex roots */
        d = b - a * x * x;
        if (d < 0.)
          {
          printf("Error: complex roots - try another initial guess\n");
           exit(1);
          }
        /* Perform iteration */
        x = pow(d, 0.2);
        dx = fabs( (x - x0) / x );
        x0 = x;
        /* Output data on iteration */
      printf("Iter = %3d    x = %8.4f    dx = %12.3e\n", count, x, dx);
     } while (dx > eps);   // Test for convergence
   return 0;
}
```

The output from the above programme is essentially identical to that from the programme iteration.c.

The while and do-while statements are particularly well suited to looping situations in which the number of passes through the loop *is not* known in advance. Conversely, situations in which the number of passes through the loop *is* known in advance are often best dealt with using a for statement. The general form of a for statement is

```
for (expression 1; expression 2; expression 3) statement
```

where expression1 is used to initialise some parameter (called an *index*) that controls the looping action, expression2 represents a condition that must be true for the loop to continue execution, and expression3 is used to alter the value of the parameter initially assigned by expression1. When a for statement is executed, expression2 is evaluated and tested at the *beginning* of each pass through the loop, whereas expression3 is evaluated at the *end* of each pass.

The programme listed below uses a for statement to evaluate the factorial of a non-negative integer.

```
/* factorial.c */
/*
   Programme to evaluate factorial of non-negative
   integer n supplied by user.
*/
#include <stdio.h>
#include <stdlib.h>
int main()
{
  int n, count;
  double fact = 1.;
  /* Read in value of n */
  printf("\nn = ");
  scanf("%d", &n);
  /* Reject negative value of n */
  if (n < 0)
    {
     printf("\nError: factorial of negative integer not defined\n");
     exit(1);
    }
  /* Calculate factorial */
  for (count = n; count > 0; --count) fact *= (double) count;
  /* Output result */
  printf("\nn = %5d      Factorial(n) = %12.3e\n", n, fact);
  return 0;
}
```

```
The typical output from the above programme is shown below:
n = 6
n =     6       Factorial(n) =      7.200e+02
%
```

The statements which occur within if-else, while, do-while, or for statements can themselves be control statements, giving rise to the possibility of nested if-else statements, conditionally executed loops, nested loops, etc. When dealing with nested control statements, it is vital to adhere religiously to the syntax rules described above, in order to avoid confusion.

Functions

We have seen that C supports the use of predefined library functions which are used to carry out a large number of commonly occurring tasks. However, C also allows programmers to define their own functions. The use of programmer-defined functions permits a large programme to be broken down into a number of smaller, self-contained units. In other words, a C programme can be *modularised* via the sensible use of programmer-defined functions. In general, modular programmes are far easier to write and debug than monolithic programmes. Furthermore, proper modularisation allows other people to grasp the logical structure of a programme with far greater ease than would otherwise be the case.

A *function* is a self-contained programme segment that carries out some specific, well-defined task. Every C programme consists of one or more functions. One of these functions must be called main. Execution of the programme always begins by carrying out the instructions contained in main. Note that if a programme contains multiple functions then their definitions may appear in *any order*. The same function can be accessed from several different places within a programme. Once the function has carried out its intended action, control is returned to the point from which the function was accessed. Generally speaking, a function processes information passed to it from the calling portion of the programme, and returns a *single value*. Some functions, however, accept information but do not return anything.

A function definition has two principal components: the *first line* (including the *argument declarations*), and the so-called *body* of the function.

The first line of a function takes the general form

```
data-type   name(type 1   arg 1,   type 2   arg 2,   ...,   type n   arg n)
```

where data-type represents the data type of the item that is returned by the function, name represents the name of the function, and type 1, type 2, ..., type n represent the data types of the arguments arg 1, arg 2, ..., arg n. The allowable data types for a function are:

```
int       for a function which returns an integer value
double    for a function which returns an floating-point value
void      for a function which does not return any value
```

The allowable data types for a function's arguments are int and double. Note that the identifiers used to reference the arguments of a function are *local*, in the sense that they are not recognised outside of the function. Thus, the argument names in a function definition *need not* be the same as those used in the segments of the programme from which the function was called. However, the corresponding data types of the arguments must always match.

The body of a function is a compound statement that defines the action to be taken by the function. Like a regular compound statement, the body can contain expression statements, control statements, other compound statements, etc. The body can even access other functions. In fact, it can even access itself—this process is known as *recursion*. In addition, however, the body must include one or more return statements in order to return a value to the calling portion of the programme.

A return statement causes the programme logic to return to the point in the programme from which the function was accessed. The general form of a return statement is:

```
return   expression;
```

This statement causes the value of expression to be returned to the calling part of the programme. Of course, the data type of expression should match the declared data type of the function. For a void function, which does not return any value, the appropriate return statement is simply:

```
return;
```

A maximum of *one* expression can be included in a return statement. Thus, a function can return a maximum of one value to the calling part of the programme. However, a function definition can include multiple return statements, each containing a different expression, which are conditionally executed, depending on the programme logic.

Note that, by convention, the main function is of type int and returns the integer value 0 to the operating system, indicating the error-free termination of the programme. In its simplest form, the main function possesses *no* arguments. The library function call exit(1), causes the execution of a programme to abort, returning the integer value 1 to the operating system, which (by convention) indicates that the programme terminated with an error status.

The programme segment listed below shows how the previous programme factorial.c can be converted into a function factorial(n) which returns the factorial (in the form of a floating-point number) of the non-negative integer n:

```
double factorial(int n)
{
  /*
      Function to evaluate factorial (in floating-point form)
      of non-negative integer n.
  */
  int count;
  double fact = 1.;
  /* Abort if n is negative integer */
  if (n < 0)
    {
     printf("\nError: factorial of negative integer not defined\n");
     exit(1);
    }
  /* Calculate factorial */
```

```
   for (count = n; count > 0; --count) fact *= (double) count;
   /* Return value of factorial */
   return fact;
}
```

A function can be accessed, or *called*, by specifying its name, followed by a list of arguments enclosed in parentheses and separated by commas. If the function call does not require any arguments then an empty pair of parentheses must follow the name of the function. The function call may be part of a simple expression, such as an assignment statement, or it may be one of the operands within a more complex expression. The arguments appearing in a function call may be expressed as constants, single variables, or more complex expressions. However, both the number and the types of the arguments must match those in the function definition.

The programme listed below (printfact.c) uses the function factorial(), described above, to print out the factorials of all the integers between 0 and 20:

```
/* printfact.c */
/*
   Programme to print factorials of all integers
   between 0 and 20
*/
#include <stdio.h>
#include <stdlib.h>
//%%%%%%%%%%%%%%%%%%%%%%%%%%%%%%%%%%%%%%%%%%%%%%%%%%%%%%%%%%%%%%%%%
double factorial(int n)
{
   /*
        Function to evaluate factorial (in floating-point form)
        of non-negative integer n.
   */
   int count;
   double fact = 1.;
   /* Abort if n is negative integer */
   if (n < 0)
     {
      printf("\nError: factorial of negative integer not defined\n");
      exit(1);
     }
```

```
    /* Calculate factorial */
    for (count = n; count > 0; —count) fact *= (double) count;
    /* Return value of factorial */
    return fact;
}
//%%%%%%%%%%%%%%%%%%%%%%%%%%%%%%%%%%%%%%%%%%%%%%%%%%%%%%%%%%%%%%
int main()
{
    int j;
    /* Print factorials of all integers between 0 and 20 */
    for (j = 0; j <= 20; ++j)
      printf("j = %3d     factorial(j) = %12.3e\n", j, factorial(j));
    return 0;
}
```

Note that the call to factorial() takes place inside a complex expression (*i.e.*, a printf() function call). Note also that the argument of factorial() has a different name (but the same data type) in the two sections of the programme (*i.e.*, in the main() function and the factorial() function). The output from the above programme looks like:

```
j =   0     factorial(j) =    1.000e+00
j =   1     factorial(j) =    1.000e+00
j =   2     factorial(j) =    2.000e+00
j =   3     factorial(j) =    6.000e+00
j =   4     factorial(j) =    2.400e+01
j =   5     factorial(j) =    1.200e+02
j =   6     factorial(j) =    7.200e+02
j =   7     factorial(j) =    5.040e+03
j =   8     factorial(j) =    4.032e+04
j =   9     factorial(j) =    3.629e+05
j =  10     factorial(j) =    3.629e+06
j =  11     factorial(j) =    3.992e+07
j =  12     factorial(j) =    4.790e+08
j =  13     factorial(j) =    6.227e+09
j =  14     factorial(j) =    8.718e+10
j =  15     factorial(j) =    1.308e+12
```

```
j  =  16     factorial(j)  =    2.092e+13
j  =  17     factorial(j)  =    3.557e+14
j  =  18     factorial(j)  =    6.402e+15
j  =  19     factorial(j)  =    1.216e+17
j  =  20     factorial(j)  =    2.433e+18
%
```

Ideally, function definitions should always *precede* the corresponding function calls in a C programme. This requirement can usually be satisfied by judicious ordering of the various functions which make up a programme, but, almost inevitably, restricts the location of the main() function to the *end* of the programme. Hence, if the order of the two functions in the above programme [*i.e.*, factorial() and main()] were *swapped* then an error message would be generated on compilation, since an attempt would be made to call factorial() prior to its definition. Unfortunately, for the sake of logical clarity, most C programmers prefer to place the main() function at the *beginning* of their programmes. After all, main() is always the first part of a programme to be executed. In such situations, function calls [within main()] are bound to precede the corresponding function definitions: fortunately, however, compilation errors can be avoided by using a construct known as a *function prototype*.

Function prototypes are conventionally placed at the beginning of a programme (*i.e.*, before the main() function) and are used to inform the compiler of the name, data type, and number and data types of the arguments, of all user-defined functions employed in the programme. The general form of a function prototype is

```
data-type   name(type 1,   type 2,   ...,   type n);
```

where data-type represents the data type of the item returned by the referenced function, name is the name of the function, and type 1, type 2,..., type n are the data types of the arguments of the function. Note that it is not necessary to specify the names of the arguments in a function prototype. Incidentally, the function prototypes for predefined library functions are contained within the associated header files which must be included at the beginning of every programme which uses these functions.

The programme listed below is a modified version of printfact.c in which the main() function is the first function to be defined:

```
/* printfact1.c */
/*
   Programme to print factorials of all integers
   between 0 and 20
*/
#include <stdio.h>
#include <stdlib.h>
/* Prototype for function factorial() */
```

```
double factorial(int);
int main()
{
   int j;
   /* Print factorials of all integers between 0 and 20 */
   for (j = 0; j <= 20; ++j)
     printf("j = %3d      factorial(j) = %12.3e\n", j, factorial(j));
   return 0;
}
//%%%%%%%%%%%%%%%%%%%%%%%%%%%%%%%%%%%%%%%%%%%%%%%%%%%%%%%%%%%%
double factorial(int n)
{
   /*
      Function to evaluate factorial (in floating-point form)
      of non-negative integer n.
   */
   int count;
   double fact = 1.;
   /* Abort if n is negative integer */
   if (n < 0)
     {
      printf("\nError: factorial of negative integer not defined\n");
      exit(1);
     }
   /* Calculate factorial */
   for (count = n; count > 0; —count) fact *= (double) count;
   /* Return value of factorial */
   return fact;
}
```

Note the presence of the function prototype for factorial() prior to the definition of main(). This is needed because the programme calls factorial() before this function has been defined. The output from the above programme is identical to that from printfact.c.

It is generally considered to be good programming practice to provide function prototypes for *all* user-defined functions accessed in a programme, whether or not they are strictly required by the compiler. The reason for this is fairly simple. If we provide a prototype for a given function then the compiler can carefully compare each use of the function, within the programme, with this prototype so as to determine whether or not we are calling the function properly. In the absence of a prototype, an incorrect call to a function (*e.g.*, using the wrong number of arguments, or arguments of the wrong data type) can give rise to run-time errors which are difficult to diagnose.

When a single value is passed to a function as an argument then the value of that argument is simply *copied* to the function. Thus, the argument's value can subsequently be altered within the function but this *will not* affect its value in the calling routine. This procedure for passing the value of an argument to a function is called *passing by value.*

Passing an argument by value has advantages and disadvantages. The advantages are that it allows a single-valued argument to be written as an expression, rather than being restricted to a single variable. Furthermore, in cases where the argument is a variable, the value of this variable is *protected* from alterations which take place within the function. The main disadvantage is that information cannot be transferred back to the calling portion of the programme via arguments. In other words, passing by value is a strictly *one-way* method of transferring information. The programme listed below, which is another modified version of printfact.c, illustrates this point:

```
/* printfact2.c */
/*
   Programme to print factorials of all integers
   between 0 and 20
*/
#include <stdio.h>
#include <stdlib.h>
/* Prototype for function factorial() */
double factorial(int);
int main()
{
   int j;
   /* Print factorials of all integers between 0 and 20 */
   for (j = 0; j <= 20; ++j)
      printf("j = %3d     factorial(j) = %12.3e\n", j, factorial(j));
   return 0;
}
//%%%%%%%%%%%%%%%%%%%%%%%%%%%%%%%%%%%%%%%%%%%%%%%%%%%%%%%%%%%%
double factorial(int n)
{
   /*
```

```
    Function to evaluate factorial (in floating-point form)
    of non-negative integer n.
  */
  double fact = 1.;
  /* Abort if n is negative integer */
  if (n < 0)
    {
     printf("\nError: factorial of negative integer not defined\n");
     exit(1);
    }
  /* Calculate factorial */
  for (; n > 0; --n) fact *= (double) n;
  /* Return value of factorial */
  return fact;
}
```

Note that the function factorial() has been modified such that its integer argument n is also used as the index in a for statement. Thus, the value of this argument is *modified* within the function. Nevertheless, the output of the above programme is identical to that from printfact.c, since the modifications to n are *not* passed back to the main part of the programme. Note, incidentally, the use of a null initialisation expression in the for statement appearing in factorial().

Pointers

One of the main characteristics of a scientific programme is that *large amounts* of numerical information are exchanged between the various functions which make up the programme. It is generally most convenient to pass this information via the *argument lists*, rather than the *names*, of the these functions. After all, only one number can be passed via a function name, whereas scientific programmes generally require far more than one number to be passed during a function call. Hence, the functions employed in scientific programmes generally return no values via their names (*i.e.*, they tend to be of data type void) but possess large strings of arguments. There is one obvious problem with this approach. Namely, a void function which passes all of its arguments by value is incapable of returning any information to the programme segment from which it was called. Fortunately, there is a way of getting around this difficulty: we can pass the arguments of a function by *reference*, rather than by *value*, using constructs known as *pointers*. This allows the *two-way* communication of information via arguments during function calls. Pointers are discussed in the following.

Suppose that v is a variable in a C programme which represents some particular data item. Of course, the programme stores this data item at some particular location in the computer's memory. The data item can thus be accessed if we know its location, or *address*, in computer memory. The address of v's memory location is determined by the expression &v, where & is a unary operator known as the *address operator*.

Suppose that we assign the address of v to another variable pv. In other words,

```
pv = &v
```

This new variable is called a *pointer* to v, since it points to the location where v is stored in memory. Remember, however, that pv represents v's address, and *not* its value.

The data item represented by v (*i.e.*, the data item stored at v's memory location) can be accessed via the expression *pv, where * is a unary operator, called the *indirection operator*, which only operates on pointer variables. Thus, *pv and v both represent the same data item. Furthermore, if we write pv = &v and u = *pv then both u and v represent the same value.

The simple programme listed below illustrates some of the points made above:

```
/* pointer.c */
/*
    Simple illustration of the action of pointers
*/
#include <stdio.h>
main()
{
   int u = 5;
   int v;
   int *pu;          // Declare pointer to an integer variable
   int *pv;          // Declare pointer to an integer variable
   pu = &u;          // Assign address of u to pu
   v = *pu;          // Assign value of u to v
   pv = &v;          // Assign address of v to pv
   printf("\nu = %d  &u = %X  pu = %X  *pu = %d", u, &u, pu, *pu);
   printf("\nv = %d  &v = %X  pv = %X  *pv = %d\n", v, &v, pv, *pv);
   return 0;
}
```

Note that pu is a pointer to u, whereas pv is a pointer to v. Incidentally, the conversion character X, which appears in the control strings of the above printf() function calls, indicates that the associated data item should be output as a *hexadecimal number*—this is the conventional method of representing an address in computer memory. Execution of the above programme yields the following output:

```
u  =  5    &u  =  BFFFFA24    pu  =  BFFFFA24    *pu  =  5
v  =  5    &v  =  BFFFFA20    pv  =  BFFFFA20    *pv  =  5
%
```

In the first line, we see that u represents the value 5, as specified in its declaration statement. The address of u is determined automatically by the compiler to be BFFFFA24 (hexadecimal). The pointer pu is assigned this value. Finally, the value to which pu points is 5, as expected. Similarly, the second line shows that v also represents the value 5. This is as expected, since we assigned the value *pu to v. The address of v is BFFFFA20. Of course, u and v have different addresses.

The unary operators & and * are members of the same precedence group as the other unary operators (*e.g.*, ++ and —). The address operator (&) can only act upon operands which possess a unique address, such as ordinary variables. Thus, the address operator *cannot* act upon arithmetic expression, such as 2 * (u + v). The indirection operator (*) can only act upon operands which are pointers.

Pointer variables, like all other variables, must be declared before they can appear in executable statements. A pointer declaration takes the general form

```
data-type    *ptvar;
```

where ptvar is the name of the pointer variable, and data-type is the data type of the data item towards which the pointer points. Note that an asterisk must always precede the name of a pointer variable in a pointer declaration.

We can now appreciate that the mysterious asterisk which appears in the declaration of an input/output stream, *e.g.*,

```
FILE    *stream;
```

is there because stream is a pointer variable (pointing towards an object of the special data type FILE). In fact, stream points towards the beginning of the associated input/output stream in memory.

Pointers are often passed to a function as arguments. This allows data items within the calling part of the programme to be accessed by the function, altered within the function, and then passed back to the calling portion of the programme in altered form. This use of pointers is referred to as passing arguments by *reference,* rather than by value.

When an argument is passed by value, the associated data item is simply copied to the function. Thus, any alteration to the data item within the function is not passed back to the calling routine. When an argument is passed by reference, however, the *address* of the associated data item is passed to the function. The contents of this address can be freely accessed by both the function and the calling routine. Furthermore, any changes made to the data item stored at this address are recognised by both the function and the calling routine. Thus, the use of a pointer as an argument allows the *two-way* communication of information between a function and its calling routine.

The programme listed below, which is yet another modified version of printfact.c, uses a pointer to pass back information from a function to its calling routine:

```
/* printfact3.c */
/*
   Programme to print factorials of all integers
   between 0 and 20
*/
#include <stdio.h>
#include <stdlib.h>
/* Prototype for function factorial() */
void factorial(int, double *);
int main()
{
   int j;
   double fact;
   /* Print factorials of all integers between 0 and 20 */
   for (j = 0; j <= 20; ++j)
    {
     factorial(j, &fact);
     printf("j = %3d      factorial(j) = %12.3e\n", j, fact);
    }
   return 0;
}
//%%%%%%%%%%%%%%%%%%%%%%%%%%%%%%%%%%%%%%%%%%%%%%%%%%%%%%%%%%%%%%%%
void factorial(int n, double *fact)
{
   /*
      Function to evaluate factorial *fact (in floating-point form)
      of non-negative integer n.
   */
   *fact = 1.;
   /* Abort if n is negative integer */
   if (n < 0)
    {
     printf("\nError: factorial of negative integer not defined\n");
```

```
        exit(1);
      }

    /* Calculate factorial */
    for (; n > 0; —n) *fact *= (double) n;
    return;
}
```

The output from this programme is again identical to that from printfact.c. Note that the function factorial() has been modified such that there is no data item associated with its name (*i.e.*, the function is of data type void). However, the argument list of this function has been extended such that there are now *two* arguments. As before, the first argument is the value of the positive integer n whose factorial is to be evaluated by the function. The second argument, fact, is a *pointer* which passes back the factorial of n (in the form of a floating-point number) to the main part of the programme. Incidentally, the compiler knows that fact is a pointer because its name is proceeded by an *asterisk* in the argument declaration for factorial(). Of course, in the body of the function, reference is made to *fact (*i.e.*, the value of the data item stored in the memory location towards which fact points) rather than fact (*i.e.*, the address of the memory location towards which fact points). Note that a void function, which returns no value, can only be called via a statement consisting of the function name followed by a list of its arguments (in parentheses and separated by commas). Thus, the function factorial() is called in the main part of the programme via the statement

```
factorial(j, &fact);
```

This statement passes the integer value j to factorial(), which, in turn, passes back the value of the factorial of j via its second argument. Note that since the second argument is passed by reference, rather than by value, it is written &fact (*i.e.*, the address of the memory location where the floating-point value fact is stored) rather than fact (*i.e.*, the value of the floating-point variable fact). Note, finally, that the function prototype for factorial() takes the form

```
void factorial(int, double *);
```

Here, the *asterisk* after double indicates that the second argument is a *pointer* to a floating-point data item.

We can now appreciate that the mysterious ampersands which must precede variable names in scanf() calls: *e.g.*,

```
scanf("%d %lf %lf", &k, &x, &y);
```

are not so mysterious, after all. scanf() is a function which returns data to its calling routine via its arguments (excluding its first argument, which is a control string). Hence, these arguments must be passed to scanf() by *reference*, rather than by *value*, otherwise they would be unable to pass information back to the calling routine. It follows that we must pass the *addresses* of variables (*e.g.*, &k) to scanf(), rather than the *values* of these variables (*e.g.*, k). Note that since the printf() function does not return any information to its calling routine via its arguments, there is no need to pass these arguments by reference—passing by value is fine. This explains why there are no ampersands in the argument list of a printf() function.

A pointer to a function can be passed to another function as an argument. This allows one function to be transferred to another, as though the first function were a variable. This is very useful in scientific programming. Imagine that we have a routine which numerically integrates a general one-dimensional function. Ideally, we would like to use this routine to integrate more than one specific function. We can achieve this by passing (to the routine) the name of the function to be integrated as an argument. Thus, for example, we can use the *same* routine to integrate a polynomial, a trigonometric function, or a logarithmic function.

Let us refer to the function whose name is passed as an argument as the *guest function*. Likewise, the function to which this name is passed is called the *host function*. A pointer to a guest function is identified in the host function definition by an entry of the form

```
data-type   (*function-name)(type 1,   type 2,  ...)
```

in the host function's argument declaration. Here, data-type is the data type of the guest function, function-name is the local name of the guest function in the host function definition, and type 1, type 2, ... are the data types of the guest function's arguments. The pointer to the guest function also requires an entry of the form

```
data-type   (*)(type 1,   type 2,   ...)
```

in the argument declaration of the host function's prototype. The guest function can be accessed within the host function definition by means of the indirection operator. To achieve this, the indirection operator must precede the guest function name, and both the indirection operator and the guest function name must be enclosed in parenthesis: i.e.,

```
(*function-name)(arg 1,   arg 2,   ...)
```

Here, arg 1, arg 2,... are the arguments passed to the guest function. Finally, the name of a guest function is passed to the host function, during a call to the latter function, via an entry like

```
function-name
```

in the host function's argument list.

The programme listed below is a rather silly example which illustrates the passing of function names as arguments to another function:

```
/* passfunction.c */

/*

  Programme to illustrate the passing of function names as

  arguments to other functions via pointers

*/

#include <stdio.h>

/* Function prototype for host fun. */

void cube(double (*)(double), double, double *);
```

```
double fun1(double);     // Function prototype for first guest function
double fun2(double);    // Function prototype for second guest function
int main()
{
   double x, res1, res2;
   /* Input value of x */
   printf("\nx = ");
   scanf("%lf", &x);
   /* Evaluate cube of value of first guest function at x */
   cube(fun1, x, &res1);
   /* Evaluate cube of value of second guest function at x */
   cube(fun2, x, &res2);
   /* Output results */
  printf("\nx = %8.4f   res1 = %8.4f   res2 = %8.4f\n", x, res1, res2);
   return 0;
}
//%%%%%%%%%%%%%%%%%%%%%%%%%%%%%%%%%%%%%%%%%%%%%%%%%%%%%%%%%%%%
void cube(double (*fun)(double), double x, double *result)
{
   /*
      Host function: accepts name of floating-point guest function
      with single floating-point argument as its first argument,
     evaluates this function at x (the value of its second argument),
     cubes the result, and returns final result via its third argument.
   */
   double y;
   y = (*fun)(x);              // Evaluate guest function at x
   *result = y * y * y;   // Cube value of guest function at x
   return;
}
//%%%%%%%%%%%%%%%%%%%%%%%%%%%%%%%%%%%%%%%%%%%%%%%%%%%%%%%%%%%%
double fun1(double z)
{
```

```
    /*
       First guest function
    */
    return 3.0 * z * z - z;
}
//%%%%%%%%%%%%%%%%%%%%%%%%%%%%%%%%%%%%%%%%%%%%%%%%%%%%%%%%%%%%%%%%
double fun2(double z)
{
    /*
       Second guest function
    */
    return 4.0 * z  - 5.0 * z * z * z;
}
```

In the above programme, the function cube() accepts the name of a guest function (with one argument) as its first argument, evaluates this function at x (the value of its second argument, which is ultimately specified by the user), cubes the result, and then passes the final result back to the main part of the programme via its third argument (which, of course, is a pointer). The two guest functions, fun1() and fun2(), whose names are passed to cube(), are both simple polynomials. The output from the above programme looks like:

```
x = 2
x =     2.0000     res1 = 1000.0000     res2 = -32768.0000
%
```

Global Variables

We have seen that a general C programme is basically a collection of functions. Furthermore, the variables used by these functions are *local* in scope: i.e., a variable defined in one function is not recognised in another. The main method of transferring data from one function to another is via the argument lists in function calls. Arguments can be passed in one of two different manners. When an argument is passed by *value* then the value of a local variable (or expression) in the calling routine is copied to a local variable in the function which is called. When an argument is passed by *reference* then a local variable in the calling routine shares the same memory location as a local variable in the function which is called: hence, a change in one variable is automatically mirrored in the other. However, there is a *third* method of transferring information from one function to another. It is possible to define variables which are *global* in extent: such variables are recognised by *all* the functions making up the programme, and have the *same* values in all of these functions.

The C compiler recognises a variable as global, as opposed to local, because its declaration is located *outside* the scope of any of the functions making up the programme. Of course, a global variable can only be used in an executable statement *after* it has been declared. Hence, the natural place to put global variable declaration statements is *before* any function definitions: i.e., right at the beginning of the programme. Global variables declarations can be used to initialise such variables, in the regular manner. However, the initial values must be expressed as constants, rather than expressions. Furthermore, the initial values are only assigned once, at the beginning of the programme.

The programme listed below is yet another version of printfact.c, in which communication between the two sections of the programme takes place entirely via global variables:

```
/* printfact4.c */
/*
   Programme to print factorials of all integers
   between 0 and 20
*/
#include <stdio.h>
#include <stdlib.h>
/* Prototype for function factorial() */
void factorial();
/* Global variable declarations */
int j;
double fact;
int main()
{
  /* Print factorials of all integers between 0 and 20 */
  for (j = 0; j <= 20; ++j)
   {
    factorial();
    printf("j = %3d      factorial(j) = %12.3e\n", j, fact);
   }
  return 0;
}
//%%%%%%%%%%%%%%%%%%%%%%%%%%%%%%%%%%%%%%%%%%%%%%%%%%%%%%%%%
void factorial()
{
  /*
     Function to evaluate factorial (in floating-point form)
     of non-negative integer j. Result stored in variable fact.
  */
```

```
    int count;
    /* Abort if j is negative integer */
    if (j < 0)
      {
        printf("\nError: factorial of negative integer not defined\n");
        exit(1);
      }
    /* Calculate factorial */
   for (count = j, fact = 1.; count > 0; —count) fact *= (double) count;
    return;
}
```

The output from the above programme is identical to that from printfact.c. Observe that the global variable j is used to pass the integer value whose factorial is to be calculated from the main part of the programme to the function factorial(), whilst the global variable fact is used to pass the calculated factorial back to the main part of the programme. Incidentally, note the use of multiple initialisation statements (separated by commas) in the for statement appearing in factorial().

Global variables should be used *sparingly* in scientific programmes (or any other type of programme), since there are inherent dangers in their employment. An alteration in the value of a global variable within a given function is carried over into all other parts of the programme. Unfortunately, such an alteration can sometimes happen inadvertently, as the side-effect of some other action. Thus, there is the possibility of the value of a global variable changing unexpectedly, resulting in a subtle programming error which can be *extremely difficult* to track down, since the offending line could be located *anywhere* in the programme. Similar errors involving local variables are much easier to debug, since the scope of local variables is far more limited than that of global variables.

Arrays

Scientific programmes very often deal with multiple data items possessing common characteristics. In such cases, it is often convenient to place the data items in question into an *array*, so that they all share a common name (*e.g.*, x). The individual data items can be either integers or floating-point numbers. However, they all must be of the *same* data type.

In C, an element of an array (*i.e.*, an individual data item) is referred to by specifying the array name followed by one or more *subscripts*, with each subscript enclosed in square brackets. All subscripts must be non-negative integers. Thus, in an n-element array called x, the array elements are x[0], x[1], ..., x[n-1]. Note that the first element of the array is x[0] and *not* x[1], as in other programming languages.

The number of subscripts determines the *dimensionality* of an array. For example, x[i] refers to an element of a one-dimensional array, x. Similarly, y[i][j] refers to an element of a two-dimensional array, y, etc.

Arrays are declared in much the same manner as ordinary variables, except that each array name must be accompanied by a size specification (which specifies the number of elements). For a one-dimensional array, the size is specified by a positive integer constant, enclosed in square brackets. The generalisation for multi-dimensional arrays is fairly obvious. Several valid array declarations are shown below:

```
int j[100];

double x[20];

double y[10][20];
```

Thus, j is a 100-element integer array, x is a 20-element floating point array, and y is a 10x20 floating-point array. Note that variable size array declarations, *e.g.*,

```
double a[n];
```

where n is an integer variable, are *illegal* in C.

It is sometimes convenient to define an array size in terms of a symbolic constant, rather than a fixed integer quantity. This makes it easier to modify a programme that utilises an array, since all references to the maximum array size can be altered by simply changing the value of the symbolic constant. This approach is used in many of the example programmes employed in this course.

Like an ordinary variable, an array can be either *local* or *global* in extent, depending on whether the associated array declaration lies inside or outside, respectively, the scope of any of the functions which constitute the programme. Both local and global arrays can be initialised via their declaration statements. For instance,

```
int j[5] = {1, 3, 5, 7, 9};
```

declares j to be a 5-element integer array whose elements have the initial values j[0]=1, j[1]=3, etc.

Single operations which involve entire arrays are *not* permitted in C. Thus, if x and y are similar arrays (*i.e.*, the same data type, dimensionality, and size) then assignment operations, comparison operations, etc. involving these two arrays must be carried out on an element by element basis. This is usually accomplished within a loop (or within nested loops, for multi-dimensional arrays).

The programme listed below is a simple illustration of the use of arrays in C. The programme reads a list of numbers, entered by the user, into a one-dimensional array, list, and then calculates the average of these numbers. The programme also calculates and outputs the deviation of each number from the average.

```
/* average.c */

/*

   Programme to calculate the average of n numbers and then

   compute the deviation of each number from the average

  Code adapted from "Programming with C", 2nd Edition, Byron Gottfreid,
```

```
   Schaum's Outline Series, (McGraw-Hill, New York NY, 1996)
*/
#include <stdio.h>
#include <stdlib.h>
#define NMAX 100
int main()
{
   int n, count;
   double avg, d, sum = 0.;
   double list[NMAX];
   /* Read in value for n */
   printf("\nHow many numbers do you want to average? ");
   scanf("%d", &n);
   /* Check that n is not too large or too small */
   if ((n > NMAX) || (n <= 0))
    {
     printf("\nError: invalid value for n\n");
     exit(1);
    }
   /* Read in the numbers and calculate their sum */
   for (count = 0; count < n; ++count)
    {
     printf("i = %d  x = ", count + 1);
     scanf("%lf", &list[count]);
     sum += list[count];
    }
   /* Calculate and display the average */
   avg = sum / (double) n;
   printf("\nThe average is %5.2f\n\n", avg);
   /* Calculate and display the deviations about the average */
   for (count = 0; count < n; ++count)
    {
     d = list[count] - avg;
```

```
    printf("i = %d  x = %5.2f  d = %5.2f\n", count + 1, list[count],
d);
    }
   return 0;
}
```

Note the use of the symbolic constant NMAX to specify the size of the array list, and, hence, the maximum number of values which can be averaged. The typical output from the above programme looks like:

```
How many numbers do you want to average? 5
i = 1  x = 4.6
i = 2  x = -2.3
i = 3  x = 8.7
i = 4  x = 0.12
i = 5  x = -2.7
The average is   1.68
i = 1  x =  4.60  d =  2.92
i = 2  x = -2.30  d = -3.98
i = 3  x =  8.70  d =  7.02
i = 4  x =  0.12  d = -1.56
i = 5  x = -2.70  d = -4.38
%
```

It is important to realise that an array name in C is essentially a *pointer* to the first element in that array. Thus, if x is a one-dimensional array then the address of the first array element can be expressed as either &x[0] or simply x. Moreover, the address of the second array element can be written as either &x[1] or (x+1). In general, the address of the (i+1)th array element can be expressed as either &x[i] or (x+i). Incidentally, it should be understood that (x+i) is a rather special type of expression, since x represents an address, whereas i represents an integer quantity. The expression (x+i) actually specifies the address of the array element which is i memory locations offset from the address of the first array element (C, of course, stores all elements of an array both contiguously and in order in the computer memory). Hence, (x+i) is a *symbolic representation* of an address, rather than an arithmetic expression.

Since &x[i] and (x+i) both represent the address of the (i+1)th element of the array x, it follows that x[i] and *(x+i) must both represent the contents of that address (*i.e.*, the value of the (i+1)th element). In fact, the latter two terms are *completely interchangeable* in C programmes.

For the moment, let us concentrate on one-dimensional arrays. An entire array can be passed to a function as an argument. To achieve this, the array name must appear by itself, without brackets or subscripts, as an argument within the function call. The corresponding argument in the function definition must be declared as an array. In order to do this, the array name is written followed by an empty pair of square brackets. The size of the array is not specified. In a function prototype, an array argument is specified by following the data type of the argument by an empty pair of square brackets.

Since, as we have seen, an array name is essentially a pointer, it is clear that when an array is passed to a function it is passed by *reference*, and not by *value*. Hence, if any of the array elements are altered within the function then these alterations *are* recognised in the calling portion of the programme. Likewise, if an array (rather than an individual array element) appears in the argument list of a scanf() function then it should *not* be preceded by the address operator (&), since an array name already is an address. The reason why arrays in C are always passed by reference is fairly obvious. In order to pass an array by value, it is necessary to copy the value of *every* element. On the other hand, to pass an array by reference it is only necessary to pass the address of the first element. Clearly, for large arrays, passing by reference is far more efficient than passing by value.

The programme listed below is yet another version of printfact.c, albeit a far more efficient one than any of those listed previously. In this version, the factorials of all the non-zero integers up to 20 are calculated in one fell swoop, by the function factorial(), using the recursion relation

$$(n+1)! = (n+1)n! \qquad ...3.1$$

The factorials are stored as elements of the array fact[], which is passed as an argument from factorial() to the main part of the programme.

```
/* printfact5.c */
/*
  Programme to print factorials of all integers
  between 0 and 20
*/
#include <stdio.h>
/* Function prototype for factorial() */
void factorial(double []);
int main()
{
  int j;
  double fact[21];               // Declaration of array fact[]
  /* Calculate factorials */
  factorial(fact);
  /* Output results */
```

```
   for (j = 0; j <= 20; ++j)
     printf("j = %3d      factorial(j) = %12.3e\n", j, fact[j]);
   return 0;
}
//%%%%%%%%%%%%%%%%%%%%%%%%%%%%%%%%%%%%%%%%%%%%%%%%%%%%%%%%%%%%%%
void factorial(double fact[])
{
   /*
     Function to calculate factorials of all integers
     between 0 and 20 (in form of floating-point
     numbers) via recursion formula
     (n+1)! = (n+1) n!
     Factorials returned in array fact[0..20]
   */
   int count;
   fact[0] = 1.;                        // Set 0! = 1
   /* Calculate 1! through 20! via recursion */
   for (count = 0; count < 20; ++count)
     fact[count+1] = (double)(count+1) * fact[count];
   return;
}
```

The output from the above programme is identical to that from printfact.c.

It is important to realise that there is *no array bound checking* in C. If an array x is declared to have 100 elements then the compiler will reserve 100 contiguous, appropriately sized, slots in computer memory on its behalf. The contents of these slots can be accessed via expressions of the form x[i], where the integer i should lie in the range 0 to 99. As we have seen, the compiler interprets x[i] to mean the contents of the memory slot which is i slots along from the beginning of the array. Unfortunately, expressions such as x[100] or x[1000] are interpreted in a like manner, leading the compiler to instruct the executable to access memory slots which lie off the end of the memory block reserved for x. Obviously, accessing elements of an array which do not exist is going to produce some sort of error. Exactly what sort of error is very difficult to say—the programme may crash, it may produce absurdly incorrect output, it may produce plausible but incorrect output, it may even produce correct output—it all depends on exactly what information is being stored in the memory locations surrounding the block of memory reserved for x. This type of error can be *extremely difficult* to debug, since it may not be immediately apparent that something has gone wrong when the programme is executed. It is, therefore, the programmer's responsibility to ensure that *all* references to array elements lie within the declared bounds of the associated arrays.

Let us now discuss multi-dimensional arrays in more detail. The elements of a multi-dimensional array are stored contiguously in a block of computer memory. In scanning across this block, from its start to its end, the order of storage is such than the last subscript of the array varies *most rapidly* whilst the first varies *least rapidly*. For instance, the elements of the two-dimensional array x[2][2] are stored in the order: x[0][0], x[0][1], x[1][0], x[1][1]. The elements of a multi-dimensional array can only be addressed if the programme is explicitly told the size of the array in its second, third, etc. dimensions. It is, therefore, not surprising to learn that when a multi-dimensional array is passed to a function, as an argument, then the associated argument declaration within the function definition *must* include *explicit size declarations* in all of the subscript positions *except the first*. The same is true for a multi-dimensional array argument appearing in a function prototype.

Character Strings

The basic C character data type is called char. A character string in C can be represented symbolically as an array of data type char. For instance, the declaration

```
char word[20] = "four";
```

initialises the 20-element character array word, and then stores the character string "four" in it. The resulting elements of word are as follows:

```
word[0] = 'f'   word[1] = 'o'   word[2] = 'u'   word[3] = 'r'   word[4]
= '\0'
```

with the remaining elements undefined. Here, 'f' represents the character ``f'', etc., and '\0' represents the so-called *null character* (ASCII code 0), which is used in C to signal the termination of a character string. The null character is automatically added to the end of any character string enclosed in double quotes. Note that, since all character strings in C must be terminated by the (invisible) null character, it takes a character array of size at least n+1 to store an n-letter string.

As with arrays of numbers, the name of a character array is essentially equivalent to a pointer to the first element of that array. Hence, word[i] and *(word + i) both refer to the same character in the character array word. Note, however, that the name of a character array is not a true pointer, since the address to which it points cannot be changed. Of course, we can always represent a character array using a true pointer. Consider the declaration

```
char *word = "four";
```

Here, word is declared to be a pointer to a char which points towards the first element of the character string 'f' 'o' 'u' 'r' '\0'. Unlike the name of a character array, a true pointer to a char can be redirected. Thus,

```
char *word =  "four";

. . .
```

```
word  =  "five";
```

is legal, whereas

```
char  word[20]  =   "four";
. . .
word  =  "five";
```

is illegal. Note that, in the former example, the addresses of the first elements of the strings "four" and "five" are probably different. Of course, the contents of a character array can always be changed, element by element—it is just the address of the first element which must remain constant. Thus,

```
char  word[20]  =   "four";
. . .
word[0]  =  'f';
word[1]  =  'i';
word[2]  =  'v';
word[3]  =  'e';
word[4]  =  '\0';
```

is perfectly legal.

Note, finally, that a character string can be printed via the printf() function by making use of a %s entry in its control string: *e.g.*,

```
printf("word  =  %s\n",  word);
```

Here, the second argument, word, can either be the name of a character array or a true pointer to the first element of a character string.

Multi-file Programmes

In a programme consisting of many different functions, it is often convenient to place each function in an individual file, and then use the make utility to compile each file separately and link them together to produce an executable.

There are a few common-sense rules associated with multi-file programmes. Since a given file is initially compiled *separately* from the rest of the programme, all symbolic constants which appear in that file must be defined at its start. Likewise, all referenced library functions must be accompanied by the appropriate references to header files. Also, any referenced user-defined functions must have their prototypes at the start of the file. Finally, all global variables used in the file must be declared at its start. This usually means that definitions for common symbolic constants, header files for common library functions, prototypes for common user-defined functions, and declarations for common global variables will appear in *multiple* files. Note that a given global variable can only be initialised in *one* of its declaration statements, which is regarded as the *true declaration* of that variable [conventionally, the true declaration appears in the file containing the function main()]. Indeed, the other declarations, which we shall term *definitions,* must be preceded by the keyword extern to distinguish them from the true declaration.

As an example, let us take the programme printfact4.c, listed previously, and break it up into multiple files, each containing a single function. The files in question are called main.c and factorial.c. The listings of the two files which make up the programme are as follows:

```
/* main.c */
/*
   Programme to print factorials of all integers
   between 0 and 20
*/
#include <stdio.h>
/* Prototype for fucntion factorial() */
void factorial();
/* Global variable declarations */
int j;
double fact;
int main()
{
   /* Print factorials of all integers between 0 and 20 */
   for (j = 0; j <= 20; ++j)
     {
       factorial();
       printf("j = %3d     factorial(j) = %12.3e\n", j, fact);
     }
   return 0;
}
```

and

```
/* factorial.c */
/*
    Function to evaluate factorial (in floating point form)
    of non-negative integer j. Result stored in variable fact.
*/
#include <stdio.h>
#include <stdlib.h>
```

```
/* Global variable definitions */
extern int j;
extern double fact;
void factorial()
{
   int count;
   /* Abort if j is negative integer */
   if (j < 0)
     {
      printf("\nError: factorial of negative integer not defined\n");
       exit(1);
     }
   /* Calculate factorial */
  for (count = j, fact = 1.; count > 0; —count) fact *= (double) count;
   return;
}
```

Note that all library functions and user-defined functions referenced in each file are declared (either via a header file or a function prototype) at the start of that file. Note, also, the distinction between the global variable *declarations* in the file main.c and the global variable *definitions* in the file factorial.c.

Command Line Parameters

The main() function may optionally possess special arguments which allow parameters to be passed to this function from the operating system. There are two such arguments, which are conventionally called argc and argv. The former argument, argc, is an integer which is set to the number of parameters passed to main(), whereas the latter argument, argv, is an array of pointers to character strings which contain these parameters. In order to pass one or more parameters to a C programme when it is executed from the operating system, the parameters must follow the programme name on the command line: *e.g.*,

```
% programme-name  parameter_1  parameter_2  parameter_3 ... parameter_n
```

The programme name will be stored in the first item in argv, followed by each of the parameters. Hence, if the programme name is followed by n parameters there will be n + 1 entries in argv, ranging from argv[0] to argv[n]. Furthermore, argc will be automatically set equal to n + 1.

The programme listed below is a simple illustration of the use of command line parameters: it simply echoes all of the parameters passed to it.

```
/* repeat.c */
/*
   Programme to read and echo data from command line
*/
int main(int argc, char *argv[])
{
   int i;
   for (i = 1; i < argc; i++) printf("%s ", argv[i]);
   printf("\n");
   return 0;
}
```

Assuming that the executable is called repeat, the typical output from this programme is as follows:

```
% repeat The quick brown fox jumped over the lazy hounds
The quick brown fox jumped over the lazy hounds
%
```

Suppose that one or more of the parameters passed to a given programme are *numbers*. As we have seen, these numbers are actually passed as character strings. Hence, before they can be employed in calculations, they must be converted into either type int or type double. This can be achieved via the use of the functions atoi() and atof() (the appropriate header file for these functions is stdlib.h). Thus, int atoi(char *ptr) converts a string pointed to by ptr into an int, whereas double atof(char *ptr) converts a string pointed to by ptr into an double. The programme listed below illustrates the use of the atof() function: it reads in a number passed as a command line parameter, interprets it as a temperature in degrees Fahrenheit, converts it to degrees Celsius, and then prints out the result.

```
/* ftoc.c */
/*
   Programme to convert temperature in Fahrenheit input
   on command line to temperature in Celsius
*/
#include <stdlib.h>
#include <stdio.h>
int main(int argc, char *argv[])
{
   double deg_f, deg_c;
```

```
    /* If no parameter passed to programme print error
       message and exit */
    if (argc < 2)
      {
        printf("Usage: ftoc temperature\n");
        exit(1);
      }
    /* Convert first command line parameter to double */
    deg_f = atof(argv[1]);
    /* Convert from Fahrenheit to Celsius */
    deg_c = (5. / 9.) * (deg_f - 32.);
    printf("%f degrees Fahrenheit equals %f degrees Celsius\n",
           deg_f, deg_c);
    return 0;
}
```

Assuming that the executable is called ftoc, the typical output from this programme is as follows:

```
% ftoc 98
98.000000 degrees Fahrenheit equals 36.666667 degrees Celsius
%
```

The Timing

The header file time.h defines a number of library functions which can be used to assess how much CPU time a C programme consumes during execution. The simplest such function is called clock(). A call to this function, with no argument, will return the amount of CPU time used so far by the calling programme. The time is returned in a special data type, clock_t, defined in time.h. This time must be divided by CLOCKS_PER_SEC, also defined in time.h, in order to covert it into seconds. The ability to measure how much CPU time a given code consumes is useful in scientific programming: *e.g.*, because it allows the effectiveness of the various available compiler optimisation flags to be determined. Optimisation usually (but not always!) speeds up the execution of a programme. However, over aggressive optimisation can often slow a programme down again.

The programme listed below illustrates the simple use of the clock() function. The programme compares the CPU time required to raise a double to the fourth power via a direct calculation and via a call to the pow() function. Actually, both operations are performed a million times and the elapsed CPU time is then divided by a million.

```
/* timing.c */
/*
   Programme to test operation of clock() function
*/
#include <time.h>
#include <math.h>
#define N_LOOP 1000000
int main()
{
   int i;
   double a = 11234567890123456.0, b;
   clock_t time_1, time_2;
   time_1 = clock();
   for (i = 0; i < N_LOOP; i++) b = a * a * a * a;
   time_2 = clock();
   printf ("CPU time needed to evaluate a*a*a*a:     %f microsecs\n",
             (double) (time_2 - time_1) / (double) CLOCKS_PER_SEC);
   time_1 = clock();
   for (i = 0; i < N_LOOP; i++) b = pow(a, 4.);
   time_2 = clock();
   printf ("CPU time needed to evaluate pow(a, 4.): %f microsecs\n",
             (double) (time_2 - time_1) / (double) CLOCKS_PER_SEC);
   return 0;
}
```

The typical output from this programme is as follows:

```
CPU time needed to evaluate a*a*a*a:     0.190000 microsecs
CPU time needed to evaluate pow(a, 4.): 1.150000 microsecs
%
```

Clearly, evaluating a fourth power using the pow() function is a lot more expensive than the direct calculation. Hence, as has already been mentioned, the pow() function should *not* be used to raise floating point quantities to small integer powers.

Random Numbers

A large class of scientific calculations (*e.g.*, so-called Monte Carlo calculations) require the use of random variables. A call to the rand() function (header file stdlib.h), with no arguments, returns a fairly good approximation to a random integer in the range 0 to RAND_MAX (defined in stdlib.h). The srand() function sets its argument, which is of type int, as the seed for a new sequence of numbers to be returned by rand(). These sequences are *repeatable* by calling srand() with the same seed value. If no seed value is provided, the rand() function is automatically seeded with the value 1. It is common practice in C programming to seed the random number generator with the number of seconds elapsed since 00:00:00 UTC, January 1st, 1970. This number is returned, as an integer, via a call to the time (NULL) function (header file: <time.h>). Seeding the generator in this manner ensures that a different set of random numbers is generated automatically each time the programme is run.

The programme listed below illustrates the use of the rand() function to construct a pseudo-random variable, x, which is uniformly distributed in the range 0 to 1. The programme calculates 10^7 values of x, and then evaluates the mean and variance of these values.

```
/* random.c */
/*
  Programme to test operation of rand() function
*/
#include <stdlib.h>
#include <stdio.h>
#include <time.h>
#define N_MAX 10000000
int main()
{
  int i, seed;
  double sum_0, sum_1, mean, var, x;
  /* Seed random number generator */
  seed = time(NULL);
  srand(seed);
  /* Calculate mean and variance of x: random number uniformly
     distributed in range 0 to 1 */
  for (i = 1, sum_0 = 0., sum_1 = 0.; i <= N_MAX; i++)
    {
      x = (double) rand() / (double) RAND_MAX;
      sum_0 += x;
      sum_1 += (x - 0.5) * (x - 0.5);
    }
  mean = sum_0 / (double) N_MAX;
```

```
    var = sum_1 / (double) N_MAX;
    printf("mean(x) = %12.10f   var(x) = %12.10f\n", mean, var);
    return 0;
}
```

The typical output from this programme is as follows:

```
mean(x) = 0.5000335261   var(x) = 0.0833193874
%
```

As is easily demonstrated, the theoretical mean and variance of x are 1/2 and 1/12, respectively. It can be seen that the values returned by the programme agree with these theoretical values to five decimal places, which is all that can be expected with only 10^7 calls.

4

C++ Extensions to C

Here, we shall briefly discuss some of the useful, non-object-orientated extensions to C introduced in the C++ language. Files containing source code which incorporates C++ elements should be distinguished from files containing plain C code via the extension .cpp.

In C, all local variables within a function must be declared *before* the first executable statement. In C++, this restriction is relaxed: a local variable can be declared (almost) *anywhere* in a function, provided that this declaration occurs prior to the variable's first use. The following code snippet illustrates the use of this new feature:

```
. . .
for (int i = 0; i < MAX; i++)
  {
    . . .
  }
. . .
```

For loop is now declared at the start of the loop, instead of at the start of the function containing the loop. This is far more convenient, and also makes the code easier to read (since we no longer have to skip back to the declarations at the start of the function to check that i is an int). Note, however, that the variable i is only defined over the extent of the loop (*i.e.*, between the curly braces). Any attempt to reference i outside the loop will result in a compilation error. In general, when a variable is declared in C++ its *scope* (*i.e.*, range of definition) extends from its declaration to the closing curly brace which terminates the current programme block. Programme blocks are functions, loops, conditionally executed compound statements, etc., and are delineated by curly braces. There are a number of restrictions to this new method of variable declaration. Variables *cannot* be declared within conditional statements, in the second or third expressions of for loops, or in function calls. In C, we have seen that in order to pass an argument to a function in such a manner that changes made to this argument within the function are passed back to the calling routine, we must actually pass a *pointer* to the argument. This procedure, which is called passing by *reference,* is illustrated in the code snippet listed below:

```
. . .
void square(double, double *);
. . .
int main()
{
   . . .
   double arg, res;
   square(arg, &res);
   . . .
   return 0;
}
. . .
void square(double x, double *y)
{
   *y = x * x;
   return;
}
```

Here, the second argument to square() is returned to main() in modified form. This argument must, therefore, be passed as a pointer. Consequently, we must write &res, rather that res, when calling square(), and we must refer to the argument as *y, rather than y, in the function itself. After a while, all these ampersands and asterisks can become a little tedious! C++ introduces a new method of passing by reference which somewhat less involved. This new method is illustrated in the following:

```
. . .
void square(double, double &);
. . .
int main()
{
   . . .
   double arg, res;
```

```
   square(arg, res);
   . . .
   return 0;
}
. . .
void square(double x, double &y)
{
   y = x * x;
   return;
}
```

Here, the second argument to square() is again passed by reference. However, this is now indicated by prepending an ampersand to the variable name in the function declaration. A corresponding ampersand appears in the associated function prototype. Note that we do not need to explicitly pass a pointer to the second argument when calling square() (this is done behind the scenes): i.e., we write res, rather than &res, when calling square(). Likewise, we do not have to explicitly deference the argument in the function itself (this is also done behind the scenes): i.e., we refer to the argument by its regular local name, y, as opposed to *y, within the function.

Functions are used in C programmes to avoid having to repeat the same block of code in many different places in the source. The use of functions also renders a code easier to read and maintain. However, there is a price to pay for the convenience of functions. When a regular function is called in an executable, the programme jumps to the block of memory in which the compiled function code is stored, and then jumps back to its original position in memory space when the function returns. Unfortunately, the large jumps in memory space associated with a function call take up a non-negligible amount of CPU time. Indeed, the overhead associated with making function calls often discourages scientific programmers from writing small functions, even when it may be desirable to do so. C++ provides a way out of this dilemma, via the use of the new keyword inline. An *inline function* looks like a normal function when it is used, but is compiled in a different manner. Calling an inline function from several different locations in a code does not result in multiple calls to a single function. Instead, the code for the inline function is inserted into the programme code by the compiler wherever the function is used.

Inline functions are only useful for *small* functions. The disadvantage of inserting the code for a large function multiple times into the code for a typical programme easily outweighs the small gain in performance obtained by the elimination of standard function calls. The break-even point for inline functions is usually about *three* executable lines.

To inline a function, a programmer adds the keyword inline at the start of the function's definition. For example:

```
inline double square(double x)
{
   return x * x;
}
```

Because the body of an inline function must be known before the compiler can insert it into the programme code, wherever the function is used, we must define such a function prior to its first use—a prototype declaration is not enough. It is common practice to define inline functions at the same location in source code files that the prototypes for regular (*i.e.*, outline) functions are placed.

Variable size array declarations of the form

```
void function(n)
{
   int n;
   double x[n];
   . . .
}
```

are *illegal* in C, which is extremely inconvenient. In C++, such declarations are enabled via the use of the new keywords new and delete. Thus, the C++ implementation of the above code snippet takes the form:

```
void function(n)
{
   . . .
   int n;
   double *x = new double[n];
   . . .
   x[i] = . . .
   . . .
   . . .
   delete x[];
   . . .
}
```

Note that x is actually declared as a pointer, rather than a standard array. The declaration new double[n] reserves a memory block which is just large enough to store n doubles, and then returns the address of the start of this block. The line delete x[] frees up the block of memory associated with the array x when it is no longer needed (note that this is not done automatically). Unfortunately, the new and delete keywords cannot be used to make variable size multi-dimensional arrays.

Complex Numbers

As we have already mentioned, the C language definition does not include complex arithmetic—presumably because the square root of minus one is not a concept which crops up very often in systems programming! Fortunately, this rather serious deficiency—at least, as far as the scientific programmer is concerned—is remedied in C++. The programme listed below illustrates the use of the C++ complex class (header file complex.h) to perform complex arithmetic using doubles:

```
/* complex.cpp */
/*
  Programme to test out C++ complex class
*/
#include <complex.h>
#include <stdio.h>
/* Define complex double data type */
typedef complex<double> dcomp;
int main()
{
  dcomp i, a, b, c, d, e, p, q, r; // Declare complex double variables
  double x, y;
  /* Set complex double variable equal to complex double constant */
  i = dcomp (0., 1.);
  printf("\ni = (%6.4f, %6.4f)\n", i);
  /* Test arithmetic operations with complex double variables */
  a = i * i;
  b = 1. / i;
  printf("\ni*i = (%6.4f, %6.4f)\n", a);
  printf("1/i = (%6.4f, %6.4f)\n", b);
  /* Test mathematical functions using complex double variables */
  c = sqrt(i);
  d = sin(i);
  e = pow(i, 0.25);
  printf("\nsqrt(i) = (%6.4f, %6.4f)\n", c);
```

```
    printf("sin(i)  =  (%6.4f,  %6.4f)\n", d);
    printf("i^0.25  =  (%6.4f,  %6.4f)\n", e);
    /* Test complex operations */
    p = conj(i);
    q = real(i);
    r = imag(i);
    printf("\nconj(i)  =  (%6.4f,  %6.4f)\n", p);
    printf("real(i)  =  %6.4f\n", q);
    printf("imag(i)  =  %6.4f\n", r);
    return 0;
}
```

The typical output from this programme is as follows:

```
i = (0.0000, 1.0000)
i*i = (-1.0000, 0.0000)
1/i = (0.0000, -1.0000)
sqrt(i) = (0.7071, 0.7071)
sin(i) = (0.0000, 1.1752)
i^0.25 = (0.9239, 0.3827)
conj(i) = (0.0000, -1.0000)
real(i) = 0.0000
imag(i) = 1.0000
%
```

The programme first of all defines the complex double type dcomp. Variables of this type are then declared, set equal to complex constants, employed in arithmetic expressions, used as the arguments of mathematical functions, etc., in much the same manner that we would perform similar operations in C with variables of type double. Note the special functions conj(), real(), and imag(), which take the complex conjugate of, find the real part of, and find the imaginary part of a complex variable, respectively.

Variable Size Multi-dimensional Arrays

Multi-dimensional arrays crop up in a wide variety of different applications in scientific programming. Moreover, it is very common for the sizes of the arrays employed in a scientific code to vary from run to run, depending on the particular values of the input parameters. For instance, in a standard fluid code the array sizes depend on the number of grid points, which, in turn, depends on the requested accuracy. Thus, a crucial test of the suitability of a programming language for scientific purposes is its ability to deal with variable size, multi-dimensional arrays in a convenient manner. Unfortunately, C fails this test rather badly, since variable size matrix declarations of the form

```
void function(a, m, n)
{
   int m, n;
   double a[m][n];
   . . .
```

are *illegal* (unlike in FORTRAN 77, where variable size matrix declarations are perfectly acceptable). Indeed, this unfortunate (and quite unnecessary!) omission in the C language definition is often put forward as a reason why C should *not* be used for scientific purposes. Fortunately, we can get around the absence of variable size multi-dimensional arrays in C by making use of a freely available C++ package called the Blitz++ library.

The programme listed below illustrates the use of the Blitz++ library. The programme adds together two matrices whose dimensions and elements are input by the user, and then prints out the result.

```
/* addmatrix.c */
/*
   Programme to add two variable dimension matrices input by user
*/
#include <stdio.h>
#include <stdlib.h>
#include <blitz/array.h>
using namespace blitz;
/* Function prototypes */
void readin(Array<double,2>);
void writeout(Array<double,2>);
void addmatrices(Array<double,2>, Array<double,2>, Array<double,2>);
int main()
{
   int n, m;
   /* Input number of rows and columns */
  printf("\nPlease input number of rows, n, and number of columns, m: ");
   scanf("%d %d", &n, &m);
   /* Check that n, m are positive integers */
```

```
    if (n <= 0 || m <= 0)
       {
          printf("\nError: invalid values for n and/or m\n");
          exit(1);
       }
    /* Array declarations */
    Array<double,2> A(n, m), B(n, m), C(n, m);
    /* Read in elements of A, row by row */
    printf("\nReading in elements of A:\n");
    readin(A);
    /* Read in elements of B, row by row */
    printf("\nReading in elements of B:\n");
    readin(B);
    /* Write out elements of A, row by row */
    printf("\nWriting out elements of A:\n");
    writeout(A);
    /* Write out elements of B, row by row */
    printf("\n\nWriting out elements of B:\n");
    writeout(B);
    /* Add matrices A and B */
    addmatrices(A, B, C);
    /* Write out matrix C = A + B, row by row */
    printf("\n\nWriting out elements of C = A + B:\n");
    writeout(C);
    printf("\n");
    return 0;
}
//%%%%%%%%%%%%%%%%%%%%%%%%%%%%%%%%%%%%%%%%%%%%%%%%%%%%%%%%%%%%%%
/*
    Read in elements of matrix M, row by row
*/
void readin(Array<double,2> M)
{
```

```
   int n = M.extent(0);
   int m = M.extent(1);
   for (int i = 0; i < n; i++)
      {
         printf("\nRow %d: ", i + 1);
         for (int j = 0; j < m; j++) scanf("%lf", &M(i, j));
      }
   return;
}
  /                                                                   /
%%%%%%%%%%%%%%%%%%%%%%%%%%%%%%%%%%%%%%%%%%%%%%%%%%%%%%%%%%%%%%%%%%%%%%
/*
   Write out elements of matrix M, row by row
*/

void writeout(Array<double,2> M)
{
   int n = M.extent(0);
   int m = M.extent(1);
   for (int i = 0; i < n; i++)
      {
         printf("\nRow %d: ", i + 1);
         for (int j = 0; j < m; j++) printf("%7.2f ", M(i, j));
      }
   return;
}
//%%%%%%%%%%%%%%%%%%%%%%%%%%%%%%%%%%%%%%%%%%%%%%%%%%%%%%
/*
      Add matrices M and N and store result in matrix P
*/
void addmatrices(Array<double,2> M, Array<double,2> N, Array<double,2> P)
{
   int n = M.extent(0);
```

```
   int m = M.extent(1);
   for (int i = 0; i < n; i++)
      for (int j = 0; j < m; j++)
         P(i, j) = M(i, j) + N(i, j);
    return;
}
```

Typical output from the programme looks like

```
Please input number of rows, n, and number of columns, m: 3 3
Reading in elements of A:
Row 1: 1 4 5
Row 2: 6 7 8
Row 3: 3 6 0
Reading in elements of B:
Row 1: 1 6 0
Row 2: 4 7 8
Row 3: 4 8 8
Writing out elements of A:
Row 1:      1.00      4.00      5.00
Row 2:      6.00      7.00      8.00
Row 3:      3.00      6.00      0.00
Writing out elements of B:
Row 1:      1.00      6.00      0.00
Row 2:      4.00      7.00      8.00
Row 3:      4.00      8.00      8.00
Writing out elements of C = A + B:
Row 1:      2.00     10.00      5.00
Row 2:     10.00     14.00     16.00
Row 3:      7.00     14.00      8.00
```

The header file for the Blitz++ library is called blitz/array.h. Moreover, any programme file which makes use of Blitz++ must include the cryptic line using namespace blitz; before the first call, otherwise compilation errors will ensue. The declaration Array<double,2> A(n, m) declares a two-dimensional n by m array of doubles. The generalisation to an array of integers, or a higher dimension array, is fairly obvious. Note that the i, j element of matrix A is refered to simply as A(i,j). The function call M.extent(0) returns the size of array M in its first dimension. Likewise, M.extent(1) returns the size of array M in its second dimension, etc. More details of the operation of Blitz++ can be found by reading the extensive documentation which accompanies this library. Unfortunately, the Blitz++ library slows down compilation considerably since it makes use of some very advanced templating features of the C++ language.

The CAM Graphics Class

There are a myriad of useful, prewritten C++ classes which are freely available on the web. In this Subsection, we shall discuss just one of these—namely, the *CAM graphics class*, whose purpose is to enable a C++ programme to generate simple line plots.

The CAM graphics class actually generates a PostScript file. PostScript is a programming language that describes the appearance of a printed page. It was developed by Adobe in 1985, and has become an industry standard for printing and imaging. All major printer manufacturers make printers that can interpret PostScript. A PostScript file is conventionally identified via a .ps suffix. The schematic code listed below illustrates the basic use of the CAM graphics class:

```
. . .
#include <gprocess.h>                    // Header file for CAM graphic class
. . .
CAMgraphicsProcess Gprocess;             // declare a graphics process
CAMpostScriptDriver Pdriver("filename.ps"); // declare a PostScript driver
Gprocess.attachDriver(Pdriver);          // attach driver to process
. . .
Gprocess.frame();                        // "frame" the first plot
. . .
Gprocess.frame();                        // "frame" the second plot
. . .
. . .
Gprocess.frame();                        // "frame" the last plot
Gprocess.detachDriver();                 // detach the driver
. . .
```

The header file for the class is called gprocess.h. The procedure for generating a plot is to first declare a graphics process, then declare a PostScript driver and attach this to a PostScript file—filename.ps, in the above example—and, finally, attach this driver to the process. A PostScript file can contain multiple pictures, or *frames*. Each frame is terminated by a call to Gprocess.frame(). Finally, the driver is detached, which has the effect of closing the PostScript file.

The programme listed below uses the CAM graphics class to plot the curve $y = \sin^2 x$ for x in the range -2π to 2π.

```
/*  camgraph1.cpp  */
/*
  Illustration of use of CAM graphics class to create simple line plot
   Programme plots y = sin^2 x versus x in range -2 PI to +2 PI
   Programme adapted from gpsmpl.cpp by Chris Anderson, UCLA 1996
*/
#include <gprocess.h>
#include <math.h>
#include <stdlib.h>
double func(double);
int main()
{
   int N_points = 400;
   double x_start = -2. * M_PI;
   double x_end = 2. * M_PI;
   double delta_x = (x_end - x_start) / ((double) N_points - 1.);
   double *x = new double[N_points];
   double *y = new double[N_points];
   for (int i = 0; i < N_points; i++)
      {
         x[i] = x_start + (double) i * delta_x;
         y[i] = func(x[i]);
         x[i] /= M_PI;
      }
   { // This brace used to limit scope of Gprocess
   CAMgraphicsProcess Gprocess;                 // declare a graphics process
   CAMpostScriptDriver Pdriver("graph1.ps");  // declare a PostScript driver
   Gprocess.attachDriver(Pdriver);            // attach driver to process
   Gprocess.setAxisRange(-2., 2., -2., 2.);   // set plotting ranges
   Gprocess.title("y = sin(x*x)");            // label the plot
   Gprocess.labelX("x / PI");
   Gprocess.labelY("y");
   Gprocess.plot(x, y, N_points);             // do the plotting
```

```
  Gprocess.frame();                // "frame" the plot
  Gprocess.detachDriver();         // detach the driver
  }                                // This brace calls the destructor
                                      for Gprocess:
                                   // without it the system() call
                                      would hang up

  delete[] x;
  delete[] y;
  system("gv graph1.ps");          // display plot on screen
  return 0;
}
double func(double x)
{
  return sin(x*x);
}
```

The command Gprocess.plot(x, y, n) plots the n values of vector y against the n values of vector x as a solid curve. The command Gprocess.setAxisRange(x_low, x_high, y_low, y_high) sets the range of plotting. Finally, the commands Gprocess.title("title"), Gprocess.labelX("x_label"), and Gprocess.labelY("y_label") label the plot, the *x*-axis, and the *y*-axis, respectively. Incidentally, the UNIX function call system("gv graph1.ps") is used to pass the command gv graph1.ps to the operating system. On execution, this command displays the contents of graph1.ps on the screen. The graph written in the file graph1.ps is shown in figure 1

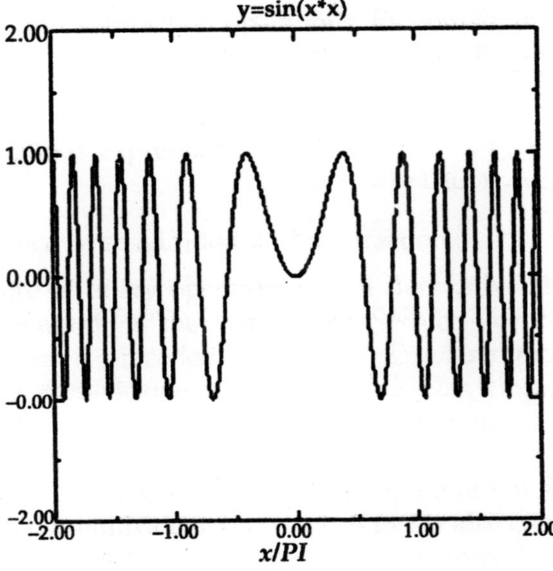

Figure 4.1: *An example plot generated by the CAM graphics class*

The programme shown below illustrates some of the more advanced features of the CAM graphics class:

```
/* camgraph2.cpp */
/*
  Illustration of use of CAMgraphics class to create more advanced line plots
   Programme plots three trigonometric functions versus x in range
     -2 PI to +2 PI using different plot styles and different
     line styles
   Programme adapted from gpsmp2.cpp by Chris Anderson, UCLA 1996
*/
#include <gprocess.h>
#include <math.h>
#include <stdlib.h>
double fun1(double);
double fun2(double);
double fun3(double);
int main()
{
   int N_points = 100;
   double x_start = -2. * M_PI;
   double x_end = 2. * M_PI;
   double delta_x = (x_end - x_start) / ((double) N_points - 1.);
   double *x  = new double[N_points];
   double *y1 = new double[N_points];
   double *y2 = new double[N_points];
   double *y3 = new double[N_points];
   for (int i = 0; i < N_points; i++)
      {
         x[i] = x_start + (double) i * delta_x;
         y1[i] = fun1(x[i]);
         y2[i] = fun2(x[i]);
         y3[i] = fun3(x[i]);
         x[i] /= M_PI;
      }
   {
```

```
CAMgraphicsProcess Gprocess;           // declare a graphics process
CAMpostScriptDriver Pdriver("graph2.ps"); // declare a PostScript driver
Gprocess.attachDriver(Pdriver);        // attach driver to process
/* First frame;  using different plot "styles" */
Gprocess.setAxisRange(-2., 2., -2., 2.); // set plotting ranges
Gprocess.title("Plots Using
Different Plot Styles");               // label the plot
Gprocess.labelX("x / PI");
Gprocess.labelY("y");
Gprocess.plot(x, y1, N_points);        // solid line (default)
Gprocess.plot(x, y2, N_points, '+');   // + markers
Gprocess.plot(x, y3, N_points, '+', 2); // + markers and solid line
Gprocess.frame();                      // "frame" the plot
/* Second frame; using different plot line "styles" */
Gprocess.setAxisRange(-2., 2., -2., 2.); // set plotting ranges
Gprocess.title("Plots Using Different Line Styles");// label the plot
Gprocess.labelX("x / PI");
Gprocess.labelY("y");
Gprocess.plot(x, y1, N_points);        // solid line (default)
Gprocess.setPlotDashPattern(1);
Gprocess.plot(x, y2, N_points);        // dashed line
Gprocess.setPlotDashPattern(4);
Gprocess.plot(x, y3, N_points);        // dashed-dot line
Gprocess.frame();                      // "frame" the plot
Gprocess.detachDriver();               // detach the driver
 }
 delete[] x;
 delete[] y1;
 delete[] y2;
 delete[] y3;
system("gv graph2.ps");                // display plots on screen
 return 0;
```

```
double fun1(double x)
{
   return sin(x);
}
double fun2(double x)
{
   return cos(x);
}
double fun3(double x)
{
   return cos(2.*x);
}
```

The command Gprocess.plot(x, y, n, '+') plots the n values of vector y against the n values of vector x as a set of points, each indicated by a '+' character. The command Gprocess.plot(x, y, n, '+', 2) does the same, but also connects the points with a solid line. The fourth argument of this command is an integer code which determines the plot style. The various options are as follows: 0 - curve; 1 - points; 2 - curve and points. The command Gprocess.setPlotDashPattern(n) sets the line style. The argument is again an integer code. The various options are: 0 - solid; 1 - dash; 2 - double-dash; 4 - dash-dot; 5 - dash-double-dot; 6 - dots.

The graphs written in the first and second frames of graph2.ps are shown in figure 2 and 3, respectively.

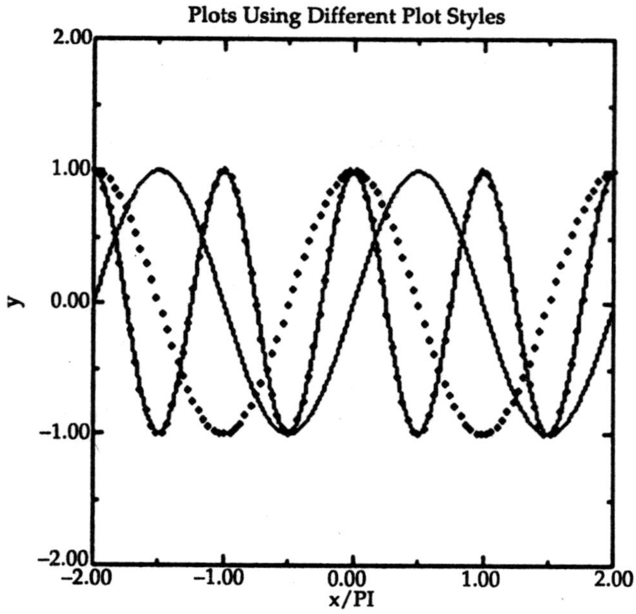

Figure 4.2: *An example plot generated by the CAM graphics class*

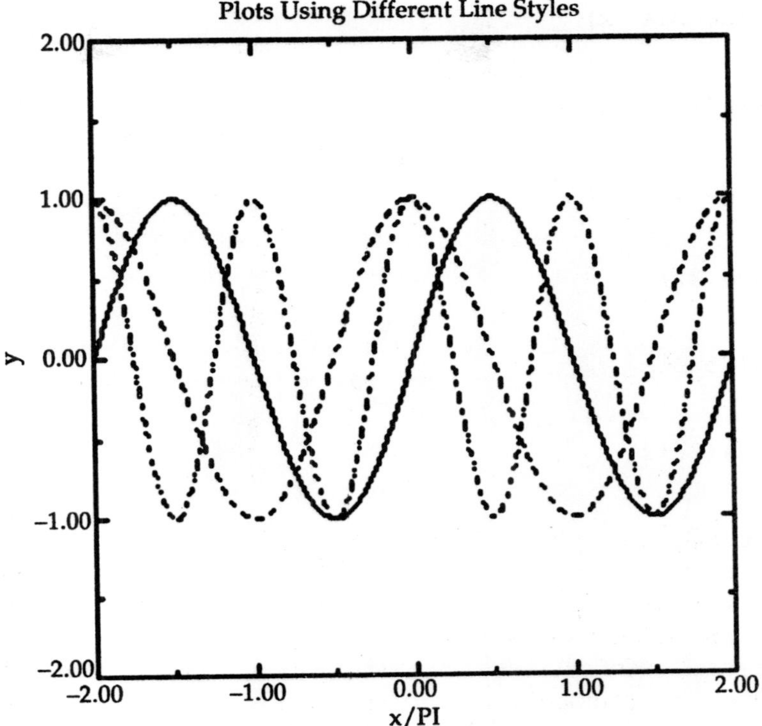

Figure 4.3: *An example plot generated by the CAM graphics class*

5

Chaotic Pendulum

We mostly deal with problems which are capable of analytic solution (so that we can easily validate our numerical solutions). Let us now investigate a problem which is quite intractable analytically, and in which meaningful progress can only be made via numerical means.

Consider a simple pendulum consisting of a point mass m, at the end of a light rigid rod of length l, attached to a fixed frictionless pivot which allows the rod (and the mass) to move freely under gravity in the vertical plane. Such a pendulum is sketched in Figure 5.1. Let us parameterise the instantaneous position of the pendulum via the angle θ the rod makes with the downward vertical. It is assumed that the pendulum is free to swing through 360°. Hence, θ and $\theta + 2\pi$ both correspond to the same pendulum position.

The angular equation of motion of the pendulum is simply

$$ml\frac{d^2\theta}{dt^2} + mg\sin\theta = 0, \qquad \text{...5.1}$$

where g is the downward acceleration due to gravity. Suppose that the pendulum is embedded in a viscous medium (*e.g.*, air). Let us assume that the viscous drag torque acting on the pendulum is governed by Stokes' law and is, thus, directly proportional to the pendulum's instantaneous velocity. It follows that, in the presence of viscous drag, the above equation generalises to

$$ml\frac{d^2\theta}{dt^2} + v\frac{d\theta}{dt} + mg\sin\theta = 0, \qquad \text{...5.2}$$

where v is a positive constant parameterising the viscosity of the medium in question. Of course, viscous damping will eventually drain all energy from the pendulum, leaving it in a stationary state. In order to maintain the motion against viscosity, it is necessary to add some external driving. For the sake of simplicity, we choose a fixed amplitude, periodic drive (which could arise, for instance, via periodic oscillations of the pendulum's pivot point). Thus, the final equation of motion of the pendulum is written

$$ml\frac{d^2\theta}{dt^2} + v\frac{d\theta}{dt} + mg\sin\theta = A\cos\omega t, \qquad \text{...5.3}$$

where A and ω are constants parameterising the amplitude and angular frequency of the external driving torque, respectively.

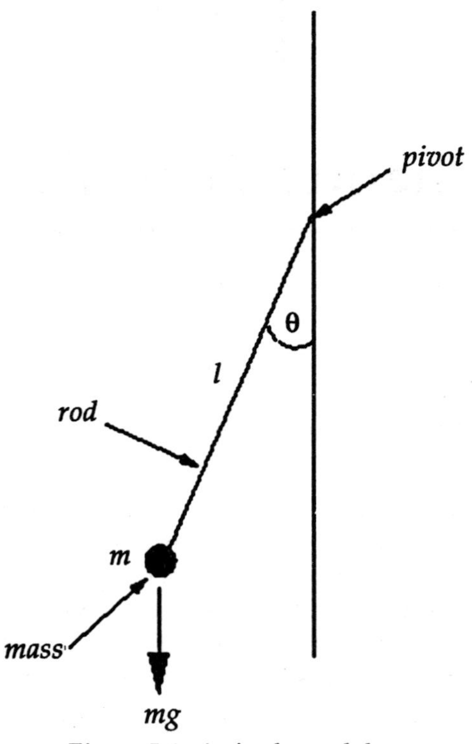

Figure 5.1: *A simple pendulum*

Let

$$\omega_0 = \sqrt{\frac{g}{l}}. \qquad \text{...5.4}$$

Of course, we recognise ω_0 as the natural (angular) frequency of small amplitude oscillations of the pendulum. We can conveniently normalise the pendulum's equation of motion by writing,

$$\hat{t} = \omega_0 t, \qquad \text{...5.5}$$

$$\hat{\omega} = \frac{\omega}{\omega_0}, \qquad \text{...5.6}$$

$$Q = \frac{mg}{\omega_0 v}, \qquad ...5.7$$

$$\hat{A} = \frac{A}{mg}, \qquad ...5.8$$

in which case equation (5.3) becomes

$$\frac{d^2\theta}{d\hat{t}^2} + \frac{1}{Q}\frac{d\theta}{d\hat{t}} + \sin\theta = \hat{A}\cos\hat{\omega}\hat{t}. \qquad ...5.9$$

From now on, the hats on normalised quantities will be omitted, for ease of notation. Note that, in normalised units, the natural frequency of small amplitude oscillations is *unity*. Moreover, Q is the familiar *quality-factor*—roughly, the number of oscillations of the undriven system which must elapse before its energy is significantly reduced via the action of viscosity. The quantity A is the amplitude of the external torque measured in units of the maximum possible gravitational torque. Finally, ω is the angular frequency of the external torque measured in units of the pendulum's natural frequency.

Equation (5.9) is clearly a second-order o.d.e. It can, therefore, also be written as two coupled first-order o.d.e.s:

$$\frac{d\theta}{dt} = v, \qquad ...5.10$$

$$\frac{dv}{dt} = -\frac{v}{Q} - \sin\theta + A\cos\omega t. \qquad ...5.11$$

Analytic Solution

Before attempting to solve the equations of motion of any dynamical system using a computer, we should, first, investigate them as thoroughly as possible via standard analytic techniques. Unfortunately, equation (5.10) and (5.11) constitute a *non-linear* dynamical system—because of the presence of the $\sin\theta$ term on the right-hand side of equation (5.11). This system, like most non-linear systems, does not possess a simple analytic solution. Fortunately, however, if we restrict our attention to *small amplitude* oscillations, such that the approximation

$$\sin\theta \simeq \theta \qquad ...5.12$$

is valid, then the system becomes *linear*, and can easily be solved analytically.

The linearised equations of motion of the pendulum take the form:

$$\frac{d\theta}{dt} = v, \qquad ...5.13$$

$$\frac{dv}{dt} = -\frac{v}{Q} - \theta + A\cos\omega t. \qquad ...5.14$$

Suppose that the pendulum's position, $\theta(0)$, and velocity, $v(0)$, are specified at time $t = 0$. As is well-known, in this case, the above equations of motion can be solved analytically to give:

$$\theta(t) = \left\{\theta(0) - \frac{A(1-\omega^2)}{\left[(1-\omega^2)^2 + \omega^2/Q^2\right]}\right\} e^{-t/2Q} \cos\omega_* t + \frac{1}{\omega_*}\left\{v(0) + \frac{\theta(0)}{2Q} - \frac{A(1-3\omega^2)/2Q}{\left[(1-\omega^2)^2 + \omega^2/Q^2\right]}\right\} e^{-t/2Q} \sin\omega_* t + \frac{A\left[(1-\omega^2)\cos\omega t + (\omega/Q)\sin\omega t\right]}{\left[(1-\omega^2)^2 + \omega^2/Q^2\right]}, \quad \text{...5.15}$$

$$v(t) = \left\{v(0) - \frac{A\omega^2/Q}{\left[(1-\omega^2)^2 + \omega^2/Q^2\right]}\right\} e^{-t/2Q} \cos\omega_* t - \frac{1}{\omega_*}\left\{\theta(0) + \frac{v(0)}{2Q} - \frac{A(1-\omega^2) - \omega^2/2Q^2}{\left[(1-\omega^2)^2 + \omega^2/Q^2\right]}\right\} e^{-t/2Q} \sin\omega_* t + \frac{\omega A\left[-(1-\omega^2)\sin\omega t + (\omega/Q)\cos\omega t\right]}{\left[(1-\omega^2)^2 + \omega^2/Q^2\right]}. \quad \text{...5.16}$$

Here,

$$\omega_* = \sqrt{1 - \frac{1}{4Q^2}}, \quad \text{...5.17}$$

and it is assumed that $Q > 1/2$.

It can be seen that the above expressions for θ and v both consist of *three* terms. The first two terms clearly represent *transients*—they depend on the initial conditions, and *decay* exponentially in time.

In fact, the e-folding time for the decay of these terms is $2Q$ (in normalised time units). The final term represents the *time-asymptotic motion* of the pendulum, and is manifestly *independent* of the initial conditions.

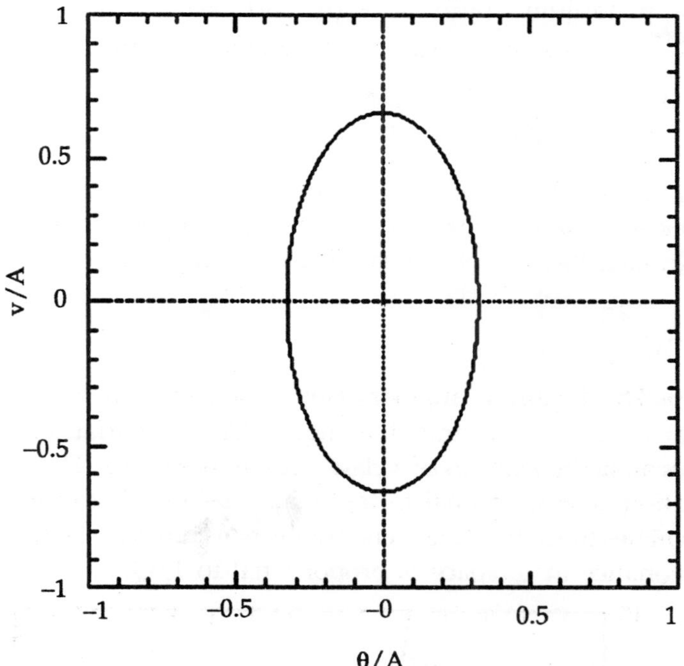

Figure 5.2: *A phase-space plot of the periodic attractor for a linear, damped, periodically driven, pendulum. Data calculated analytically for* $Q=4$ *and* $\omega = 2$

It is often convenient to *visualise* the motion of a dynamical system as an orbit, or trajectory, in *phase-space*, which is defined as the space of all of the dynamical variables required to specify the instantaneous state of the system. For the case in hand, there are two dynamical variables, v and θ, and so phase-space corresponds to the $\theta - v$ plane. Note that each different point in this plane corresponds to a unique instantaneous state of the pendulum. [Strictly speaking, we should also consider to be a dynamical variable, since it appears explicitly on the right-hand side of equation (5.11).]

It is clear, from equation (5.15) and (5.16), that if we wait long enough for all of the transients to decay away then the motion of the pendulum settles down to the following simple orbit in phase-space:

$$\theta(t) = \frac{A\left[\left(1-\omega^2\right)\cos\omega t + (\omega/Q)\sin\omega t\right]}{\left[\left(1-\omega^2\right)^2 + \omega^2/Q^2\right]}, \qquad \text{...5.18}$$

$$v(t) = \frac{\omega A\left[-\left(1-\omega^2\right)\sin\omega t + (\omega/Q)\cos\omega t\right]}{\left[\left(1-\omega^2\right)^2 + \omega^2/Q^2\right]}. \qquad \text{...5.19}$$

This orbit traces out the closed curve

$$\left(\frac{\theta}{\tilde{A}}\right)^2 + \left(\frac{v}{\omega\tilde{A}}\right)^2 = 1, \qquad \text{...5.20}$$

in phase-space, where

$$\tilde{A} = \frac{A}{\sqrt{\left(1-\omega^2\right)^2 + \omega^2/Q^2}}. \qquad \text{...5.21}$$

As illustrated in figure 5.2, this curve is an *ellipse* whose principal axes are aligned with the v and θ axes. Observe that the curve is *closed*, which suggests that the associated motion is *periodic* in time. In fact, the motion repeats itself exactly every

$$\tau = \frac{2\pi}{\omega} \qquad \text{...5.22}$$

normalised time units. The maximum angular displacement of the pendulum from its undriven rest position ($\theta = 0$) is $\tilde{A}$. As illustrated in figure 5.3, the variation of $\tilde{A}$ with driving frequency ω displays all of the features of a classic resonance curve. The maximum amplitude of the driven oscillation is proportional to the quality-factor, Q, and is achieved when the driving frequency matches the natural frequency of the pendulum (*i.e.*, when $|\omega| = 1$). Moreover, the width of the resonance in ω-space is proportional to $1/Q$.

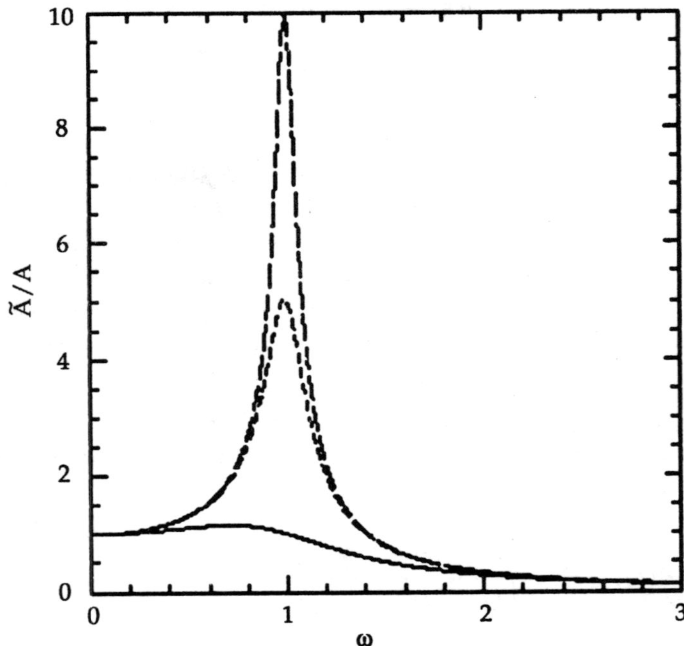

Figure 5.3: The maximum angular displacement of a linear, damped, periodically driven, pendulum as a function of driving frequency. The solid curve corresponds to $Q=1$. The short-dashed curve corresponds to $Q=5$. The long-dashed curve corresponds to $Q=10$. Analytic data

The phase-space curve shown in figure 5.2 is called a *periodic attractor*. It is termed an "attractor" because, irrespective of the initial conditions, the trajectory of the system in phase-space tends asymptotically to—in other words, is attracted to—this curve as $t \to \infty$. This gravitation of phase-space trajectories towards the attractor is illustrated in figure 5.4 and 5.5.

Of course, the attractor is termed "periodic" because it corresponds to motion which is periodic in time.

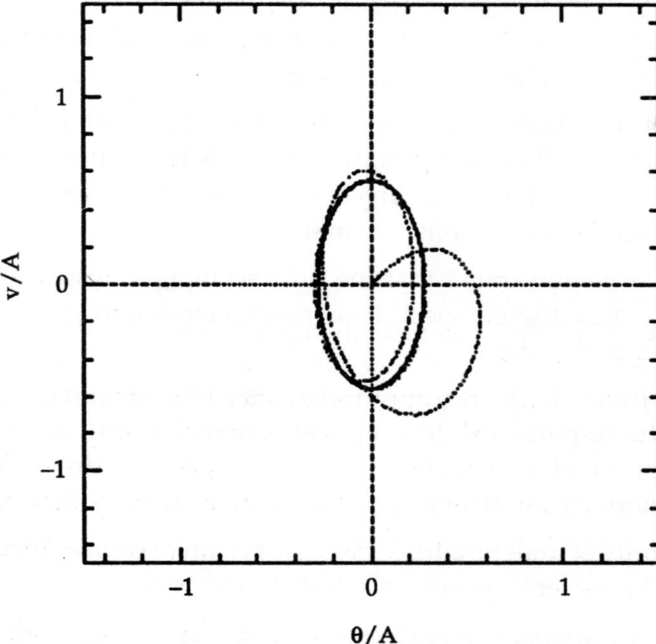

Figure 5.4: *The phase-space trajectory of a linear, damped, periodically driven, pendulum. Data calculated analytically for* $Q=1$ *and* $\omega=2$. *Here,* $v(0)/A=0$ *and* $\theta(0)/A=0$

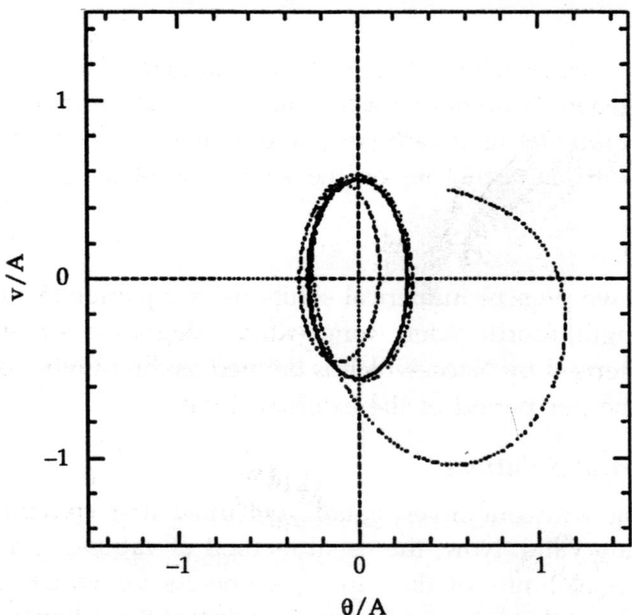

Figure 5.5: *The phase-space trajectory of a linear, damped, periodically driven, pendulum. Data calculated analytically for* $Q=1$ *and* $\omega=2$. *Here,* $v(0)A=0.5$ *and* $\theta(0)/A=0.5$

Let us summarise our findings, so far. We have discovered that if a damped pendulum is subject to a low amplitude, periodic, drive then its *time-asymptotic* response (*i.e.*, its response after any transients have died away) is *periodic*, with the same period as the driving torque.

Moreover, the response exhibits *resonant* behaviour as the driving frequency approaches the natural frequency of oscillation of the pendulum.

The amplitude of the resonant response, as well as the width of the resonant window, is governed by the amount of damping in the system. After a little reflection, we can easily appreciate that all of these results are a direct consequence of the *linearity* of the pendulum's equations of motion in the low amplitude limit.

In fact, it is easily demonstrated that the time-asymptotic response of *any* intrinsically stable linear system (with a discrete spectrum of normal modes) to a periodic drive is periodic, with the same period as the drive.

Moreover, if the driving frequency approaches one of the natural frequencies of oscillation of the system then the response exhibits resonant behaviour. But, is this the only allowable time-asymptotic response of a dynamical system to a periodic drive? Most undergraduate students might be forgiven for answering this question in the affirmative.

After all, the majority of undergraduate classical dynamics courses focus almost exclusively on linear systems. The correct answer, as we shall see, is no.

The response of a *non-linear* system to a periodic drive is generally far more rich and diverse than simple periodic motion. Since the majority of naturally occurring dynamical systems are non-linear, it is clearly important that we gain a basic understanding of this phenomenon.

Unfortunately, we cannot achieve this goal via a standard analytic approach—non-linear equations of motion generally do not possess simple analytic solutions. Instead, we must use *computers*. As an example, let us investigate the dynamics of a damped pendulum, subject to a periodic drive, with *no restrictions* on the amplitude of the pendulum's motion.

Numerical Solution

In the following, we present numerical solutions of equation (5.10) and (5.11) obtained using a fixed step-length, fourth-order, Runge-Kutta integration scheme. The step-length is conveniently parameterised by Nacc, which is defined as the number of time-steps taken by the integration scheme per period of the external drive.

Validation of Numerical Solutions

Before proceeding with our investigation, we must first convince ourselves that our numerical solutions are valid. Now, the usual method of validating a numerical solution is to look for some special limits of the input parameters for which analytic solutions are available, and then to test the numerical solution in one of these limits against the associated analytic solution.

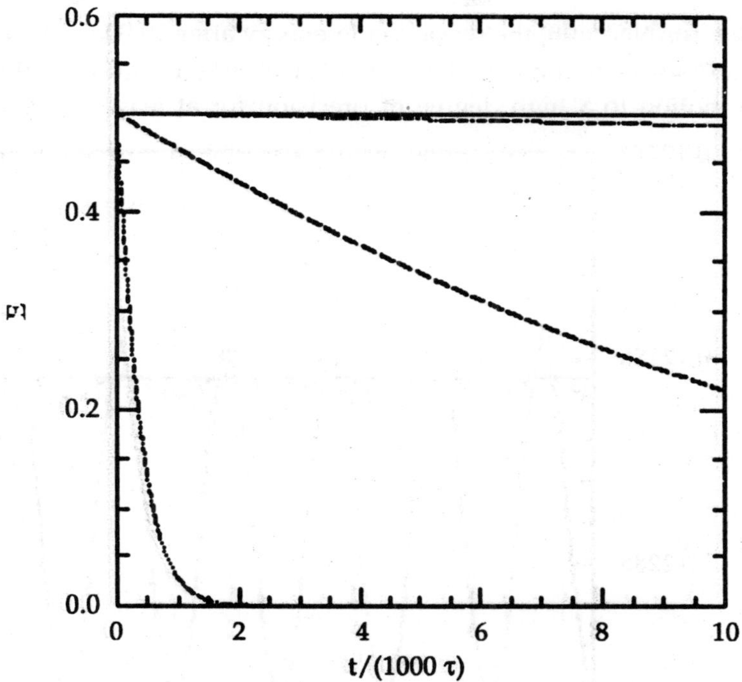

Figure 5.6: *The normalised energy* ε *of an undamped, undriven, pendulum versus time (measured in natural periods of oscillation* τ*). Data calculated numerically for* $Q=10^{16}$, $A=10^{-16}$, $\omega=1$, $\theta(0)=0$, *and* $v(0)=1$. *The dotted curve shows data for Nacc=12. The dashed curve shows data for Nacc=24. The dot-dashed curve shows data for Nacc=48. Finally, the solid curve shows data for Nacc=96*

One special limit of equation (5.10) and (5.11) occurs when there is no viscous damping (*i.e.*, $Q \rightarrow \infty$) and no external driving (*i.e.*, $A \rightarrow 0$). In this case, we expect the normalised energy of the pendulum

$$\varepsilon = 1 + \frac{v^2}{2} - \cos\theta \qquad \text{...5.23}$$

to be a *constant* of the motion. Note that ε is defined such that the energy is zero when the pendulum is in its stable equilibrium state (*i.e.*, at rest, pointing vertically downwards). Figure 5.6 shows ε versus time, calculated numerically for an undamped, undriven, pendulum. Curves are plotted for various values of the parameter Nacc, which, in this special case, measures the number of time-steps taken by the integrator per (low amplitude) natural period of oscillation of the pendulum.

It can be seen that for Nacc=12 there is a strong spurious loss of energy, due to truncation error in the numerical integration scheme, which eventually drains all energy from the pendulum after about 2000 oscillations. For Nacc=24, the spurious energy loss is less severe, but, nevertheless, still causes a more than 50 per cent reduction in pendulum energy after 10,000 oscillations. For Nacc=48, the reduction in energy after 10,000 oscillations is only about

1 per cent. Finally, for Nacc=96, the reduction in energy after 10,000 oscillation is completely negligible. This test seems to indicate that when Nacc $\geq$ 100 our numerical solution describes the pendulum's motion to a high degree of precision for at least 10,000 oscillations.

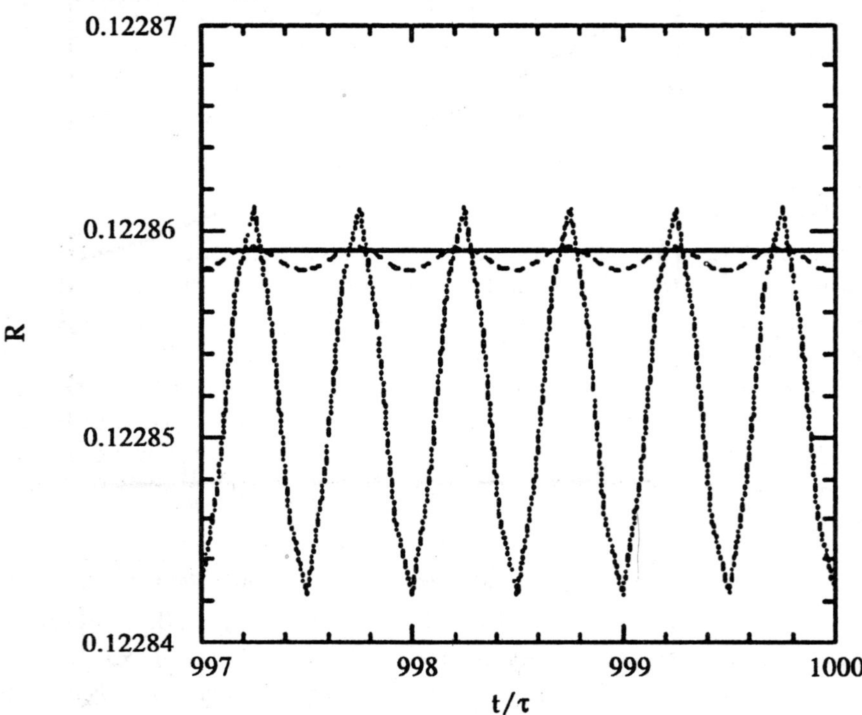

Figure 5.7: *The parameter R associated with a linearised, damped, periodically driven, pendulum versus time (measured in units of the period of oscillation of the external drive). Data calculated numerically for* $Q=2$, $A=1$, $\omega=3$, $\theta(0)=0$, *and* $v(0)=0$. *The dotted curve shows data for Nacc=12. The dashed curve shows data for Nacc=24. The solid curve shows data for Nacc=48*

Another special limit of equation (5.10) and (5.11) occurs when these equations are *linearised* to give equation (5.13) and (5.14). In this case, we expect

$$R=\sqrt{\theta^2+(v/\omega)^2} \qquad \text{...5.24}$$

to be a constant of the motion, *after* all transients have died away. Figure 5.7 shows R versus time, calculated numerically, for a linearised, damped, periodically driven, pendulum. Curves are plotted for various values of the parameter Nacc, which measures the number of time-steps taken by the integrator per period of oscillation of the external drive.

As Nacc increases, it can be seen that the amplitude of the spurious oscillations in R, which are due to truncation error in the numerical integration scheme, decreases rapidly. Indeed, for Nacc $\geq$ 48 these oscillations become effectively undetectable. According to the analysis in Sect. 4.2, the parameter R should take the value

$$R = \frac{A}{\sqrt{\left(1-\omega^2\right)^2 + \omega^2 / Q^2}}. \qquad ...5.25$$

Thus, for the case in hand (*i.e.*, $Q=2, A=1, \omega=3$), we expect R=0.122859. It can be seen that this prediction is borne out very accurately in figure 5.7. The above test essentially confirms our previous conclusion that when Nacc$\geq$100 our numerical solution matches pendulum's actual motion to a high degree of accuracy for many thousands of oscillation periods.

Poincaré Section

For the sake of definiteness, let us fix the normalised amplitude and frequency of the external drive to be $A=1.5$ and $\omega=2/3$, respectively. Furthermore, let us investigate any changes which may develop in the nature of the pendulum's time-asymptotic motion as the quality-factor Q is varied. Of course, if Q is made sufficiently small (*i.e.*, if the pendulum is embedded in a sufficiently viscous medium) then we expect the amplitude of the pendulum's time-asymptotic motion to become low enough that the linear analysis outlined in Sect. 4.2 remains valid. Indeed, we expect non-linear effects to manifest themselves as Q is gradually made larger, and the amplitude of the pendulum's motion consequently increases to such an extent that the small angle approximation breaks down.

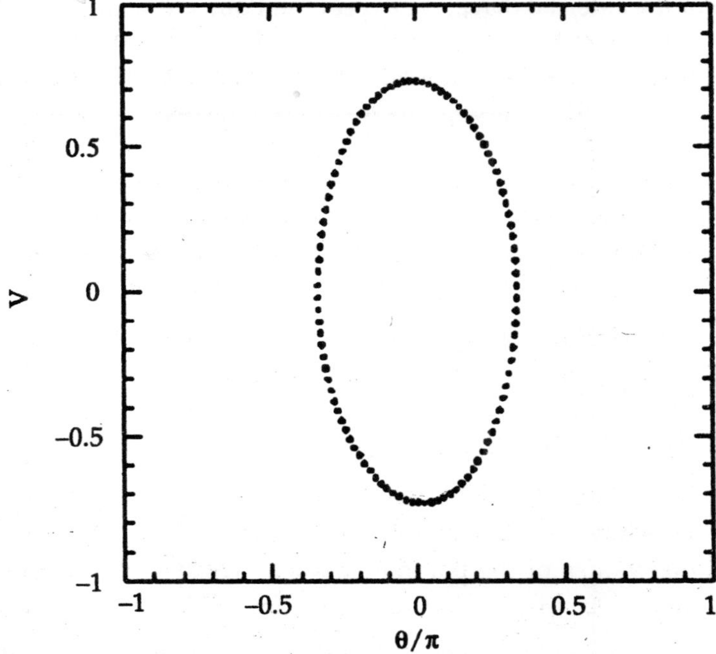

Figure 5.8: *Equally spaced (in time) points on a time-asymptotic orbit in phase-space. Data calculated numerically for* $Q=0.5$, $A=1.5$, $\omega=2/3$, $\theta(0)=0$, $v(0)=0$, *and* $Nacc=100$

Figure 5.8 shows a time-asymptotic orbit in phase-space calculated numerically for a case where Q is sufficiently small (*i.e.*, $Q=1/2$) that the small angle approximation holds reasonably well. Not surprisingly, the orbit is very similar to the analytic orbits described in Sect. 4.2. The fact that the orbit consists of a *single* loop, and forms a *closed* curve in phase-space, strongly suggests that the corresponding motion is periodic with the same period as the external drive—we term this type of motion *period-1* motion. More generally, period-n motion consists of motion which repeats itself exactly every n periods of the external drive (and, obviously, does not repeat itself on any time-scale less than n periods). Of course, period-1 motion is the only allowed time-asymptotic motion in the small angle limit.

It would certainly be helpful to possess a graphical test for period-n motion. In fact, such a test was developed more than a hundred years ago by the French mathematician Henry Poincaré—nowadays, it is called a *Poincaré section* in his honour. The idea of a Poincaré section, as applied to a periodically driven pendulum, is very simple. As before, we calculate the time-asymptotic motion of the pendulum, and visualise it as a series of points in $\theta - v$ phase-space. However, we only plot *one point per period* of the external drive. To be more exact, we only plot a point when

$$\omega t = \phi + k2\pi \qquad \text{...5.26}$$

where k is any integer, and ϕ is referred to as the *Poincaré phase*. For period-1 motion, in which the motion repeats itself exactly every period of the external drive, we expect the Poincaré section to consist of only *one* point in phase-space (*i.e.*, we expect all of the points to plot on top of one another). Likewise, for period-2 motion, in which the motion repeats itself exactly every two periods of the external drive, we expect the Poincaré section to consist of *two* points in phase-space (*i.e.*, we expect alternating points to plot on top of one another). Finally, for period-n motion we expect the Poincaré section to consist of n points in phase-space.

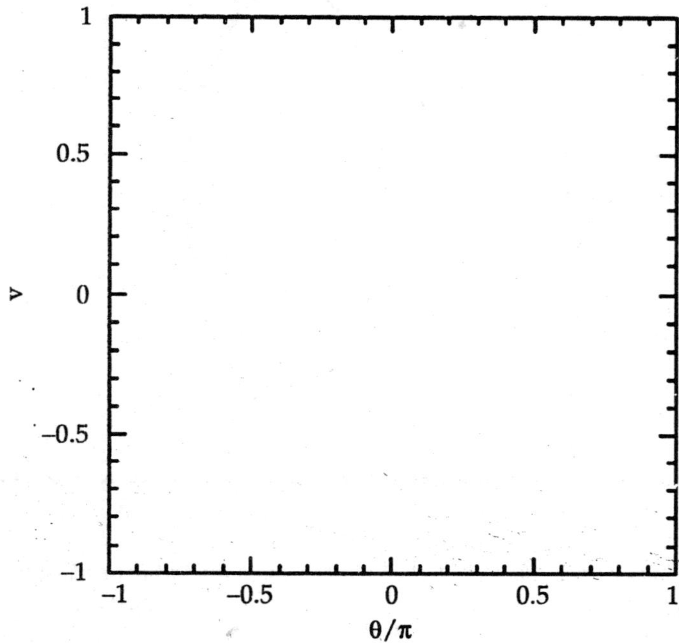

Figure 5.9: The Poincaré section of a time-asymptotic orbit. Data calculated numerically for $Q = 0.5$, $A = 1.5$, $\omega = 2/3$, $\theta(0) = 0$, $v(0) = 0$, $Nacc = 100$, *and* $\phi = 0$

Figure 5.9 displays the Poincaré section of the orbit shown in figure 5.8. The fact that the section consists of a single point confirms that the motion displayed in figure 5.8 is indeed period-1 motion.

Spatial Symmetry Breaking

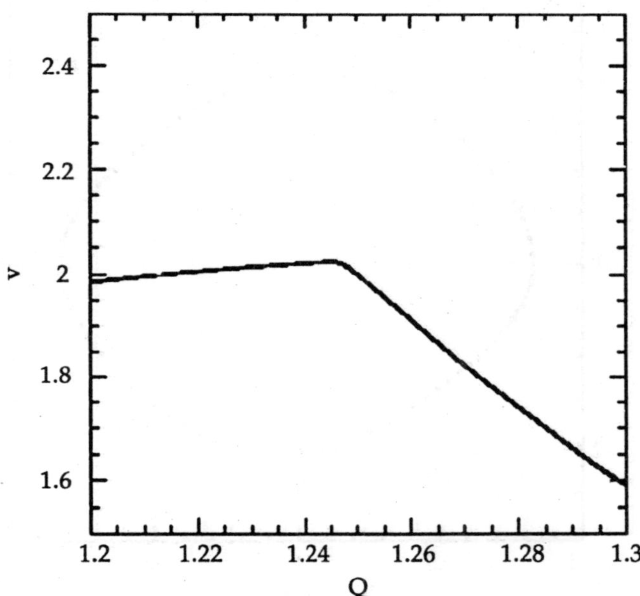

Figure 5.10: *The v-coordinate of the Poincaré section of a time-asymptotic orbit plotted against the quality-factor Q. Data calculated numerically for* $A = 1.5$, $\omega = 2/3$, $\theta(0) = 0$, $v(0) = 0$, $Nacc = 100$, *and* $\phi = 0$

Suppose that we now gradually increase the quality-factor Q. What happens to the simple orbit shown in figure 5.8? It turns out that, at first, nothing particularly exciting happens. The size of the orbit gradually increases, indicating a corresponding increase in the amplitude of the pendulum's motion, but the general nature of the motion remains unchanged. However, something interesting does occur when Q is increased beyond about 1.2. Figure 5.10 shows the v-coordinate of the orbit's Poincaré section plotted against Q in the range 1.2 and 1.3. Note the sharp downturn in the curve at $Q \simeq 1.245$.

What does this signify? Well, figure 5.11 shows the time-asymptotic phase-space orbit just before the downturn (*i.e.*, at $Q = 1.24$), and figure 5.12 shows the orbit somewhat after the downturn (*i.e.*, at $Q = 1.30$). It is clear that the downturn is associated with a sudden change in the nature of the pendulum's time-asymptotic phase-space orbit. Prior to the downturn, the orbit spends as much time in the region $\theta < 0$ as in the region $\theta > 0$. However, after the downturn the orbit spends the majority of its time in the region $\theta < 0$. In other words, after the downturn the pendulum bob favours the region to the *left* of the pendulum's vertical. This is somewhat surprising, since there is nothing in the pendulum's equations of motion which differentiates between the regions to the left and to the right of the vertical. We refer to a solution of this type—which fails to realise the full symmetry of the dynamical system in question—as a *symmetry breaking* solution. In this case, because the particular symmetry which is broken is a *spatial* symmetry, we refer to the process by which the symmetry breaking solution suddenly appears, as the control parameter Q is adjusted, as *spatial symmetry breaking*. Needless to say, spatial symmetry breaking is an intrinsically *non-linear* process—it cannot take place in dynamical systems possessing linear equations of motion.

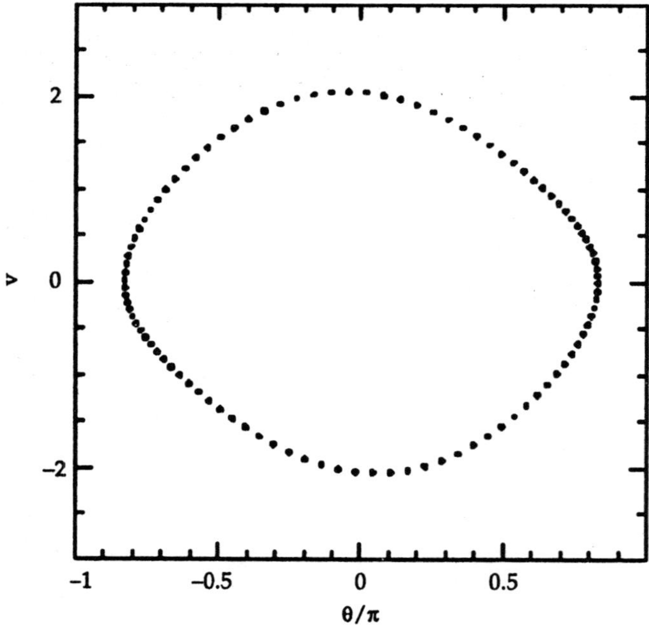

***Figure 5.11:** Equally spaced (in time) points on a time-asymptotic orbit in phase-space. Data calculated numerically for* $Q = 1.24$, $A = 1.5$, $\omega = 2/3$, $\theta(0) = 0$, $v(0) = 0$, *and* $Nacc = 100$

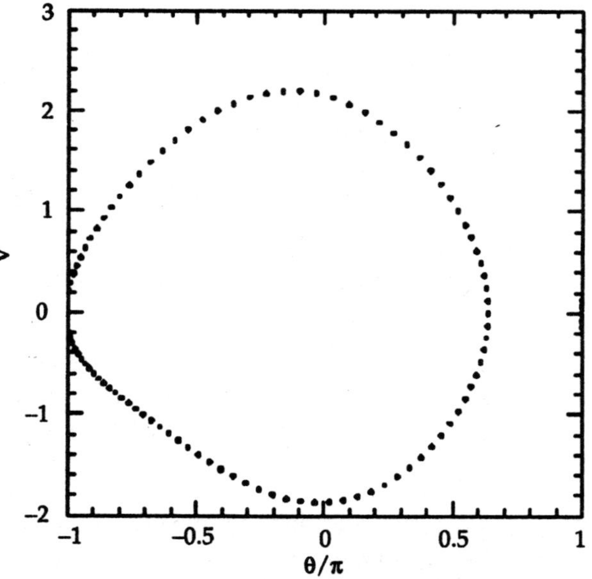

***Figure 5.12:** Equally spaced (in time) points on a time-asymptotic orbit in phase-space. Data calculated numerically for* $Q = 1.30$, $A = 1.5$, $\omega = 2/3$, $\theta(0) = 0$, $v(0) = 0$, *and* $Nacc = 100$

It stands to reason that since the pendulum's equations of motion favour neither the left nor the right then the left-favouring orbit pictured in figure 5.12 must be accompanied by a mirror image right-favouring orbit.

How do we obtain this mirror image orbit? It turns out that all we have to do is keep the physics parameters Q, A, and ω fixed, but *change the initial conditions* $\theta(0)$ and $v(0)$. Figure 5.13 shows a time-asymptotic phase-space orbit calculated with the same physics parameters used in figure 5.12, but with the initial conditions $\theta(0)=0$ and $v(0)=-3$, instead of $\theta(0)=0$ and $v(0)=0$.

It can be seen that the orbit is indeed the mirror image of that pictured in figure 5.12.

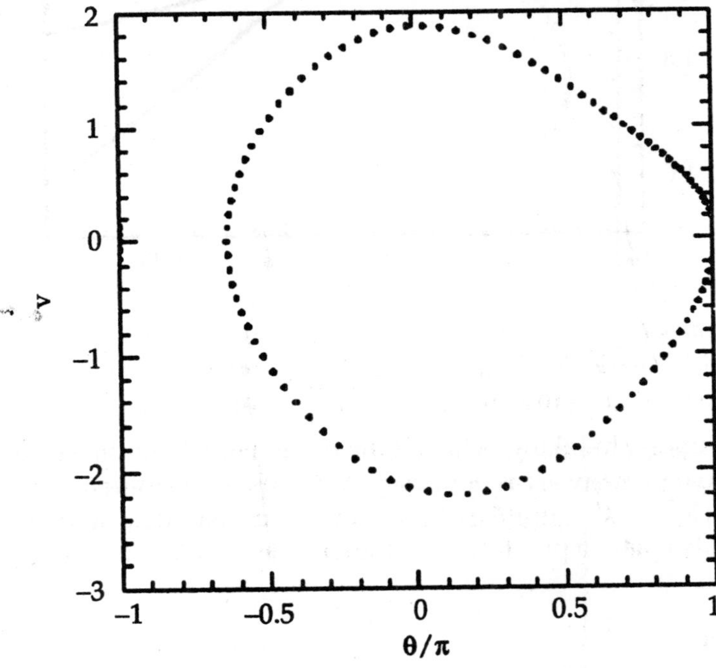

Figure 5.13: *Equally spaced (in time) points on a time-asymptotic orbit in phase-space. Data calculated numerically for* $Q=1.30$, $A=1.5$, $\omega=2/3$, $\theta(0)=0$, $v(0)=-3$, *and* $Nacc=100$

Figure 5.14 shows the v-coordinate of the Poincaré section of a time-asymptotic orbit, calculated with the same physics parameters used in figure 5.10, versus Q in the range 1.2 and 1.3. The figure is interpreted as follows.

When Q is less than a critical value, which is about 1.245, then the two sets of initial conditions lead to motions which converge on the *same*, left-right symmetric, period-1 attractor. However, once Q exceeds the critical value then the attractor *bifurcates* into two asymmetric, mirror image, period-1 attractors.

Obviously, the bifurcation is indicated by the forking of the curve shown in figure 5.14. The lower and upper branches correspond to the left- and right-favouring attractors, respectively.

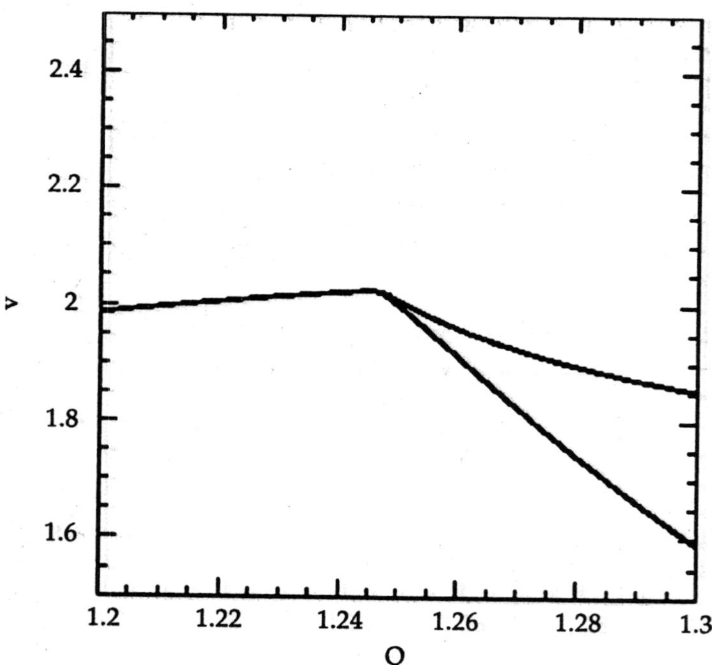

Figure 5.14: *The v-coordinate of the Poincaré section of a time-asymptotic orbit plotted against the quality-factor Q. Data calculated numerically for* $A = 1.5$, $\omega = 2/3$, *and Nacc=100. Data is shown for two sets of initial conditions:* $\theta(0) = 0$ *and* $v(0) = 0$ *(lower branch); and* $\theta(0) = 0$ *and* $v(0) = -3$ *(upper branch)*

Spontaneous symmetry breaking, which is the fundamental non-linear process illustrated, plays an important role in many areas of physics. For instance, symmetry breaking gives mass to elementary particles in the unified theory of electromagnetic and weak interactions. Symmetry breaking also plays a pivotal role in the so-called "inflation" theory of the expansion of the early universe.

Basins of Attraction

We have seen that when $Q = 1.3$, $A = 1.5$, and $\omega = 2/3$ there are two co-existing period-1 attractors in $\theta - v$ phase-space. The time-asymptotic trajectory of the pendulum through phase-space converges on one or other of these attractors depending on the initial conditions: i.e., depending on the values of $\theta(0)$ and $v(0)$.

Let us define the *basin of attraction* of a given attractor as the locus of all points in the $\theta(0)$-$v(0)$ plane which lead to motion which ultimately converges on that attractor. We have seen that in the low-amplitude (*i.e.,* linear) limit there is only a single period-1 attractor in phase-space, and all possible initial conditions lead to motion which converges on this attractor.

In other words, the basin of attraction for the low-amplitude attractor constitutes the entire $\theta(0)$-$v(0)$ plane. The present case, in which there are *two* co-existing attractors in phase-space, is somewhat more complicated.

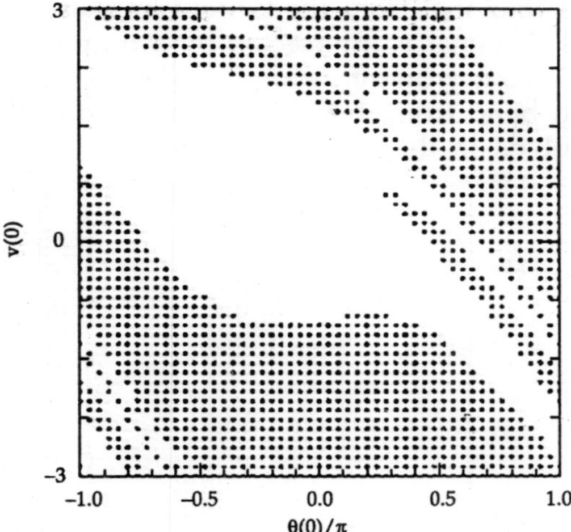

Figure 5.15: *The basins of attraction for the asymmetric, mirror image, attractors pictured in figure 5.12 and 5.13. Regions of $\theta(0)-v(0)$ space which lead to motion converging on the left-favouring attractor shown in figure 5.12 are coloured white: regions of $\theta(0)-v(0)$ space which lead to motion converging on the right-favouring attractor shown in figure 5.13 are coloured black. Data calculated numerically for* $Q=1.3$, $A=1.5$, $\omega=2/3$, $Nacc=100$, *and* $\phi=0$

Figure 5.15 shows the basins of attraction, in $\theta(0)-v(0)$ space, of the asymmetric, mirror image, attractors pictured in figure 5.12 and 5.13. The basin of attraction of the left-favouring attractor shown in figure 5.12 is coloured black, whereas the basin of attraction of the right-favouring attractor shown in figure 5.13 is coloured white. It can be seen that the two basins form a complicated interlocking pattern. Since we can identify the angles π and $-\pi$, the right-hand edge of the pattern connects smoothly with its left-hand edge. In fact, we can think of the pattern as existing on the surface of a *cylinder*.

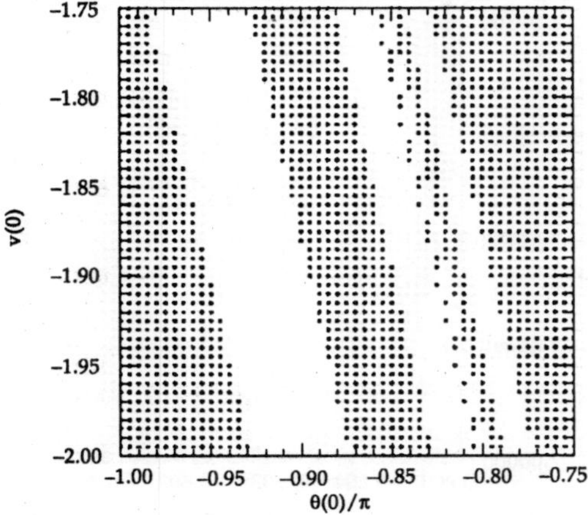

Figure 5.16: *Detail of the basins of attraction for the asymmetric, mirror image, attractors pictured in figure 5.12 and 5.13. Regions of $\theta(0)-v(0)$ space which lead to motion converging on the left-favouring attractor shown in figure 5.12 are coloured white: regions of $\theta(0)-v(0)$ space which lead to motion converging on the right-favouring attractor shown in figure 5.13 are coloured black. Data calculated numerically for* $Q=1.3$, $A=1.5$, $\omega=2/3$, *Nacc=100, and* $\phi=0$

Suppose that we take a diagonal from the bottom left-hand corner of figure 5.15 to its top right-hand corner. This diagonal is intersected by a number of black bands of varying thickness. Observe that the two narrowest bands (*i.e.*, the fourth band from the bottom left-hand corner and the second band from the upper right-hand corner) both exhibit structure which is not very well resolved in the present picture.

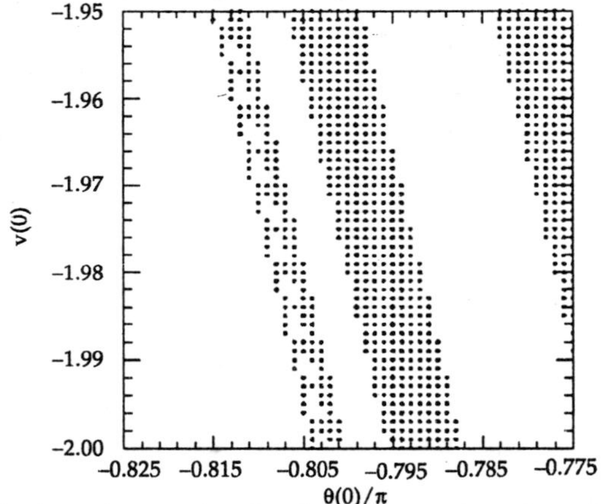

Figure 5.17: *Detail of the basins of attraction for the asymmetric, mirror image, attractors pictured in figure 5.12 and 5.13. Regions of* $\theta(0)$ - $v(0)$ *space which lead to motion converging on the left-favouring attractor shown in figure 5.12 are coloured white: regions of* $\theta(0)$ - $v(0)$ *space which lead to motion converging on the right-favouring attractor shown in figure 5.13 are coloured black. Data calculated numerically for* $Q=1.3$, $A=1.5$, $\omega=2/3$, $Nacc=100$, *and* $\phi=0$

Figure 5.16 is a blow-up of a region close to the lower left-hand corner of figure 5.15. It can be seen that the unresolved band in the latter figure (*i.e.*, the second and third bands from the right-hand side in the former figure) actually consists of a closely spaced *pair* of bands. Note, however, that the narrower of these two bands exhibits structure which is not very well resolved in the present picture.

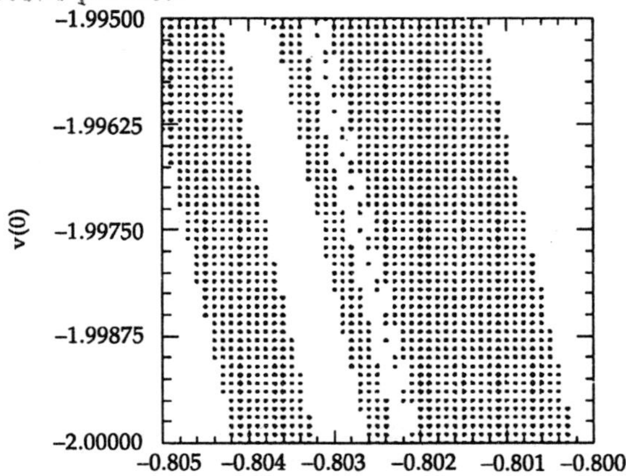

Figure 5.18: *Detail of the basins of attraction for the asymmetric, mirror image, attractors pictured in figure 5.12 and 5.13. Regions of* $\theta(0)$ - $v(0)$ *space which lead to motion converging on the left-favouring attractor shown in figure 5.12 are coloured white: regions of* $\theta(0)$ - $v(0)$ *space which lead to motion converging on the right-favouring attractor shown in figure 5.13 are coloured black. Data calculated numerically for* $Q=1.3$, $A=1.5$, $\omega=2/3$, $Nacc=100$, *and* $\phi=0$

Figure 5.17 is a blow-up of a region of figure 5.16. It can be seen that the unresolved band in the latter figure (*i.e.*, the first and second bands from the left-hand side in the former figure) actually consists of a closely spaced *pair* of bands. Note, however, that the broader of these two bands exhibits structure which is not very well resolved in the present picture.

Figure 5.18 is a blow-up of a region of figure 5.17. It can be seen that the unresolved band in the latter figure (*i.e.*, the first, second, and third bands from the right-hand side in the former figure) actually consists of a closely spaced *triplet* of bands. Note, however, that the narrowest of these bands exhibits structure which is not very well resolved in the present picture.

It should be clear, by this stage, that no matter how closely we look at figure 5.15 we are going to find structure which we cannot resolve. In other words, the separatrix between the two basins of attraction shown in this figure is a curve which exhibits structure *at all scales*. Mathematicians have a special term for such a curve—they call it a *fractal*.

Many people think of fractals as mathematical toys whose principal use is the generation of pretty pictures. However, it turns out that there is a close connection between fractals and the dynamics of non-linear systems—particularly systems which exhibit chaotic dynamics. We have just seen an example in which the boundary between the basins of attraction of two co-existing attractors in phase-space is a fractal curve.

This turns out to be a fairly general result: i.e., when multiple attractors exist in phase-space the separatrix between their various basins of attraction is invariably fractal. What is this telling us about the nature of non-linear dynamics? Well, returning to figure 5.15, we can see that in the region of phase-space in which the fractal behaviour of the separatrix manifests itself most strongly (*i.e.*, the region where the light and dark bands fragment) the system exhibits abnormal sensitivity to initial conditions. In other words, we only have to change the initial conditions slightly (*i.e.*, so as to move from a dark to a light band, or *vice versa*) in order to significantly alter the time-asymptotic motion of the pendulum (*i.e.*, to cause the system to converge to a left-favouring instead of a right-favouring attractor, or *vice versa*). Fractals and extreme sensitivity to initial conditions are themes which will recur in our investigation of non-linear dynamics.

Period-doubling Bifurcations

Let us now return to figure 5.10. Recall, that as the quality-factor Q is gradually increased, the time-asymptotic orbit of the pendulum through phase-space undergoes a sudden transition, at $Q \simeq 1.245$, from a left-right symmetric, period-1 orbit to a left-favouring, period-1 orbit. What happens if we continue to increase Q? Figure 5.19 is basically a continuation of figure 5.10.

It can be seen that as Q is increased the left-favouring, period-1 orbit gradually evolves until a critical value of Q, which is about 1.348, is reached. When Q exceeds this critical value the nature of the orbit undergoes another sudden change: this time from a left-favouring, period-1 orbit to a left-favouring, *period*-2 orbit.

Obviously, the change is indicated by the forking of the curve in figure 5.19. This type of transition is termed a *period-doubling bifurcation*, since it involves a sudden doubling of the repetition period of the pendulum's time-asymptotic motion.

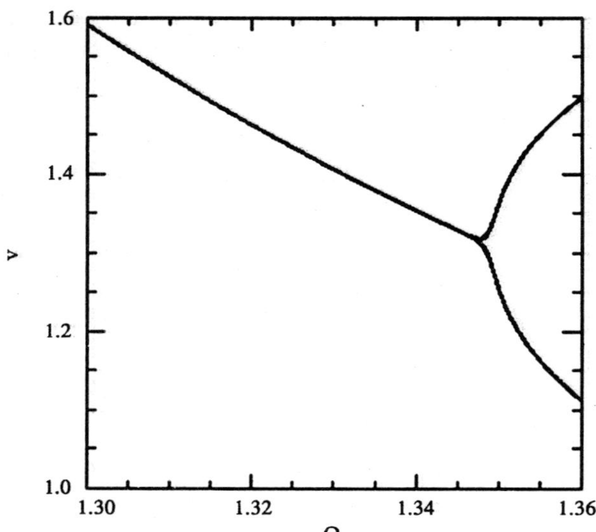

Figure 5.19: *The v-coordinate of the Poincaré section of a time-asymptotic orbit plotted against the quality-factor Q. Data calculated numerically for* $A = 1.5$, $\omega = 2/3$, $\theta(0) = 0$, $v(0) = 0$, $Nacc = 100$, *and* $\phi = 0$

We can represent period-1 motion schematically as $AAAAAA\cdots$, where A represents a pattern of motion which is repeated every period of the external drive. Likewise, we can represent period-2 motion as $ABABAB\cdots$, where A and B represent *distinguishable* patterns of motion which are repeated every alternate period of the external drive. A period-doubling bifurcation is represented: $AAAAAA\cdots \to ABABAB\cdots$. Clearly, all that happens in such a bifurcation is that the pendulum suddenly decides to do something slightly different in alternate periods of the external drive.

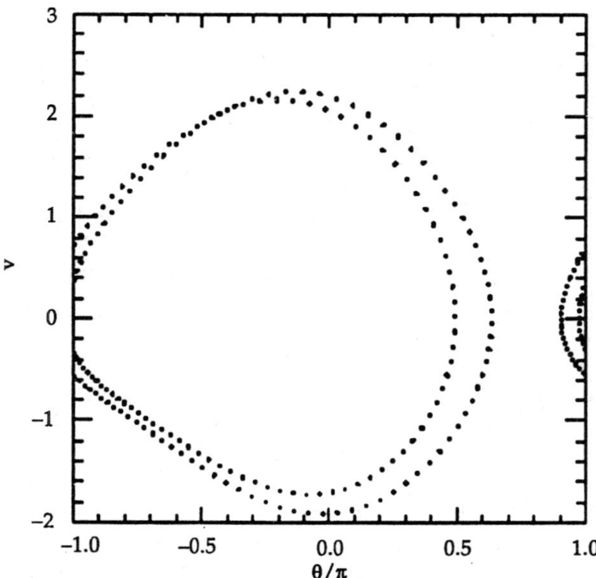

Figure 5.20: *Equally spaced (in time) points on a time-asymptotic orbit in phase-space. Data calculated numerically for* $Q = 1.36$, $A = 1.5$, $\omega = 2/3$, $\theta(0) = 0$, $v(0) = -3$, *and* $Nacc = 100$

Figure 5.20 shows the time-asymptotic phase-space orbit of the pendulum calculated for a value of Q somewhat higher than that required to trigger the above mentioned period-doubling bifurcation.

It can be seen that the orbit is left-favouring (*i.e.*, it spends the majority of its time on the left-hand side of the plot), and takes the form of a closed curve consisting of *two* interlocked loops in phase-space. Recall that for period-1 orbits there was only a single closed loop in phase-space.

Figure 5.21 shows the Poincaré section of the orbit shown in figure 5.20. The fact that the section consists of *two* points confirms that the orbit does indeed correspond to period-2 motion.

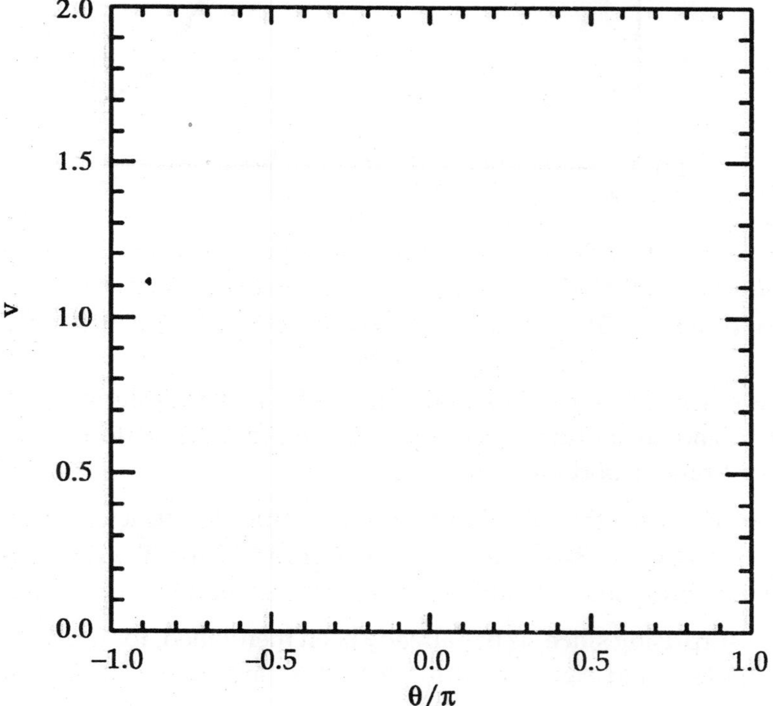

Figure 5.21: The Poincaré section of a time-asymptotic orbit. Data calculated numerically for $Q = 1.36$, $A = 1.5$, $\omega = 2/3$, $\theta(0) = 0$, $v(0) = 0$, *Nacc=100, and* $\phi = 0$

A period-doubling bifurcation is an example of *temporal symmetry breaking*. The equations of motion of the pendulum are invariant under the transformation $t \to t + \tau$, where τ is the period of the external drive. In the low amplitude (*i.e.*, linear) limit the time-asymptotic motion of the pendulum always respects this symmetry.

However, as we have just seen, in the non-linear regime it is possible to obtain solutions which spontaneously break this symmetry. Obviously, motion which repeats itself every two periods of the external drive is not as temporally symmetric as motion which repeats every period of the drive.

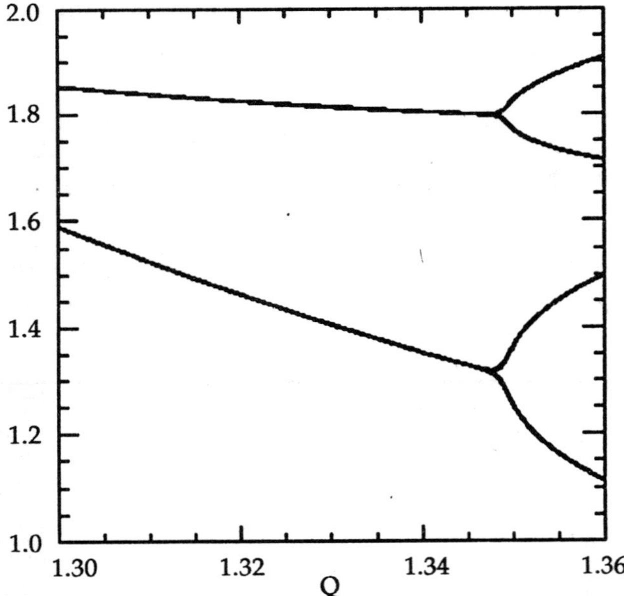

Figure 5.22: The v-coordinate of the Poincaré section of a time-asymptotic orbit plotted against the quality-factor Q. Data calculated numerically for $A=1.5$, $\omega=2/3$, Nacc=100, and $\phi=0$. Data is shown for two sets of initial conditions: $\theta(0)=0$ and $v(0)=0$ (lower branch); and $\theta(0)=0$ and $v(0)=-2$ (upper branch)

Figure 5.22 is essentially a continuation of figure 5.14. Data is shown for two sets of initial conditions which lead to motions converging on left-favouring (lower branch) and right-favouring (upper branch) periodic attractors.

We have already seen that the left-favouring periodic attractor undergoes a period-doubling bifurcation at $Q=1.348$. It is clear from figure 5.22 that the right-favouring attractor undergoes a similar bifurcation at almost exactly the same Q-value.

This is hardly surprising since, as has already been mentioned, for fixed physics parameters (*i.e.*, Q, A, ω), the left- and right-favouring attractors are mirror-images of one another.

Route to Chaos

Let us return to figure 5.19, which tracks the evolution of a left-favouring periodic attractor as the quality-factor Q is gradually increased. Recall that when Q exceeds a critical value, which is about 1.348, then the attractor undergoes a period-doubling bifurcation which converts it from a period-1 to a period-2 attractor.

This bifurcation is indicated by the forking of the curve in figure 5.19. Let us now investigate what happens as we continue to increase Q. figure 5.23 is basically a continuation of figure 5.19.

It can be seen that, as Q is gradually increased, the attractor undergoes a period-doubling bifurcation at $Q=1.348$, as before, but then undergoes a *second* period-doubling bifurcation (indicated by the second forking of the curves) at $Q \simeq 1.370$, and a *third* bifurcation at $Q \simeq 1.375$. Obviously, the second bifurcation converts a period-2 attractor into a period-4 attractor (hence, two curves split apart to give four curves).

Likewise, the third bifurcation converts a period-4 attractor into a period-8 attractor (hence, four curves split into eight curves).

Shortly after the third bifurcation, the various curves in the figure seem to expand explosively and merge together to produce an area of almost solid black. As we shall see, this behaviour is indicative of the onset of *chaos*.

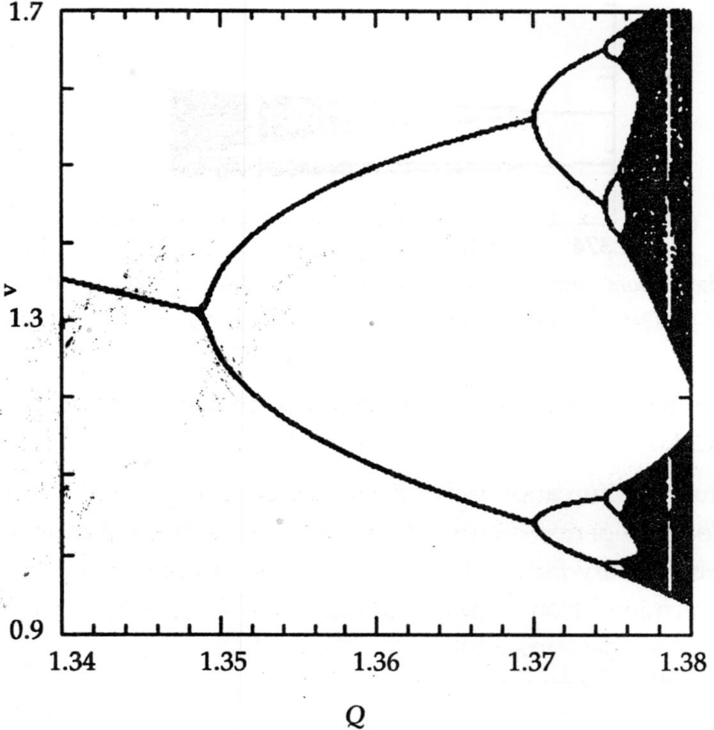

Figure 5.23: *The v-coordinate of the Poincaré section of a time-asymptotic orbit plotted against the quality-factor Q. Data calculated numerically for* $A = 1.5$, $\omega = 2/3$, $\theta(0) = 0$, $v(0) = 0$, *Nacc=100, and* $\phi = 0$

Figure 5.24 is a blow-up of figure 5.23, showing more details of the onset of chaos. The period-4 to period-8 bifurcation can be seen quite clearly. However, we can also see a period-8 to period-16 bifurcation, at $Q \simeq 1.3755$.

Finally, if we look carefully, we can see a hint of a period-16 to period-32 bifurcation, just before the start of the solid black region. Figures 5.23 and 5.24 seem to suggest that the onset of chaos is triggered by an *infinite series* of period-doubling bifurcations.

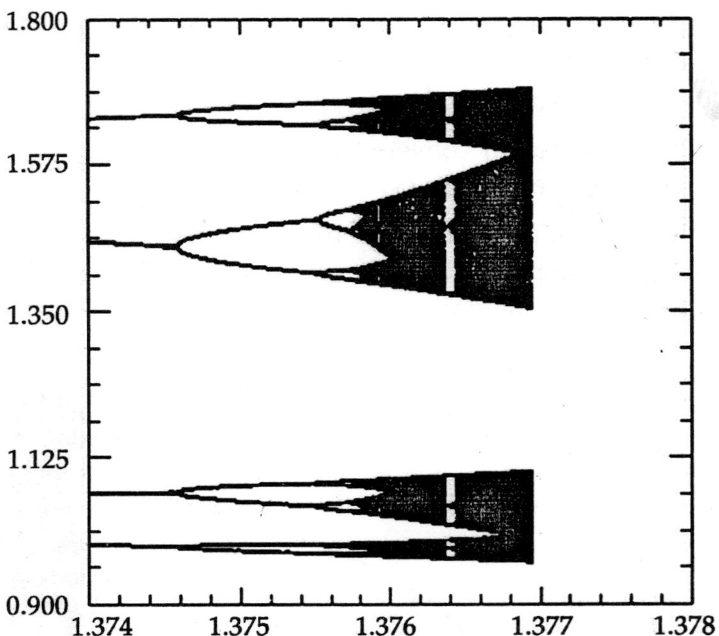

Figure 5.24: *The v-coordinate of the Poincaré section of a time-asymptotic orbit plotted against the quality-factor Q. Data calculated numerically for* $A = 1.5$, $\omega = 2/3$, $\theta(0) = 0$, $v(0) = 0$, *Nacc=100, and* $\phi = 0$

Table 5.1 gives some details of the sequence of period-doubling bifurcations shown in figure 5.23 and 5.24.

Let us introduce a bifurcation index *n*: the period-1 to period-2 bifurcation corresponds to $n = 1$; the period-2 to period-4 bifurcation corresponds to $n = 2$; and so on. Let Q_n be the critical value of the quality-factor *Q* above which the *n*th bifurcation is triggered. Table 5.1 shows the Q_n, determined from figure 5.23 and 5.24, for $n = 1$ to 5. Also shown is the ratio

$$F_n = \frac{Q_{n-1} - Q_{n-2}}{Q_n - Q_{n-1}} \qquad \text{...5.27}$$

for $n = 3$ to 5. It can be seen that Tab. 2 offers reasonably convincing evidence that this ratio takes the *constant* value $F = 4.69$. It follows that we can estimate the critical *Q*-value required to trigger the th bifurcation via the following formula:

$$Q_n = Q_1 + (Q_2 - Q_1) \sum_{j=0}^{n-2} \frac{1}{F^j}, \qquad \text{...5.28}$$

for $n > 1$. Note that the distance (in *Q*) between bifurcations decreases rapidly as increases. In fact, the above formula predicts an *accumulation* of period-doubling bifurcations at $Q = Q_\infty$ where

$$Q_\infty = Q_1 + (Q_2 - Q_1)\sum_{j=0}^{\infty}\frac{1}{F^j} \equiv Q_1 + (Q_2 - Q_1)\frac{F}{F-1} = 1.3758. \qquad ...5.29$$

Note that our calculated accumulation point corresponds almost exactly to the onset of the solid black region in figure 5.24. By the time that Q exceeds Q_∞, we expect the attractor to have been converted into a *period-infinity* attractor via an infinite series of period-doubling bifurcations.

A period-infinity attractor is one whose corresponding motion *never* repeats itself, no matter how long we wait. In dynamics, such bounded aperiodic motion is generally referred to as *chaos*.

Hence, a period-infinity attractor is sometimes called a *chaotic attractor*. Now, period-n motion is represented by n separate curves in figure 5.24. It is, therefore, not surprising that chaos (*i.e.*, period-infinity motion) is represented by an infinite number of curves which merge together to form a region of solid black.

Table 5.1: The period-doubling cascade.

Bifurcation	n	Q_n	$Q_n - Q_{n-1}$	F_n
period-1 → period-2	1	1.34870	–	–
period-2 → period-4	2	1.37003	0.02133	–
period-4 → period-8	3	1.37458	0.00455	4.69 ± 0.01
period-8 → period-16	4	1.37555	0.00097	4.69 ± 0.04
period-16 → period-32	5	1.37575	0.00020	4.9 ± 0.20

Let us examine the onset of chaos in a little more detail. Figures 5.25-5.28 show details of the pendulum's time-asymptotic motion at various stages on the period-doubling cascade. Figure 5.25 shows period-4 motion: note that the Poincaré section consists of four points, and the associated sequence of net rotations per period of the pendulum repeats itself every four periods.

Figure 5.26 shows period-8 motion: now the Poincaré section consists of eight points, and the rotation sequence repeats itself every eight periods. Figure 5.27 shows period-16 motion: as expected, the Poincaré section consists of sixteen points, and the rotation sequence repeats itself every sixteen periods.

Finally, figure 5.28 shows chaotic motion. Note that the Poincaré section now consists of a set of four *continuous* line segments, which are, presumably, made up of an infinite number of points (corresponding to the infinite period of chaotic motion).

Note, also, that the associated sequence of net rotations per period shows no obvious sign of ever repeating itself. In fact, this sequence looks rather like one of the previously shown periodic sequences with the addition of a small random component.

The generation of *apparently random* motion from equations of motion, such as equation (5.10) and (5.11), which contain no overtly random elements is one of the most surprising features of non-linear dynamics.

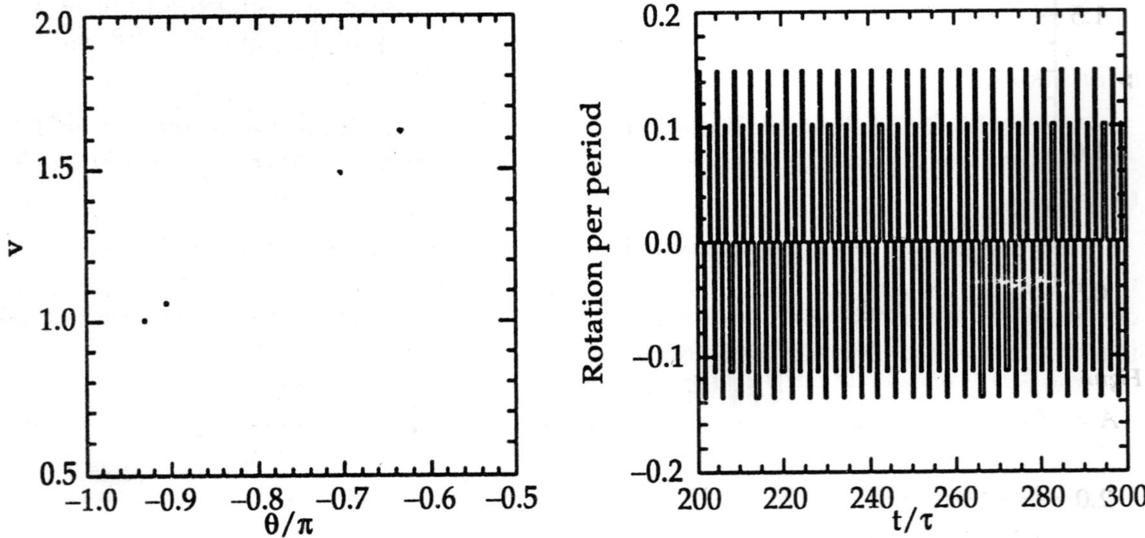

Figure 5.25: The Poincaré section of a time-asymptotic orbit. Data calculated numerically for $Q = 1.372$, $A = 1.5$, $\omega = 2/3$, $\theta(0) = 0$, $v(0) = 0$, *Nacc=100, and* $\phi = 0$. *Also, shown is the net rotation per period,* $\Delta\theta/2\pi$, *calculated at the Poincaré phase* $\phi = 0$

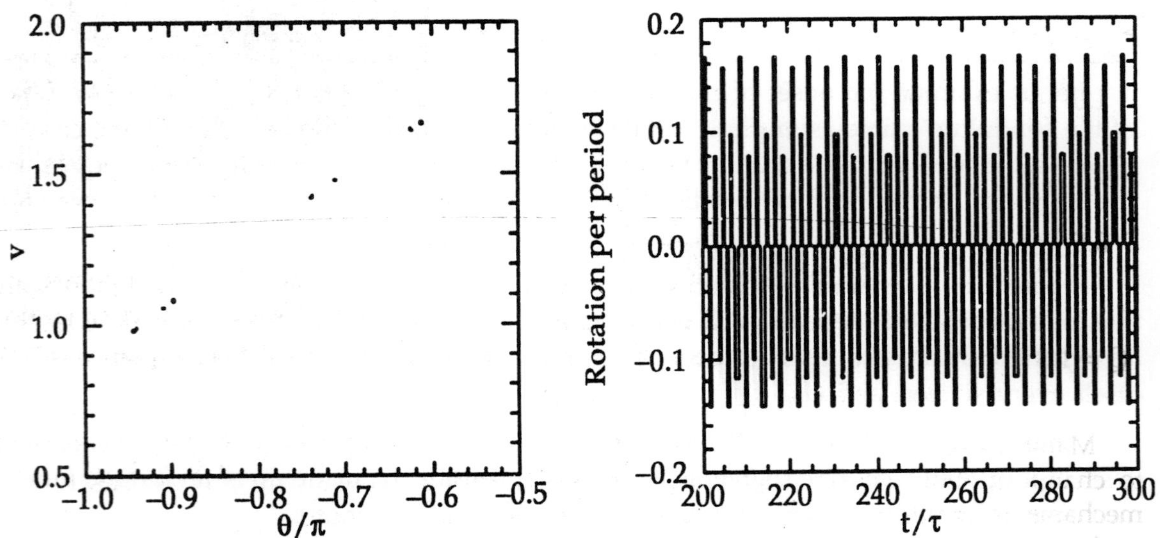

Figure 5.26: The Poincaré section of a time-asymptotic orbit. Data calculated numerically for $Q = 1.375$, $A = 1.5$, $\omega = 2/3$, $\theta(0) = 0$, $v(0) = 0$, *Nacc=100, and* $\phi = 0$. *Also, shown is the net rotation per period,* $\Delta\theta/2\pi$, *calculated at the Poincaré phase* $\phi = 0$

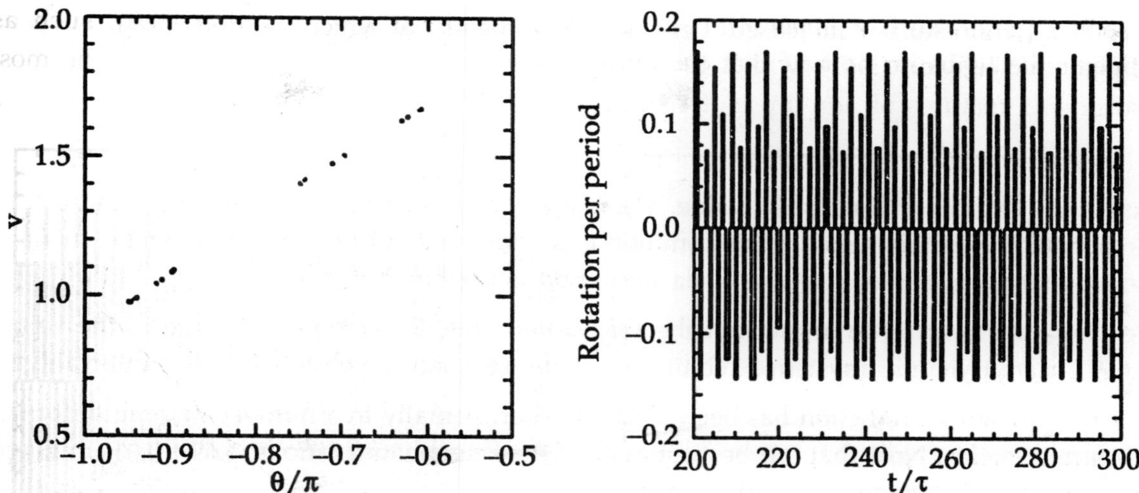

Figure 5.27: *The Poincaré section of a time-asymptotic orbit. Data calculated numerically for* $Q = 1.3757$, $A = 1.5$, $\omega = 2/3$, $\theta(0) = 0$, $v(0) = 0$, *Nacc=100, and* $\phi = 0$. *Also, shown is the net rotation per period,* $\Delta\theta/2\pi$, *calculated at the Poincaré phase* $\phi = 0$

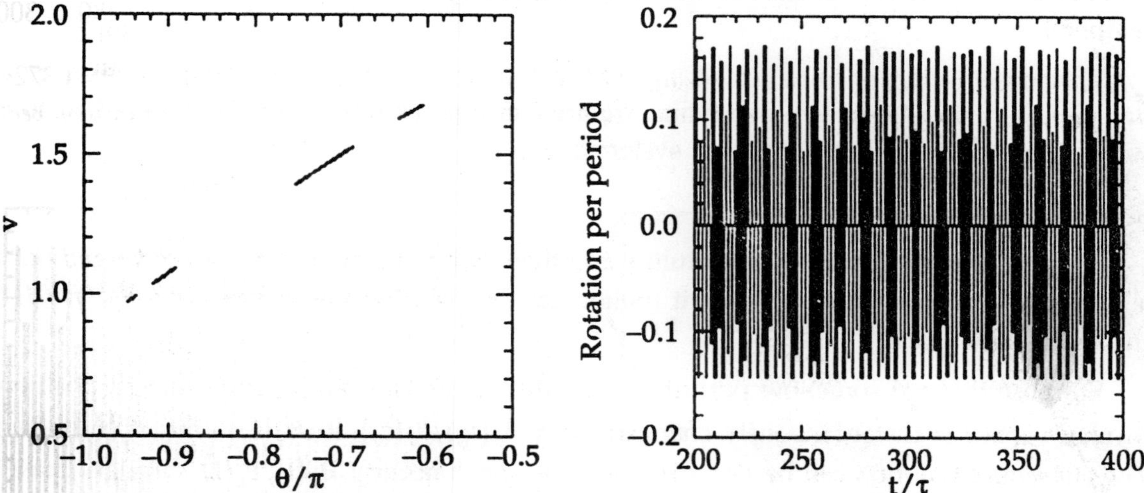

Figure 5.28: *The Poincaré section of a time-asymptotic orbit. Data calculated numerically for* $Q = 1.376$, $A = 1.5$, $\omega = 2/3$, $\theta(0) = 0$, $v(0) = 0$, *Nacc=100, and* $\phi = 0$. *Also, shown is the net rotation per period,* $\Delta\theta/2\pi$, *calculated at the Poincaré phase* $\phi = 0$

Many non-linear dynamical systems, found in nature, exhibit a transition from periodic to chaotic motion as some control parameter is varied. Now, there are various known mechanisms by which chaotic motion can arise from periodic motion. However, a transition to chaos via an infinite series of period-doubling bifurcations, as illustrated above, is certainly amongst the most commonly occurring of these mechanisms. Around 1975, the physicist Mitchell Feigenbaum was investigating a simple mathematical model, known as the *logistic map*, which exhibits a transition to chaos, via a sequence of period-doubling bifurcations, as a control parameter is r increased. Let r_n be the value of r at which the first 2^n-period cycle appears. Feigenbaum noticed that the ratio

$$F_n = \frac{r_{n-1} - r_{n-2}}{r_n - r_{n-1}} \qquad \text{...5.30}$$

converges rapidly to a constant value, $F = 4.669...$, as n increases. Feigenbaum was able to demonstrate that this value of F is common to a wide range of different mathematic models which exhibit transitions to chaos via period-doubling bifurcations.

Feigenbaum went on to argue that the *Feigenbaum ratio*, F_n, should converge to the value 4.669... in *any* dynamical system exhibiting a transition to chaos via period-doubling bifurcations.

This amazing prediction has been verified experimentally in a number of quite different physical systems. Note that our best estimate of the Feigenbaum ratio is 4.69 ± 0.01, in good agreement with Feigenbaum's prediction.

The existence of a universal ratio characterising the transition to chaos via period-doubling bifurcations is one of many pieces of evidence indicating that chaos is a *universal* phenomenon (*i.e.*, the onset and nature of chaotic motion in different dynamical systems has many common features).

This observation encourages us to believe that in studying the chaotic motion of a damped, periodically driven, pendulum we are learning lessons which can be applied to a wide range of non-linear dynamical systems.

Sensitivity to Initial Conditions

Suppose that we launch our pendulum and then wait until its motion has converged onto a particular attractor. The subsequent motion can be visualised as a trajectory $\theta_0(t), v_0(t)$ through phase-space.

Suppose that we somehow perturb the pendulum, at time $t = t_0$, such that its position in phase-space is instantaneously changed from $\theta_0(t_0), v_0(t_0)$ to $\theta_0(t_0) + \delta\theta_0, v_0(t_0) + \delta v_0$. The subsequent motion can be visualised as a second trajectory $\theta_1(t), v_1(t)$ through phase-space. What is the relationship between the original trajectory $\theta_0(t), v_0(t)$ and the perturbed trajectory $\theta_1(t), v_1(t)$?

In other words, does the phase-space separation between the two trajectories, whose components are

$$\delta\theta(\Delta t) = \theta_1(t_0 + \Delta t) - \theta_0(t_0 + \Delta t), \qquad \text{...5.31}$$

$$\delta v(\Delta t) = v_1(t_0 + \Delta t) - v_0(t_0 + \Delta t), \qquad \text{...5.32}$$

grow in time, decay in time, or stay more or less the same? What we are really investigating is how sensitive the time-asymptotic motion of the pendulum is to initial conditions.

According to the linear analysis,

$$\delta\theta(\Delta t) = \delta\theta_0 \cos(\omega_* \Delta t) e^{-\Delta t/2Q} + \frac{1}{\omega_*}\left\{\delta v_0 + \frac{\delta\theta_0}{2Q}\right\} \sin(\omega_* \Delta t) e^{-\Delta t/2Q}, \quad \text{...5.33}$$

$$\delta v(\Delta t) = \delta v_0 \cos(\omega_* \Delta t) e^{-\Delta t/2Q} - \frac{1}{\omega_*}\left\{\delta\theta_0 + \frac{\delta v_0}{2Q}\right\} \sin(\omega_* \Delta t) e^{-\Delta t/2Q}, \quad \text{...5.34}$$

assuming $\sin(\omega_* t_0) = 0$. It is clear that in the linear regime, at least, the pendulum's time-asymptotic motion is not particularly sensitive to initial conditions.

In fact, if we move the pendulum's phase-space trajectory slightly off the linear attractor, as described above, then the perturbed trajectory decays back to the attractor *exponentially* in time. In other words, if we wait long enough then the perturbed and unperturbed motions of the pendulum become effectively indistinguishable. Let us now investigate whether this insensitivity to initial conditions carries over into the non-linear regime.

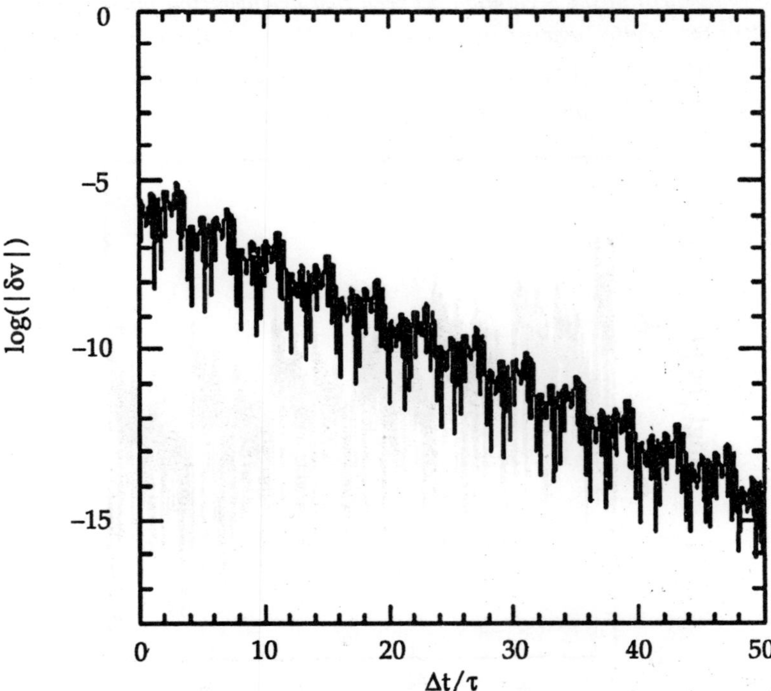

Figure 5.29: *The -component of the separation between two neighbouring phase-space trajectories (one of which lies on an attractor) plotted against normalised time. Data calculated numerically for* $Q = 1.372$, $A = 1.5$, $\omega = 2/3$, $\theta(0) = 0$, $v(0) = 0$, *and Nacc=100. The separation between the two trajectories is initialised to* $\delta\theta_0 = \delta v_0 = 10^{-6}$ *at* $\Delta t = 0$

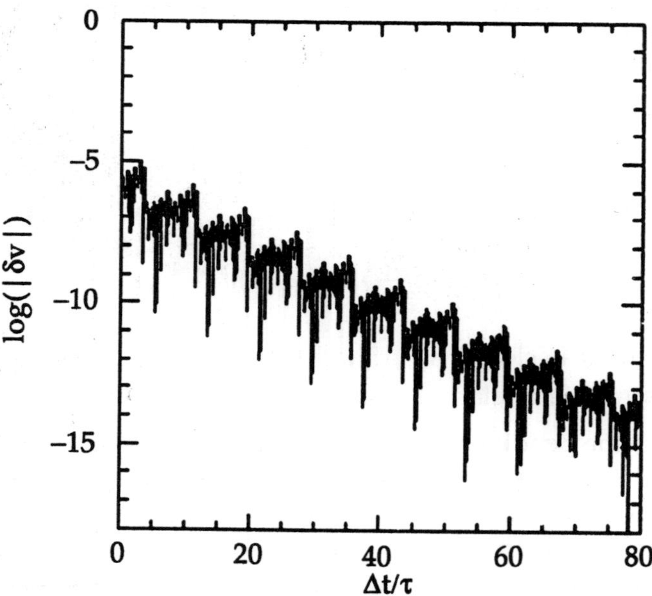

Figure 5.30: *The v-component of the separation between two neighbouring phase-space trajectories (one of which lies on an attractor) plotted against normalised time. Data calculated numerically for* $Q = 1.375$, $A = 1.5$, $\omega = 2/3$, $\theta(0) = 0$, $v(0) = 0$, *and Nacc=100. The separation between the two trajectories is initialised to* $\delta\theta_0 = \delta v_0 = 10^{-6}$ *at* $\Delta t = 0$

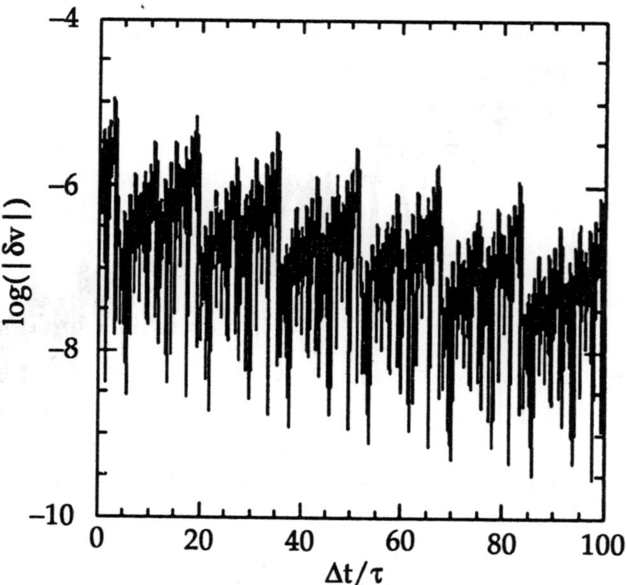

Figure 5.31: *The v-component of the separation between two neighbouring phase-space trajectories (one of which lies on an attractor) plotted against normalised time. Data calculated numerically for* $Q = 1.3757$, $A = 1.5$, $\omega = 2/3$, $\theta(0) = 0$, $v(0) = 0$, *and Nacc=100. The separation between the two trajectories is initialised to* $\delta\theta_0 = \delta v_0 = 10^{-6}$ *at* $\Delta t = 0$

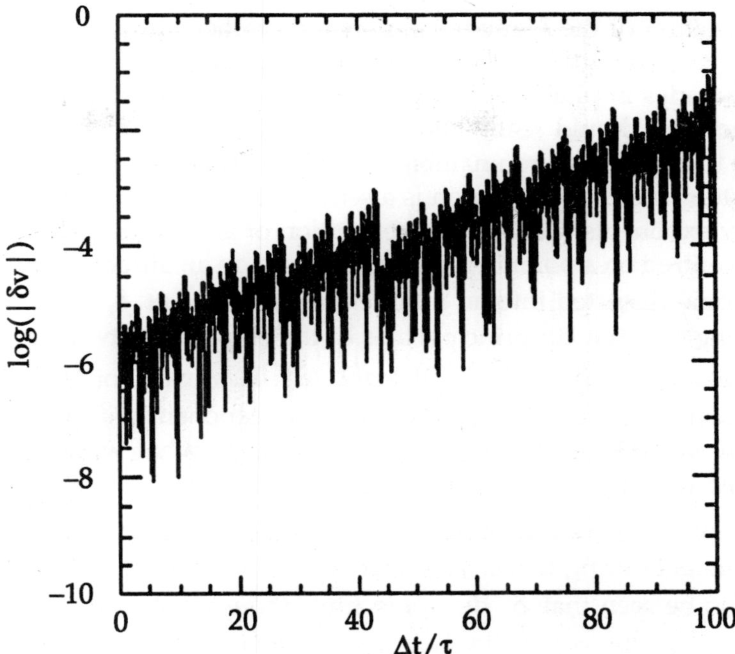

Figure 5.32: *The v-component of the separation between two neighbouring phase-space trajectories (one of which lies on an attractor) plotted against normalised time. Data calculated numerically for* $Q = 1.376$, $A = 1.5$, $\omega = 2/3$, $\theta(0) = 0$, $v(0) = 0$, *and Nacc=100. The separation between the two trajectories is initialised to* $\delta\theta_0 = \delta v_0 = 10^{-6}$ *at* $\Delta t = 0$

Figures 5.29-5.32 show the results of the experiment described above, in which the pendulum's phase-space trajectory is moved slightly off an attractor and the phase-space separation between the perturbed and unperturbed trajectories is then monitored as a function of time, at various stages on the period-doubling cascade. To be more exact, the figures show the logarithm of the absolute magnitude of the v-component of the phase-space separation between the perturbed and unperturbed trajectories as a function of normalised time.

Figure 5.29 shows the time evolution of the v-component of the phase-space separation, δv, between two neighbouring trajectories, one of which is the period-4 attractor illustrated in figure 5.25. It can be seen that δv decays rapidly in time. In fact, the graph of $\log(|\delta v|)$ versus Δt can be plausibly represented as a *straight-line* of negative gradient λ. In other words,

$$|\delta v(\Delta t)| = \delta v_0 e^{\lambda \Delta t}, \qquad \text{...5.35}$$

where the quantity λ is known as the *Liapunov exponent*. Clearly, in this case, λ measures the strength of the exponential convergence of the two trajectories in phase-space. Of course, the graph of $\log(|\delta v|)$ versus Δt is not exactly a straight-line. There are deviations due to δv the fact that oscillates, as well as decays, in time. There are also deviations because the strength of the exponential convergence between the two trajectories varies along the attractor.

The above definition of the Liapunov exponent is rather inexact, for two main reasons. In the first place, the strength of the exponential convergence/divergence between two neighbouring trajectories in phase-space, one of which is an attractor, generally *varies* along the attractor. Hence, we should really take formula (5.35) and somehow *average* it over the attractor, in order to obtain a more unambiguous definition of λ. In the second place, since the dynamical system under investigation is a second-order system, it actually possesses *two* different Liapunov exponents. Consider the evolution of an infinitesimal circle of perturbed initial conditions, centred on a point in phase-space lying on an attractor. During its evolution, the circle will become distorted into an infinitesimal ellipse. Let δ_k, where $k = 1,2$, denote the phase-space length of the kth principal axis of the ellipse. The two Liapunov exponents, λ_1 and λ_2, are defined via $\delta_k(\Delta t) \simeq \delta_k(0)\exp(\lambda_k \Delta t)$. However, for large Δt the diameter of the ellipse is effectively controlled by the Liapunov exponent with the most positive real part. Hence, when we refer to *the* Liapunov exponent, λ, what we generally mean is the Liapunov exponent with the most positive real part.

Figure 5.30 shows the time evolution of the v-component of the phase-space separation, δ_v, between two neighbouring trajectories, one of which is the period-8 attractor illustrated in figure 5.26. It can be seen that δ_v decays in time, though not as rapidly as in figure 5.29. Another way of saying this is that the Liapunov exponent of the periodic attractor shown in figure 5.26 is negative (*i.e.*, it has a negative real part), though not as negative as that of the periodic attractor shown in figure 5.25.

Figure 5.31 shows the time evolution of the v-component of the phase-space separation, δ_v, between two neighbouring trajectories, one of which is the period-16 attractor illustrated in figure 5.27. It can be seen that δ_v decays weakly in time. In other words, the Liapunov exponent of the periodic attractor shown in figure 5.27 is small and negative.

Finally, figure 5.32 shows the time evolution of the v-component of the phase-space separation, δ_v, between two neighbouring trajectories, one of which is the chaotic attractor illustrated in figure 5.28. It can be seen that δ_v *increases* in time. In other words, the Liapunov exponent of the chaotic attractor shown in figure 5.28 is *positive*. Further investigation reveals that, as the control parameter Q is gradually increased, the Liapunov exponent changes sign and becomes positive at exactly the same point that chaos ensues in Fig 5.24.

Periodic attractors are characterised by negative Liapunov exponents, whereas chaotic attractors are characterised by positive exponents. But, how can an attractor have a positive Liapunov exponent? Surely, a positive exponent necessarily implies that neighbouring phase-space trajectories *diverge* from the attractor (and, hence, that the attractor is not a true attractor)? It turns out that this is not the case. The chaotic attractor shown in figure 5.28 is a true attractor, in the sense that neighbouring trajectories rapidly converge onto it—*i.e.*, after a few periods of the external drive their Poincaré sections plot out the same four-line segment shown in figure 5.28. Thus, the exponential divergence of neighbouring trajectories, characteristic of chaotic attractors, takes place *within the attractor* itself. Obviously, this exponential divergence must come to an end when the phase-space separation of the trajectories becomes comparable to the extent of the attractor.

A dynamical system characterised by a positive Liapunov exponent, λ, has a *time horizon* beyond which regular deterministic prediction breaks down. Suppose that we measure the initial conditions of an experimental system very accurately. Obviously, no measurement is perfect: there is always some error δ_0 between our estimate and the true initial state. After a time *t*, the discrepancy grows to $\delta(t) \sim \delta_0 \exp(\lambda t)$. Let be a measure of our tolerance: i.e., a prediction within a of the true state is considered acceptable. It follows that our prediction becomes unacceptable when $\delta \gg a$, which occurs when

$$t > t_h \sim \frac{1}{\lambda} \ln\left(\frac{a}{\delta_0}\right). \qquad ...5.36$$

Note the *logarithmic* dependence on δ_0. This ensures that, in practice, no matter how hard we work to reduce our initial measurement error, we cannot predict the behaviour of the system for longer than a few multiples of $1/\lambda$.

It follows the chaotic attractors are associated with motion which is essentially *unpredictable.* In other words, if we attempt to integrate the equations of motion of a chaotic system then even the slightest error made in the initial conditions will be amplified exponentially over time and will rapidly destroy the accuracy of our prediction. Eventually, all that we will be able to say is that the motion lies somewhere on the chaotic attractor in phase-space, but exactly where it lies on the attractor at any given time will be unknown to us.

The hyper-sensitivity of chaotic systems to initial conditions is sometimes called the *butterfly effect*. The idea is that a butterfly flapping its wings in a South American rain-forest could, in principle, affect the weather in Texas (since the atmosphere exhibits chaotic dynamics). This idea was first publicised by the meteorologist Edward Lorenz, who constructed a very crude model of the convection of the atmosphere when it is heated from below by the ground. Lorenz discovered, much to his surprise, that his model atmosphere exhibited chaotic motion—which, at that time, was virtually unknown to physics. In fact, Lorenz was essentially the first scientist to fully understand the nature and ramifications of chaotic motion in physical systems. In particular, Lorenz realised that the chaotic dynamics of the atmosphere spells the doom of long-term weather forecasting: the best one can hope to achieve is to predict the weather a few days in advance ($1/\lambda$ for the atmosphere is of order a few days).

Definition of Chaos

There is no universally agreed definition of chaos. However, most people would accept the following working definition:

Chaos is aperiodic time-asymptotic behaviour in a deterministic system which exhibits sensitive dependence on initial conditions.

This definition contains three main elements:

1. *Aperiodic time-asymptotic behaviour*—this implies the existence of phase-space trajectories which do not settle down to fixed points or periodic orbits. For practical reasons, we insist that these trajectories are not too rare. We also require the trajectories to be *bounded*: i.e., they should not go off to infinity.

2. *Deterministic*—this implies that the equations of motion of the system possess no random inputs. In other words, the irregular behaviour of the system arises from non-linear dynamics and not from noisy driving forces.
3. *Sensitive dependence on initial conditions*—this implies that nearby trajectories in phase-space separate exponentially fast in time: i.e., the system has a positive Liapunov exponent.

Periodic Windows

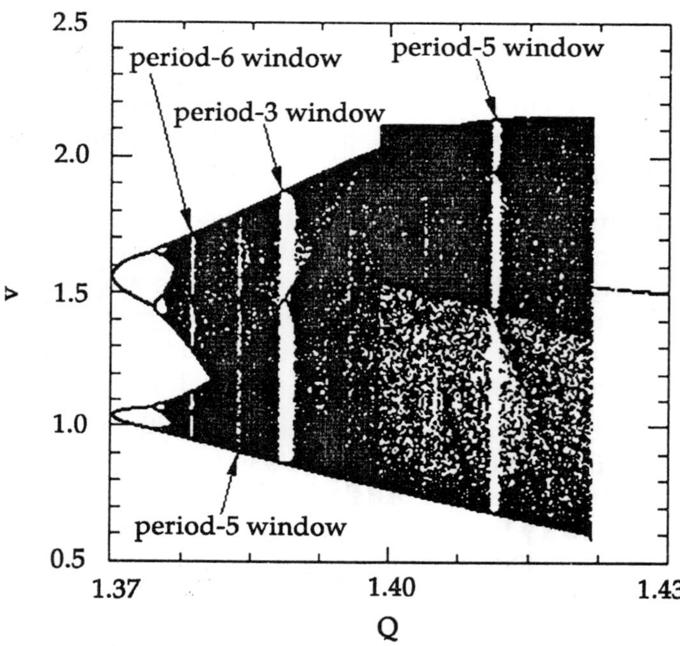

Figure 5.33: *The v-coordinate of the Poincaré section of a time-asymptotic orbit plotted against the quality-factor Q. Data calculated numerically for* $A = 1.5$, $\omega = 2/3$, $\theta(0) = 0$, $v(0) = -0.75$, *Nacc=100, and* $\phi = 0$

Let us return to figure 5.24. Recall, that this figure shows the onset of chaos, via a cascade of period-doubling bifurcations, as the quality-factor Q is gradually increased.

Figure 5.33 is essentially a continuation of figure 5.24 which shows the full extent of the chaotic region (in Q-v space). It can be seen that the chaotic region ends abruptly when Q exceeds a critical value, which is about 1.4215.

Beyond this critical value, the time-asymptotic motion appears to revert to period-1 motion (*i.e.*, the solid black region collapses to a single curve).

It can also be seen that the chaotic region contains many narrow windows in which chaos reverts to periodic motion (*i.e.*, the solid black region collapses to n curves, where n is the period of the motion) for a short interval in Q. The four widest windows are indicated on the figure.

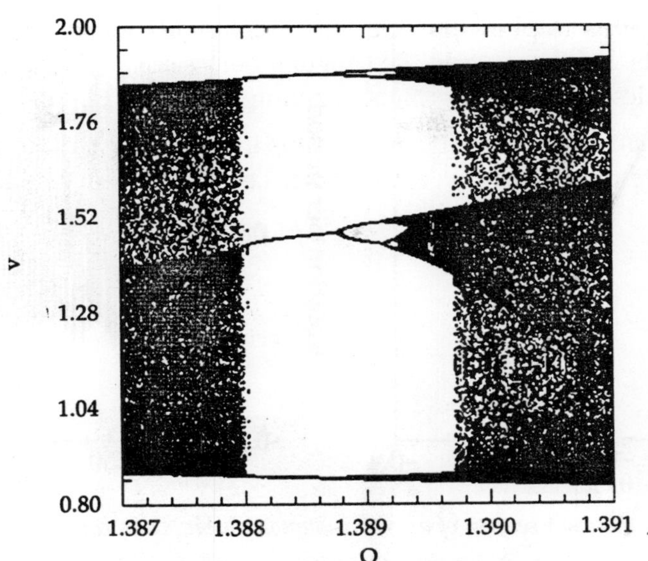

Figure 5.34: *The v-coordinate of the Poincaré section of a time-asymptotic orbit plotted against the quality-factor Q. Data calculated numerically for* $A = 1.5$, $\omega = 2/3$, $\theta(0) = 0$, $v(0) = -0.75$, *Nacc=100, and* $\phi = 0$

Figure 5.34 is a blow-up of the period-3 window shown in figure 5.33. It can be seen that the window appears "out of the blue" as Q is gradually increased. However, it can also be seen that, as Q is further increased, the window breaks down, and eventually disappears, due to the action of a cascade of period-doubling bifurcations. The same basic mechanism operates here as in the original period-doubling cascade, except that now the orbits are of period 3.2^n, instead of 2.2^n. Note that all of the other periodic windows seen in figure 5.33 break down in an analogous manner, as Q is increased.

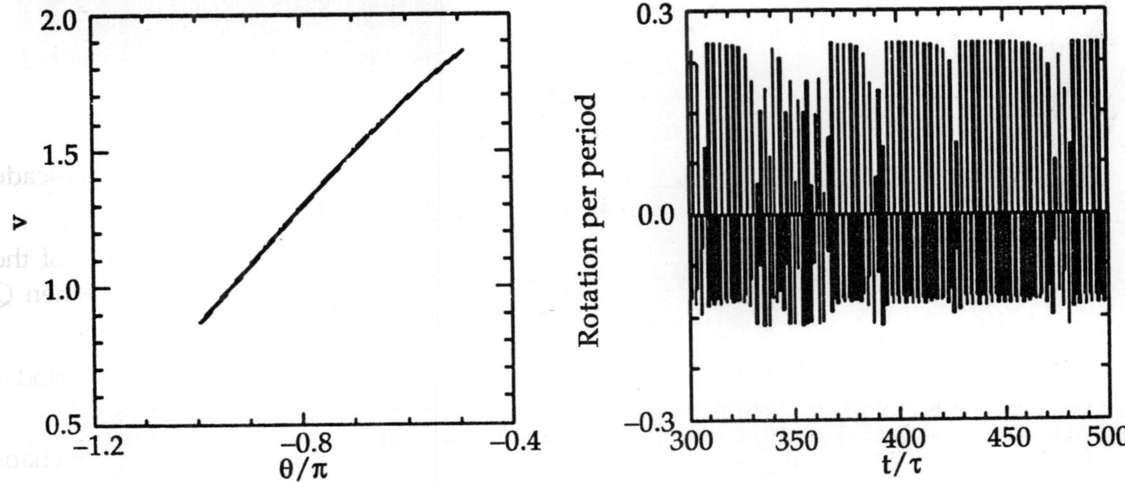

Figure 5.35: *The Poincaré section of a time-asymptotic orbit. Data calculated numerically for* $Q = 1.387976$, $A = 1.5$, $\omega = 2/3$, $\theta(0) = 0$, $v(0) = -0.75$, *Nacc=100, and* $\phi = 0$. *Also, shown is the net rotation per period,* $\Delta\theta / 2\pi$, *calculated at the Poincaré phase* $\phi = 0$

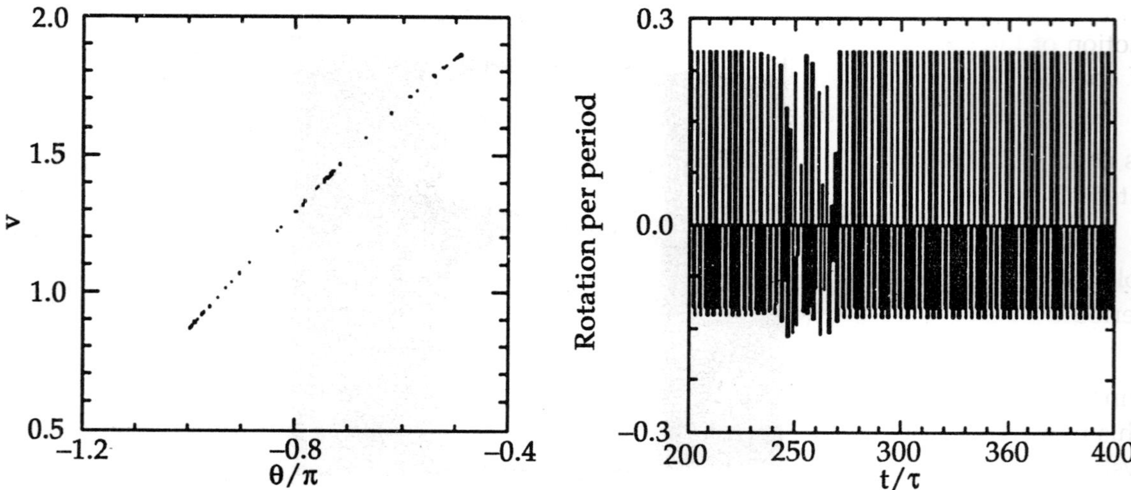

Figure 5.36: *The Poincaré section of a time-asymptotic orbit. Data calculated numerically for* $Q = 1.387977$, $A = 1.5$, $\omega = 2/3$, $\theta(0) = 0$, $v(0) = -0.75$, *Nacc=100, and* $\phi = 0$. *Also, shown is the net rotation per period,* $\Delta\theta/2\pi$, *calculated at the Poincaré phase* $\phi = 0$

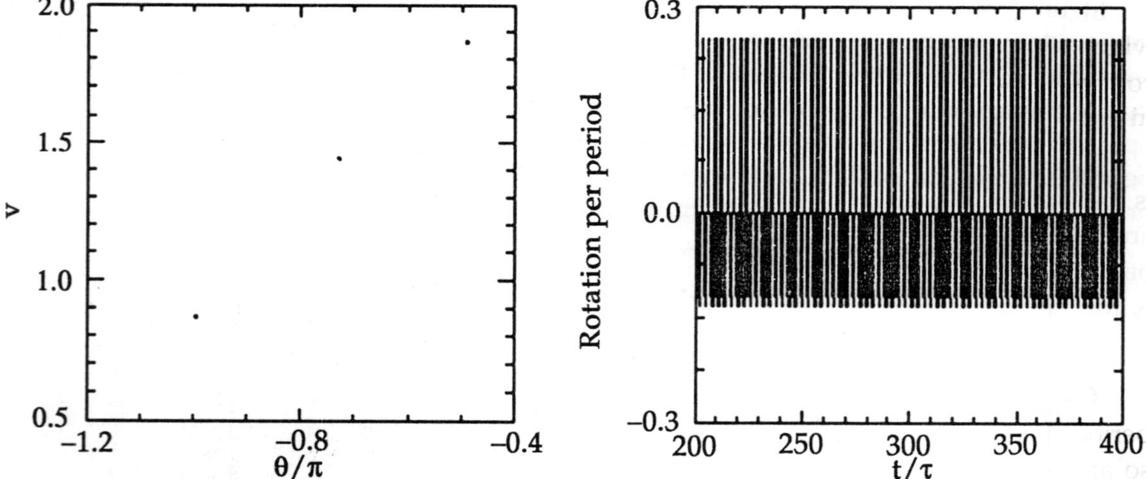

Figure 5.37: *The Poincaré section of a time-asymptotic orbit. Data calculated numerically for* $Q = 1.387978$, $A = 1.5$, $\omega = 2/3$, $\theta(0) = 0$, $v(0) = -0.75$, *Nacc=100, and* $\phi = 0$. *Also, shown is the net rotation per period,* $\Delta\theta/2\pi$, *calculated at the Poincaré phase* $\phi = 0$

We now understand how periodic windows break down. But, how do they appear in the first place? Figures 5.35-5.37 show details of the pendulum's time-asymptotic motion calculated *just before* the appearance of the period-3 window (shown in figure 5.34), *just at* the appearance of the window, and *just after* the appearance of the window, respectively. It can be seen, from figure 5.35, that just before the appearance of the window the attractor is chaotic (*i.e.,* its Poincaré section consists of a line, rather than a discrete set of points), and the time-asymptotic motion of the pendulum consists of intervals of period-3 motion interspersed with intervals of chaotic motion.

Figure 5.36 shows that just at the appearance of the window the attractor loses much of its chaotic nature (*i.e.*, its Poincaré section breaks up into a series of points), and the chaotic intervals become shorter and much less frequent.

Finally, figure 5.37 shows that just after the appearance of the window the attractor collapses to a period-3 attractor, and the chaotic intervals cease altogether. All of the other periodic windows seen in figure 5.33 appear in an analogous manner to that just described.

The typical time-asymptotic motion seen just prior to the appearance of a period-n window consists of intervals of period-n motion interspersed with intervals of chaotic motion. This type of behaviour is called *intermittency*, and is observed in a wide variety of non-linear systems.

As we move away from the window, in parameter space, the intervals of periodic motion become gradually shorter and more infrequent. Eventually, they cease altogether. Likewise, as we move towards the window, the intervals of periodic motion become gradually longer and more frequent. Eventually, the whole motion becomes periodic.

In 1973, Metropolis and co-workers investigated a class of simple mathematical models which all exhibit a transition to chaos, via a cascade of period-doubling bifurcations, as some control parameter r is increased. They were able to demonstrate that, for these maps, the order in which stable periodic orbits occur as r is increased is fixed.

That is, *stable periodic attractors always occur in the same sequence* as r is varied. This sequence is called the universal or *U-sequence*. It is possible to make a fairly convincing argument that any physical system which exhibits a transition to chaos via a sequence of period-doubling bifurcations should also exhibit the U-sequence of stable periodic attractors. Up to period-6, the U-sequence is

$$1,2,2\times 2,6,5,3,2\times 3,5,6,4,6,5,6.$$

The beginning of this sequence is familiar: periods 1, 2, 2×2 are the first stages of the period-doubling cascade. (The later period-doublings give rise to periods greater than 6, and so are omitted here).

The next periods, 6, 5, 3 correspond to the first three periodic windows shown in figure 5.33. Period 2×3 is the first component of the period-doubling cascade which breaks up the period-3 window. The next period, 5, corresponds to the last periodic window shown in figure 5.33. The remaining periods, 6, 4, 6, 5, 6, correspond to tiny periodic windows, which, in practice, are virtually impossible to observe.

It follows that our driven pendulum system exhibits the U-sequence of stable periodic orbits fairly convincingly. This sequence has also been observed experimentally in other, quite different, dynamical systems. The existence of a universal sequence of stable periodic orbits in dynamical systems which exhibit a transition to chaos via a cascade of period-doubling bifurcations is another indication that chaos is a universal phenomenon.

Further Investigation

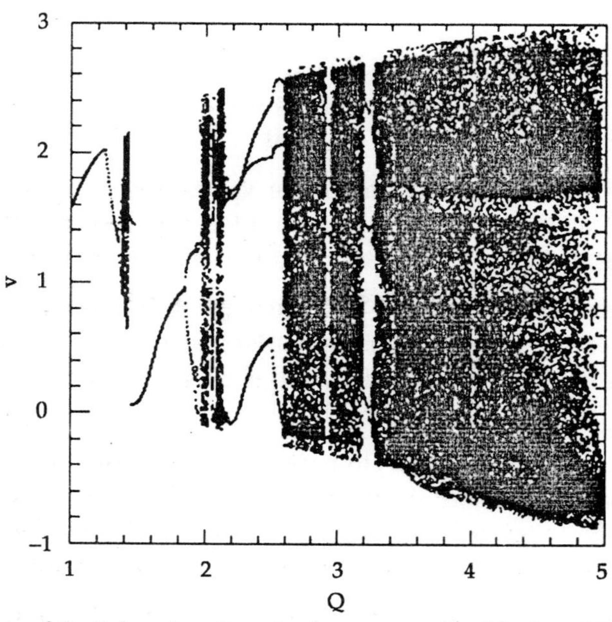

Figure 5.38: The -coordinate of the Poincaré section of a time-asymptotic orbit plotted against the quality-factor Q. Data calculated numerically for $A = 1.5$, $\omega = 2/3$, $\theta(0) = 0$, $v(0) = 0$, *Nacc=100, and* $\phi = 0$

Figure 5.38 shows the complete *bifurcation diagram* for the damped, periodically driven, pendulum (with $A = 1.5$ and $\omega = 2/3$). It can be seen that the chaotic region investigated in the previous section is, in fact, the first, and least extensive, of *three* different chaotic regions.

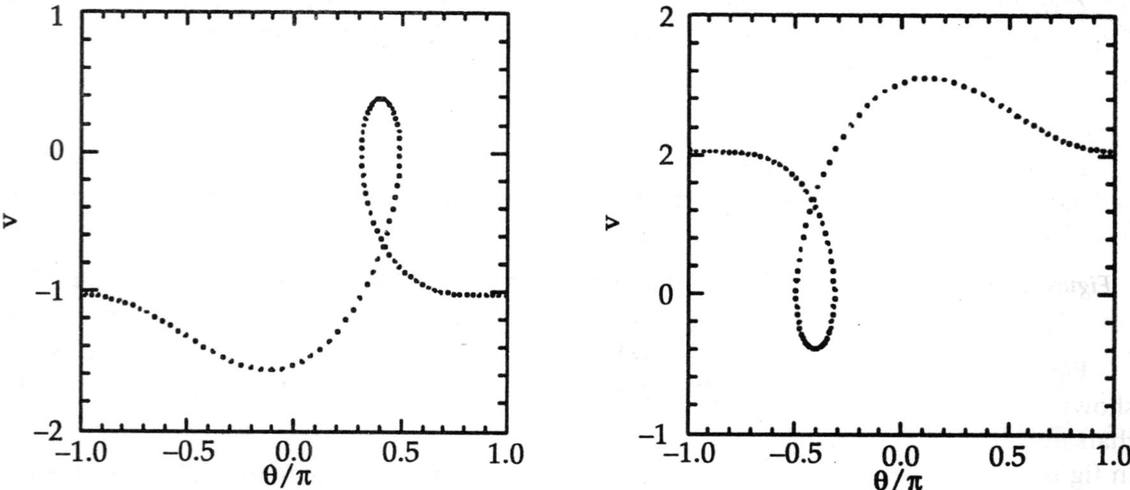

Figure 5.39: Equally spaced (in time) points on a time-asymptotic orbit in phase-space. Data calculated numerically for $Q = 1.5$, $A = 1.5$, $\omega = 2/3$, $\theta(0) = 0$, $v(0) = 0$, *and Nacc=100. Also shown is the time-asymptotic orbit calculated for the modified initial conditions* $\theta(0) = 0$, *and* $v(0) = -1$

The interval between the first and second chaotic regions is occupied by the period-1 orbits shown in figure 5.39. Note that these orbits differ somewhat from previously encountered period-1 orbits, because the pendulum executes a complete rotation (either to the left or to the right) every period of the external drive. Now, an n, l periodic orbit is defined such that

$$\theta(t+n\tau)=\theta(t)+2\pi l$$

for all t(after the transients have died away). It follows that all of the periodic orbits which we encountered in previous sections were $l=0$ orbits: i.e., their associated motions did not involve a net rotation of the pendulum. The orbits show in figure 5.39 are $n=1, l=-1$ and $n=1, l=+1$ orbits, respectively. The existence of periodic orbits in which the pendulum undergoes a net rotation, either to the left or to the right, is another example of spatial symmetry breaking—there is nothing in the pendulum's equations of motion which distinguishes between the two possible directions of rotation.

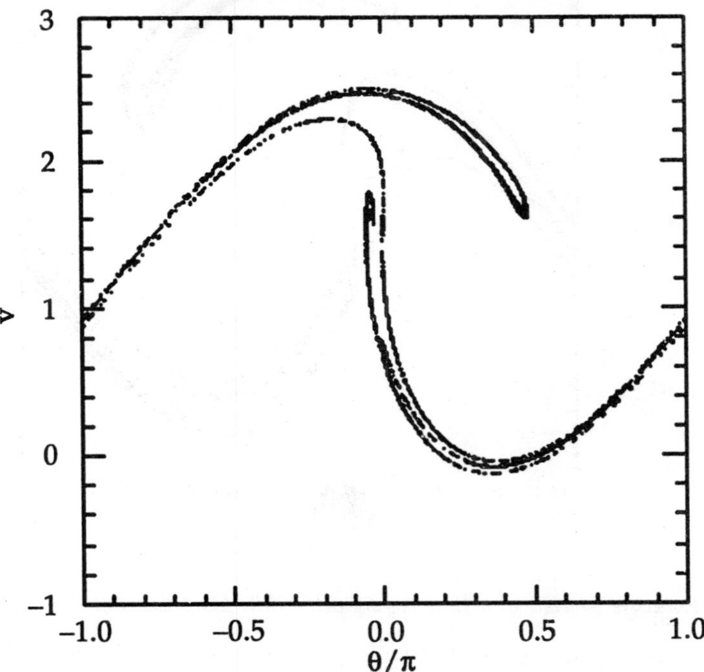

Figure 5.40: *The Poincaré section of a time-asymptotic orbit. Data calculated numerically for* Q=2.13, $A=1.5$, $\omega=2/3$, $\theta(0)=0$, $v(0)=0$, *Nacc=100, and* $\phi=0$

Figure 5.40 shows the Poincare section of a typical attractor in the second chaotic region shown in figure 5.38. It can be seen that this attractor is far more convoluted and extensive than the simple 4-line chaotic attractor pictured in figure 5.28. In fact, the attractor shown in figure 5.40 is clearly a *fractal* curve. It turns out that virtually all chaotic attractors exhibit fractal structure.

The interval between the second and third chaotic regions shown in figure 5.38 is occupied by $n=3$, l=0 periodic orbits. Figure 5.41 shows the Poincare section of a typical attractor in the third chaotic region. It can be seen that this attractor is even more overtly fractal in nature than that pictured in the previous figure. Note that the fractal nature of chaotic attractors is closely associated with some of their unusual properties. Trajectories on a chaotic attractor remain confined to a bounded region of phase-space, and yet they separate from their neighbours exponentially fast (at least, initially). How can trajectories diverge endlessly and still stay bounded? The basic mechanism is described below. If we imagine a blob of initial conditions in phase-space, then these undergo a series of repeated *stretching and folding* episodes, as the chaotic motion unfolds. The stretching is what gives rise to the divergence of neighbouring trajectories. The folding is what ensures that the trajectories remain bounded. The net result is a phase-space structure which looks a bit like filo dough—in other words, a fractal structure.

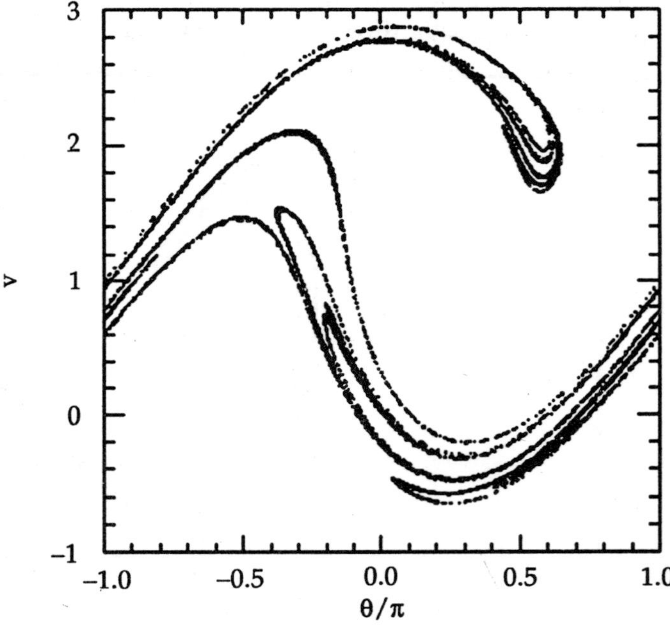

Figure 5.41: *The Poincaré section of a time-asymptotic orbit. Data calculated numerically for* $Q = 3.9$, $A = 1.5$, $\omega = 2/3$, $\theta(0) = 0$, $v(0) = 0$, *Nacc=100, and* $\phi = 0$

6

Matrix Algebra

A matrix (plural matrices, or less commonly matrixes) is a rectangular array of numbers, such as

$$\begin{bmatrix} 1 & 9 & 13 \\ 20 & 55 & 4 \end{bmatrix}.$$

An item in a matrix is called an entry or an element. The example has entries 1, 9, 13, 20, 55, and 4. Entries are often denoted by a variable with two subscripts, as shown on the right. Matrices of the same size can be added and subtracted entrywise and matrices of compatible sizes can be multiplied. These operations have many of the properties of ordinary arithmetic, except that matrix multiplication is not commutative, that is, *AB* and *BA* are not equal in general. Matrices consisting of only one column or row define the components of vectors, while higher-dimensional (e.g., three-dimensional) arrays of numbers define the components of a generalisation of a vector called a tensor.

Matrices are a key tool in linear algebra. One use of matrices is to represent linear transformations, which are higher-dimensional analogues of linear functions of the form $f(x) = cx$, where c is a constant; matrix multiplication corresponds to composition of linear transformations. Matrices can also keep track of the coefficients in a system of linear equations. For a square matrix, the determinant and inverse matrix (when it exists) govern the behaviour of solutions to the corresponding system of linear equations, and eigenvalues and eigenvectors provide insight into the geometry of the associated linear transformation.

Matrices find many applications. Physics makes use of matrices in various domains, for example in geometrical optics and matrix mechanics; the latter led to studying in more detail matrices with an infinite number of rows and columns. Graph theory uses matrices to keep track of distances between pairs of vertices in a graph. Computer graphics uses matrices to project 3-dimensional space onto a 2-dimensional screen. Matrix calculus generalises classical analytical notions such as derivatives of functions or exponentials to matrices. The latter is a recurring need in solving ordinary differential equations. Serialism and dodecaphonism are musical movements of the 20th century that use a square mathematical matrix to determine the pattern of music intervals.

A major branch of numerical analysis is devoted to the development of efficient algorithms for matrix computations, a subject that is centuries old but still an active area of research. Matrix decomposition methods simplify computations, both theoretically and practically. For sparse matrices, specifically tailored algorithms can provide speedups; such matrices arise in the finite element method, for example.

Nearly every scientific problem which is solvable on a computer can be represented by matrices. However the ease of solution can depend crucially on the types of matrices involved. There are 3 main classes of problems which we might want to solve:

1. Trivial Algebraic Manipulation such as addition, $A+B$ or multiplication, AB, of matrices.
2. *Systems of Equations:* $Ax = b$ where A and b are known and we have to find x. This also includes the case of finding the inverse, A^{-1}. The standard example of such a problem is Poisson's Equation.
3. *Eigenvalue Problems:* $Ax = \alpha x$. This also includes the generalised eigenvalue problem: $Ax = \alpha Bx$. Here we will consider the time-independent Schrödinger equation.

In most cases there is no point in writing your own routine to solve such problems. There are many computer libraries, such as Numerical Algorithms Group (n.d.), Lapack Numerical Library (n.d.) (for linear algebra problems and eigenvalue problems) which contain well tried routines. In addition vendors of machines with specialised architectures often provide libraries of such routines as part of the basic software.

Types of Matrices

There are several ways of classifying matrices depending on symmetry, sparsity, etc. Here we provide a list of types of matrices and the situation in which they may arise in physics.

- *Hermitian Matrices:* Many Hamiltonians have this property especially those containing magnetic fields: $A_{ji} = A_{ij}^{\dagger}$ where at least some elements are complex.
- *Real Symmetric Matrices:* These are the commonest matrices in physics as most Hamiltonians can be represented this way: $A_{ji} = A_{ij}$ and all A_{ij} are real. This is a special case of Hermitian matrices.
- *Positive Definite Matrices:* A special sort of Hermitian matrix in which all the eigenvalues are positive. The overlap matrices used in tight-binding electronic structure calculations are like this. Sometimes matrices are non-negative definite and zero eigenvalues are also allowed. An example is the dynamical matrix describing vibrations of the atoms of a molecule or crystal, where $\omega^2 \geq 0$.

- *Unitary Matrices:* The product of the matrix and its Hermitian conjugate is a unit matrix, $U^{\dagger}U = 1$. A matrix whose columns are the eigenvectors of a Hermitian matrix is unitary; the unitarity is a consequence of the orthogonality of the eigenvectors. A scattering (S) matrix is unitary; in this case a consequence of current conservation.
- *Diagonal Matrices:* All matrix elements are zero except the diagonal elements, $A_{ij} = 0$ when $i \neq j$. The matrix of eigenvalues of a matrix has this form. Finding the eigenvalues is equivalent to diagonalisation.
- *Tridiagonal Matrices:* All matrix elements are zero except the diagonal and first off diagonal elements, $A_{i,i} \neq 0$, $A_{i,i\pm1} \neq 0$. Such matrices often occur in 1 dimensional problems and at an intermediate stage in the process of diagonalisation.
- *Upper and Lower Triangular Matrices:* In Upper Triangular Matrices all the matrix elements below the diagonal are zero, $A_{ij} = 0$ for $i > j$. A Lower Triangular Matrix is the other way round, $A_{ij} = 0$ for $i < j$. These occur at an intermediate stage in solving systems of equations and inverting matrices.
- *Sparse Matrices:* Matrices in which most of the elements are zero according to some pattern. In general sparsity is only useful if the number of non-zero matrix elements of an $N \times N$ matrix is proportional to N rather than N^2. In this case it may be possible to write a function which will multiply a given vector x by the matrix A to give Ax without ever having to store all the elements of A. Such matrices commonly occur as a result of simple discretisation of partial differential equations, and in simple models to describe many physical phenomena.
- *General Matrices:* Any matrix which doesn't fit into any of the above categories, especially non-square matrices.

There are a few extra types which arise less often:

- *Complex Symmetric Matrices:* Not generally a useful symmetry. There are however two related situations in which these occur in theoretical physics: Green's functions and scattering (S) matrices. In both these cases the real and imaginary parts commute with each other, but this is not true for a general complex symmetric matrix.
- *Symplectic Matrices:* This designation is used in 2 distinct situations:
  - The eigenvalues occur in pairs which are reciprocals of one another. A common example is a Transfer Matrix.
  - Matrices whose elements are Quaternions, which are 2×2 matrices like

$$\begin{pmatrix} a & b \\ -b^* & a^* \end{pmatrix}. \qquad \text{...6.1}$$

  - Such matrices describe systems involving spin-orbit coupling. All eigenvalues are doubly degenerate (Kramers degeneracy).

Simple Matrix Problems

Addition and Subtraction

In programming routines to add or subtract matrices it pays to remember how the matrix is stored in the computer. Unfortunately this varies from language to language: in C(++) and Pascal the first index varies slowest and the matrix is stored one complete row after another; whereas in FORTRAN the first index varies fastest and the matrix is stored one complete column after another. It is generally most efficient to access the matrix elements in the order in which they are stored. Hence the simple matrix algebra, $A = B + C$, should be written in C as

```
const int N =??;
int i, j;
double A[N][N], B[N][N], C[N][N];
for( i = 0; i < N; i++)
    for( j = 0; j < N; j++)
        A[i][j] = B[i][j] + C[i][j];
```

or its equivalent using pointers. In FORTRAN90 this should be written

```
integer :: i,j
integer, parameter :: N = ??
real, double precision :: A(N,N), B(N,N), C(N,N)
do i = 1,N
    do j=1,N
        A(j,i) = B(j,i) + C(j,i)
    end do
end do
```

Note the different ordering of the loops in these 2 examples. This is intended to optimise the order in which the matrix elements are accessed. It is perhaps worth noting at this stage that in both C++ and FORTRAN90 it is possible to define matrix type variables (classes) so that the programmes could be reduced to

```
matrix A(N,N), B(N,N), C(N,N);
A = B + C;
```

The time taken to add or subtract 2 $M \times N$ matrices is generally proportional to the total number of matrix elements, MN, although this may be modified by parallel architecture.

Multiplication of Matrices

Unlike addition and subtraction of matrices it is difficult to give a general machine independent rule for the optimum algorithm for the operation. In fact matrix multiplication is a classic example of an operation which is very dependent on the details of the architecture of the machine. We quote here a general purpose example but it should be noted that this does not necessarily represent the optimum ordering of the loops.

```
const int L = ??, M = ??, N = ??;
int i, j, k;
double A[L][N], B[L][M], C[M][N], sum;
for ( j = 0; j < N; i++)
    for ( i = 0; i < L; j++)
    {
        for ( sum = 0, k = 0; k < M; k++)
            sum += B[i][k] * C[k][j];
        A[i][j] = sum;
    }
```

in C. The time taken to multiply 2 $N \times N$ matrices is generally proportional to N^3, although this may be modified by parallel processing.

Elliptic Equations — Poisson's Equation

At this point we digress to discuss Elliptic Equations such as Poisson's equation. In general these equations are subject to boundary conditions at the outer boundary of the range; there are no initial conditions, such as we would expect for the Wave or Diffusion equations. Hence they cannot be solved by adapting the methods for simple differential equations.

One Dimension

We start by considering the one dimensional Poisson's equation

$$\frac{d^2V}{dx^2} = f(x). \qquad ...6.2$$

The 2nd derivative may be discretised in the usual way to give

$$V_{n-1} - 2V_n + V_{n+1} = \delta x^2 \cdot f_n \qquad ...6.3$$

where we define $f_n = f(x = x_n = n\delta x)$.

The boundary conditions are usually of the form $V(x) = V_0$ at $x = x_0$ and $V(x) = V_{N+1}$ at $x = x_{N+1}$, although sometimes the condition is on the first derivative. Since V_0 and V_{N+1} are both known the $n = 1$ and $n = N$ equations (6.3) may be written as

$$-2V_1 + V_2 = \delta x^2 \cdot f_1 - V_0 \qquad ...6.4$$

$$V_{N-1} - 2V_N = \delta x^2 \cdot f_N - V_{N+1}. \qquad ...6.5$$

This may seem trivial but it maintains the convention that all the terms on the left contain unknowns and everything on the right is known. It also allows us to rewrite the (6.3) in matrix form as

$$\begin{bmatrix} -2 & 1 & & & & & & \\ 1 & -2 & 1 & & & & & \\ & 1 & -2 & 1 & & & & \\ & & \ddots & \ddots & \ddots & & & \\ & & & 1 & -2 & 1 & & \\ & & & & \ddots & \ddots & \ddots & \\ & & & & & 1 & -2 & 1 \\ & & & & & & 1 & -2 \end{bmatrix} \begin{bmatrix} V_1 \\ V_2 \\ V_3 \\ \vdots \\ V_n \\ \vdots \\ V_{N-1} \\ V_N \end{bmatrix} = \begin{bmatrix} \delta x^2 \cdot f_1 - V_0 \\ \delta x^2 \cdot f_2 \\ \delta x^2 \cdot f_3 \\ \vdots \\ \delta x^2 \cdot f_n \\ \vdots \\ \delta x^2 \cdot f_{N-1} \\ \delta x^2 \cdot f_N - V_{N+1} \end{bmatrix} \quad \text{...6.6}$$

which is a simple matrix equation of the form

$$Ax = b \quad \text{...6.7}$$

in which A is tridiagonal. Such equations can be solved by methods which we shall consider below. For the moment it should suffice to note that the tridiagonal form can be solved particularly efficiently and that functions for this purpose can be found in most libraries of numerical functions.

There are several points which are worthy of note.

- We could only write the equation in this matrix form because the boundary conditions allowed us to eliminate a term from the 1st and last lines.
- Periodic boundary conditions, such as $V(x+L) = V(x)$ can be implemented, but they have the effect of adding a non-zero element to the top right and bottom left of the matrix, A_{1N} and A_{N1}, so that the tridiagonal form is lost.
- It is sometimes more efficient to solve Poisson's or Laplace's equation using Fast Fourier Transformation (FFT). Again there are efficient library routines available (Numerical Algorithms Group, n.d.). This is especially true in machines with vector processors.

2 or more Dimensions

Poisson's and Laplace's equations can be solved in 2 or more dimensions by simple generalisations of the schemes discussed for 1D. However the resulting matrix will not in general be tridiagonal. The discretised form of the equation takes the form

$$V_{m,n-1} + V_{m,n+1} + V_{m-1,n} + V_{m+1,n} - 4V_{m,n} = \delta x^2 \cdot f_{m,n}. \quad \text{...6.8}$$

The 2 dimensional index pairs $\{m,n\}$ may be mapped on to one dimension for the purpose of setting up the matrix. A common choice is so-called dictionary order,

$$(1,1), (1,2),\ldots(1,N),(2,1),(2,2),\ldots(2,N),\ldots(N,1),(N,2),\ldots(N,N)$$

Alternatively Fourier transformation can be used either for all dimensions or to reduce the problem to tridiagonal form.

Systems of Equations and Matrix Inversion

We now move on to look for solutions of problems of the form

$$AX = B \qquad \text{...6.9}$$

where A is an $M \times M$ matrix and X and B are $M \times N$ matrices. Matrix inversion is simply the special case where B is an $M \times M$ unit matrix.

Exact Methods

Most library routines, for example those in the NAG (Numerical Algorithms Group, n.d.) or LaPack (Lapack Numerical Library, n.d.) libraries are based on some variation of Gaussian elimination. The standard procedure is to call a first function which performs an *LU* decomposition of the matrix A,

$$A = LU \qquad \text{...6.10}$$

where L and U are lower and upper triangular matrices respectively, followed by a function which performs the operations

$$Y = L^{-1}B \qquad \text{...6.11}$$

$$X = U^{-1}Y \qquad \text{...6.12}$$

on each column of B. The procedure is sometimes described as Gaussian Elimination. A common variation on this procedure is partial pivoting. This is a trick to avoid numerical instability in the Gaussian Elimination (or *LU* decomposition) by sorting the rows of A to avoid dividing by small numbers. There are several aspects of this procedure which should be noted:

- The *LU* decomposition is usually done in place, so that the matrix A is overwritten by the matrices L and U.
- The matrix B is often overwritten by the solution X.
- A well written *LU* decomposition routine should be able to spot when A is singular and return a flag to tell you so.
- Often the *LU* decomposition routine will also return the determinant of A.
- Conventionally the lower triangular matrix L is defined such that all its diagonal elements are unity. This is what makes it possible to replace A with both L and U.
- Some libraries provide separate routines for the Gaussian Elimination and the Back-Substitution steps. Often the Back-Substitution must be run separately for each column of B.
- Routines are provided for a wide range of different types of matrices. The symmetry of the matrix is not often used however.

The time taken to solve N equations in N unknowns is generally proportional to N^3 for the Gaussian-Elimination (LU-decomposition) step. The Back-Substitution step goes as N^2 for each column of B. As before this may be modified by parallelism.

Iterative Methods

As an alternative to the above library routines, especially when large sparse matrices are involved, it is possible to solve the equations iteratively.

The Jacobi Method

We first divide the matrix A into 3 parts

$$A = D + L + U \qquad \text{...6.13}$$

where D is a diagonal matrix (i.e. $D_{ij} = 0$ for $i \neq j$) and L and U are strict lower and upper triangular matrices respectively (i.e. $L_{ii} = U_{ii} = 0$ for all i). We now write the Jacobi or Gauss-Jacobi iterative procedure to solve our system of equations as

$$X^{n+1} = D^{-1}\left[B - (L+U)X^n\right] \qquad \text{...6.14}$$

where the superscripts n refer to the iteration number. Note that in practice this procedure requires the storage of the diagonal matrix, D, and a function to multiply the vector, X^n by $L+U$. This algorithm resembles the iterative solution of hyperbolic or parabolic partial differential equations, and can be analysed in the same spirit. In particular care must be taken that the method is stable. Simple C code to implement this for a 1D Poisson's equation is given below.

```
int i;
const int N = ??;    // incomplete code
double xa[N], xb[N], b[N];
while ( ... ) // incomplete code
{
    for ( i = 0; i < N; i++ )
        xa[i] = (b[i] - xb[i-1] - xb[i+1]) * 0.5;
    for ( i = 0; i < N; i++ )
        xb[i] = xa[i];
}
```

Note that 2 arrays are required for X and that the matrices, D, L and U don't appear explicitly.

The Gauss-Seidel Method

Any implementation of the Jacobi Method above will involve a loop over the matrix elements in each column of X^{n+1}. Instead of calculating a completely new matrix X^{n+1} at each iteration and then replacing X^n with it before the next iteration, as in the above code, it might seem sensible to replace the elements of X^n with those of X^{n+1} as soon as they have been calculated. Naively we would expect faster convergence. This is equivalent to rewriting the Jacobi equation as

$$X^{n+1} = (D+L)^{-1}\left[B-UX^n\right]. \quad ...6.15$$

This Gauss-Seidel method has better convergence than the Jacobi method, but only marginally so. As before we consider the example of the solution of the 1*D* Poisson's equation. As *C* programmes the basic structure might be something like

```
int i;
const int N = ??;    //incomplete code
double x[N], b[N];
while ( ... ) //incomplete code
{
    for ( i = 0; i < N; i++ )
        x[i] = (b[i] - x[i-1] - x[i+1]) * 0.5;
}
```

Note that only one array is now required to store X, whereas the Jacobi method needed 2. The time required for each iteration is proportional to *N* for each column of *B*, assuming *A* is sparse. The number of iterations required depends on the details of the problem, on the quality of the initial guess for X, and on the accuracy of the required solution.

Matrix Eigenvalue Problems

Schrödinger's Equation

In dimensionless form the time-independent Schrödinger equation can be written as

$$-\nabla^2\psi + V(r)\psi = E\psi. \quad ...6.16$$

The Laplacian, ∇^2, can be represented in discrete form as in the case of Laplace's or Poisson's equations. For example, in 1*D* (6.16) becomes

$$-\psi_{j-1} + (2+\delta x V_j)\psi_j - \psi_{j+1} = E\psi_j \quad ...6.17$$

which can in turn be written in terms of a tridiagonal matrix *H* as

$$H\underline{\psi} = E\underline{\psi}. \quad ...6.18$$

An alternative and more common procedure is to represent the eigenfunction in terms of a linear combination of basis functions so that we have

$$\psi(r) = \sum_{\beta} a_\beta \phi_\beta(r). \quad ...6.19$$

The basis functions are usually chosen for convenience and as some approximate analytical solution of the problem. Thus in chemistry it is common to choose the ϕ_β to be known atomic orbitals. In solid state physics often plane waves are chosen. Inserting (6.19) into (6.16) gives

$$\sum_\beta a_\beta(-\nabla^2 + V(r))\phi_\beta(r) = E\sum_\beta a_\beta\phi_\beta(r). \quad ...6.20$$

Multiplying this by one of the ϕ's and integrating gives

$$\sum_\beta \int dr\phi_\alpha^*(r)(-\nabla^2 + V(r))\phi_\beta(r)a_\beta = E\sum_\beta \int dr\phi_\alpha^*(r)\phi_\beta(r)a_\beta. \quad ...6.21$$

We now define 2 matrices

$$H \equiv H_{\alpha\beta} \quad = \quad \int dr\phi_\alpha^*(r)(-\nabla^2 + V(r))\phi_\beta(r) \quad ...6.22$$

$$S \equiv S_{\alpha\beta} \quad = \quad \int dr\phi_\alpha^*(r)\phi_\beta(r) \quad ...6.23$$

so that the whole problem can be written concisely as

$$\sum_\beta H_{\alpha\beta}a_\beta \quad = \quad E\sum_\beta S_{\alpha\beta}a_\beta \quad ...6.24$$

$$H_a \quad = \quad ESa \quad ...6.25$$

which has the form of the generalised eigenvalue problem. Often the ϕ's are chosen to be orthogonal so that $S_{\alpha\beta} = \delta_{\alpha\beta}$ and the matrix S is eliminated from the problem.

General Principles

The usual form of the eigenvalue problem is written

$$Ax = ax \quad ...6.26$$

where A is a square matrix x is an eigenvector and α is an eigenvalue. Sometimes the eigenvalue and eigenvector are called latent root and latent vector respectively. An $N \times N$ matrix usually has N distinct eigenvalue/eigenvector pairs. The full solution of the eigenvalue problem can then be written in the form

$$AU_r \quad = \quad U_r\underline{\underline{\alpha}} \quad ...6.27$$

$$U_lA \quad = \quad \underline{\underline{\alpha}}U_l \quad ...6.28$$

where $\underline{\underline{\alpha}}$ is a diagonal matrix of eigenvalues and $U_r(U_l)$ are matrices whose columns (rows) are the corresponding eigenvectors. U_l and U_r are the left and right handed eigenvectors respectively, and $U_l = U_r^{-1}$.

- For Hermitian matrices U_l and U_r are unitary and are therefore Hermitian transposes of each other: $U_l^\dagger = U_r$.
- For Real Symmetric matrices U_l and U_r are also real. Real unitary matrices are sometimes called orthogonal.

Full Diagonalisation

Routines are available to diagonalise real symmetric, Hermitian, tridiagonal and general matrices. In the first 2 cases this is usually a 2 step process in which the matrix is first tridiagonalised (transformed to tridiagonal form) and then passed to a routine for diagonalising a tridiagonal matrix. Routines are available which find only the eigenvalues or both eigenvalues and eigenvectors.

The former are usually much faster than the latter. Usually the eigenvalues of a Hermitian matrix are returned sorted into ascending order, but this is not always the case (check the description of the routine). Also the eigenvectors are usually normalised to unity. For non-Hermitian matrices only the right-handed eigenvectors are returned and are not normalised. In fact it is not always clear what normalisation means in the general case. Some older FORTRAN and all C (not C++) and PASCAL routines for complex matrices store the real and imaginary parts as separate arrays. The eigenvalues and eigenvectors may also be returned in this form. This is due to 2 facts

- The original routines from Wilkinson and Reinsch were written in ALGOL, which had no complex type.
- Many FORTRAN compilers (even recent ones) handle complex numbers very inefficiently, in that the use a function even for complex addition rather than inline code. In C++ it is worthwhile checking the complex header file to see how this is implemented.

The Generalised Eigenvalue Problem

A common generalisation of the simple eigenvalue problem involves 2 matrices

$$Ax = \alpha Bx. \qquad \text{...6.29}$$

This can easily be transformed into a simple eigenvalue problem by multiplying both sides by the inverse of either A or B. This has the disadvantage however that if both matrices are Hermitian $B^{-1}A$ is not, and the advantages of the symmetry are lost, together, possibly, with some important physics. There is actually a more efficient way of handling the transformation. Using Cholesky factorisation an *LU* decomposition of a positive definite matrix can be carried out such that

$$B = LL^{\dagger} \qquad \text{...6.30}$$

which can be interpreted as a sort of square root of B. Using this we can transform the problem into the form

$$\left[L^{-1}A(L^{\dagger})^{-1}\right]\left[L^{\dagger}x\right] = \alpha\left[L^{\dagger}x\right] \qquad \text{...6.31}$$

$$A'y = \alpha y. \qquad \text{...6.32}$$

Most libraries contain routines for solving the generalised eigenvalue problem for Hermitian and Real Symmetric matrices using Cholesky Factorisation followed by a standard routine. Problem 6 contains a simple and informative example.

Partial Diagonalisation

Often only a fraction of the eigenvalues are required, sometimes only the largest or smallest. Generally if more than 10% are required there is nothing to be gained by not calculating all of them, as the algorithms for partial diagonalisation are much less efficient per eigenvalue than those for full diagonalisation. Routines are available to calculate the largest or smallest eigenvalues and also all eigenvalues in a particular range.

Sturm Sequence

The Sturm sequence is a very nice algorithm found in most libraries. It finds all the eigenvalues in a given range, $\alpha_{min} < \alpha < \alpha_{max}$, and the corresponding eigenvectors. It is also able to find the number of such eigenvalues very quickly and will return a message if insufficient storage has been allocated for the eigenvectors. It does require a tridiagonalisation beforehand and is often combined with the Lanczos algorithm, to deal with sparse matrices.

Sparse Matrices and the Lanczos Algorithm

None of the above diagonalisation procedures make any use of sparsity. A very useful algorithm for tridiagonalising a sparse matrix is the Lanczos algorithm. This algorithm was developed into a very useful tool by the Cambridge group of Volker Heine (including (e.g.) Roger Haydock, Mike Kelly, John Pendry) in the late 60's and early 70's. A suite of programmes based on the Lanczos algorithm can be obtained by anonymous ftp from the HENSA archive.

Oscillations of a Crane

A crane manufacturer wants to know how his newly designed crane is going to behave in a high wind. He wants to know the oscillation frequencies of a steel wire with and without masses on the end. This project is to write a programme which calculates the modes of oscillation and their frequencies.

Analysis

The difference equations for the oscillating wire can be derived by dividing it into N segments (10 or 20 should be sufficient) each of which can be considered rigid. This would be the case if the wire consisted of a series of rods connected together.

Let x_n be the displacement of the bottom of the nth segment. Let θ_n be the angle it makes with the vertical. Let T_n be the tension in the nth segment. Assume that the mass of the wire is located at the joints of the segments, Δm at each joint. $\Delta m = \rho\delta z$ where δz is the length of each segment, and ρ is the mass per unit length of the wire. The equation of motion for the mass at the bottom of the nth segment is,

$$\Delta m \frac{d^2 x_n}{dt^2} = T_{n+1}\sin(\theta_{n+1}) - T_n \sin(\theta_n). \qquad ...6.33$$

Assume small oscillations so that $\sin(\theta_n) = (x_n - x_{n-1})/\delta z$.

$$\Delta m \frac{d^2 x_n}{dt^2} = \frac{1}{\delta z}\left[T_{n+1}x_{n+1} - (T_{n+1} + T_n)x_n + T_n x_{n-1}\right]. \qquad ...6.34$$

The mass associated with the end of the wire will be only $\Delta m/2$ since there is no contribution from the (N+1)th segment. Consequently, the equation of motion for this point is,

$$(M + \frac{1}{2}\Delta m)\frac{d^2 x_N}{dt^2} = -T_N \sin(\theta_N) = -T_N (x_N - x_{N-1})/\delta z \,, \qquad \text{...6.35}$$

where M is any mass carried by the crane. In addition the displacement of the top of the wire is zero, so that $x_0 = 0$ in the equation for $n = 1$.

The modes of oscillation are calculated by seeking solutions of the form $x_n(t) = y_n \exp(i\omega t)$. Substituting this into the equations of motion gives,

$$-\Delta m \omega^2 y_n = (T_{n+1} y_{n+1} - (T_n + T_{n+1}) y_n + T_n y_{n-1})/\delta z \qquad \text{...6.36}$$

$$-(M + \frac{1}{2}\Delta m)\omega^2 y_N = T_N (y_{N-1} - y_N)/\delta z \qquad \text{...6.37}$$

The specification of the equations is completed by noting that, from the equilibrium conditions,

$$T_N = (M + \frac{1}{2}\Delta m)g, \quad T_n = T_{n+1} + g\Delta m \,, \qquad \text{...6.38}$$

where g is the acceleration due to gravity.

The equations can be organised in the form,

$$Ay = -M\omega^2 y \,, \qquad \text{...6.39}$$

where y is the column vector of displacements, $(y_1, y_2, \ldots, y_N)$, A is a symmetric tridiagonal matrix and M is a diagonal matrix. The problem becomes one of finding the eigenvalues, $-\omega^2$, and eigenvectors, y, of a generalised eigenvalue problem. The eigenvectors show how the wire distorts when oscillating in each mode and the eigenvalues give the corresponding oscillation frequencies. Low frequency modes are more important than high frequency modes to the crane manufacturer.

The problem can be solved most easily by using a LaPack routine which finds the eigenvalues and eigenvectors directly. However, before doing so it is necessary to eliminate the matrix M using the same method.

Phonons in a Quasicrystal

Until a few years ago, it was thought that there were only two different kinds of solids: crystals, in which the atoms are arranged in a regular pattern with translational symmetry (there may be defects, of course); and amorphous solids, in which there is no long range order, although there is some correlation between the positions of nearby atoms. It was also known that it was impossible for a crystal to have five fold rotational symmetry, since this is incompatible with translational order.

This was how things stood until 1984, when Shechtman *et al.*, were measuring the X ray diffraction pattern of an alloy of Al and Mn and got a sharp pattern with clear five fold symmetry. The sharpness of the pattern meant that there had to be long range order, but the five fold symmetry meant that the solid could not be crystalline. Shechtman called the material a "quasicrystal".

One possible explanation (although this has still not been conclusively established) is that quasicrystals are three dimensional analogues of Penrose tilings (Scientific American, January 1977 — Penrose tilings were known as a mathematical curiosity before quasicrystals were discovered). Penrose found that you could put together two (or more) different shapes in certain well defined ways so that they "tiled" the plane perfectly, but with a pattern that never repeated itself. Sure enough, some of these tilings do have five fold symmetries; and sure enough, there is perfect long range order (although no translational symmetry) so that the diffraction pattern from a Penrose lattice would be sharp.

The Fibonacci Lattice

The mathematical theory of Penrose tilings gets quite high brow and abstruse, but everything is very simple in one dimension. Then the two shapes are lines of different lengths, which we shall call *A* and *C*, for Adult and Child (Fibonacci actually studied the dynamics of rabbit population). Every year each adult has one child and each child becomes an adult. Let us start with a single child

$$C \qquad \text{...6.40}$$

and then repeatedly apply the "generation rule,"

$$C \mapsto A, \quad A \mapsto AC, \qquad \text{...6.41}$$

to obtain longer and longer sequences. The first few sequences generated are,

$$C \quad A \quad AC \quad ACA \quad ACAAC \quad ACAACACA. \qquad \text{...6.42}$$

Note the interesting property that each generation is the "sum" of the 2 previous generations: $ACAAC = ACA \oplus AC$

In a one dimensional Fibonacci quasicrystal, the longs and shorts could represent the interatomic distances; or the strengths of the bonds between the atoms; or which of two different types of atom is at that position in the chain.

The Model

This project is to write a programme to work out the phonons (normal mode vibrations) of a Fibonacci quasicrystal with two different sorts of atom. The "adults" and "children" are the masses, *M* and *m*, of the two kinds of atom. Since we are only interested in small vibrations, we may represent the forces between the atoms as simple springs (spring constant *K*) and we will assume that all the springs are identical (*K* is independent of the types of the atoms at the ends of the spring). The equation of motion of such a system may be written as

$$m_n \frac{d^2 x_n}{dt^2} = -m_n \omega^2 x_n = K(x_{n+1} - 2x_n + x_{n-1}), \qquad \text{...6.43}$$

where x_n and m_n are the displacement and mass of the *n*th atom. If the chain contains *N* atoms, you will then have a set of *N* coupled second order ordinary differential equations.

The choice of boundary conditions for the atoms at either end of the chain is up to you and should not make much difference when the chain is long enough: you could fix the atoms at either end to immovable walls, $x_0 = x_{N+1} = 0$, or you could leave them free by removing the springs at the 2 ends, $K_{0,1} = K_{N,N+1} = 0$. (Periodic boundary conditions — when the chain of atoms is looped around and joined up in a ring — are convenient for analytic work but not so good for numerical work in this case. Why?)

The equations (6.43) are linear algebraic equations which may be cast as a tridiagonal matrix eigenproblem. The eigenvalues of this problem are ω^2 and so give the vibrational frequencies, and the eigenvectors give the corresponding normal mode coordinates, $x_n, n = 1, N$.

Note, however, that the presence of the masses m_n in (6.43) means that the problem is in the generalised form $Ax = \omega^2 Bx$. This can be transformed into the normal eigenvalue problem by making the substitution $x_n = m_n^{-1/2} y_n$ and multiplying the nth equation by $m_n^{-1/2}$. I suggest you solve the eigenproblem by using a NAG or LaPack routine for the eigenvalues only. There should be no problem in dealing with chains of several hundred atoms or more.

The idea is to investigate the spectrum (the distribution of the eigenvalues) as the ratio, m/M, of the two masses changes. Your programme should list the eigenvalues in ascending order and then plot a graph of eigenvalue against position in the list. When $m = M$, the crystal is "perfect", the graph is smooth, and you should be able to work out all the eigenvalues analytically (easiest when using periodic boundary conditions). But when m and M begin to differ, the graph becomes a "devil's staircase" with all sorts of fascinating fractal structure. Try to understand the behaviour at small ω (when the wavelength is long and the waves are "acoustic") and the limits as $m/M \to 0$ and $m/M \to \infty$.

Another interesting thing to do is to plot values of m/M (on the y axis) against the vibrational frequencies (on the x axis). Choose a value of m/M, work out all the frequencies, and put a point on the graph for each. The graph now has a line of points parallel to the x axis at the given y value. Do this for a number of values of m/M and see how the spectrum develops as m/M changes. Again, you should try to understand the behaviour when the frequency tends to zero, and the large and small mass ratio limits.

If you have time it is interesting to investigate the fractal structure by focusing in on a single peak for a short sequence and investigating how it splits when you add another "generation". You should find that the behaviour is independent of the number of generations at which you start.

7

Ordinary Differential Equations

Types of Differential Equation

Here, we will consider the methods of solution of the sorts of ordinary differential equations (ODEs) which occur very commonly in physics. By ODEs we mean equations involving derivatives with respect to a single variable, usually time. Although we will formulate the discussion in terms of linear ODEs for which we know the analytical solution, this is simply to enable us to make comparisons between the numerical and analytical solutions and does not imply any restriction on the sorts of problems to which the methods can be applied. In the practical work you will encounter examples which do not fit neatly into these categories. We consider 3 basic differential equations:

$$\frac{dy}{dt}+\alpha y = 0 \quad\Rightarrow\quad y = y_0\exp(-\alpha t) \qquad \text{The } \textit{Decay} \text{ equation} \qquad ...7.1$$

$$\frac{dy}{dt}-\alpha y = 0 \quad\Rightarrow\quad y = y_0\exp(+\alpha t) \qquad \text{The } \textit{Growth} \text{ equation} \qquad ...7.2$$

$$\frac{dy}{dt}\pm i\omega y = 0 \quad\Rightarrow\quad y = y_0\exp(\mp i\omega t) \qquad \text{The } \textit{Oscillation} \text{ equation} \qquad ...7.3$$

which are representative of most more complex cases. Higher order differential equations can be reduced to 1st order by appropriate choice of additional variables. The simplest such choice is to define new variables to represent all but the highest order derivative. For example, the damped harmonic oscillator equation, usually written as

$$m\frac{d^2y}{dt^2}+\eta\frac{dy}{dt}+\kappa y = 0 \qquad ...7.4$$

can be rewritten in terms of y and velocity $v = dy/dt$ in the form of a pair of 1st order ODEs

$$\frac{dv}{dt} \qquad + \frac{\eta}{m} v + \frac{\kappa}{m} y \qquad = 0 \qquad \text{...7.5}$$

$$\frac{dy}{dt} \qquad - v \qquad = 0 \qquad \text{...7.6}$$

Similarly any nth order differential equation can be reduced to n1st order equations. Such systems of ODEs can be written in a very concise notation by defining a vector, y say, whose elements are the unknowns, such as y and v in (7.1). Any ODE in n unknowns can then be written in the general form

$$\frac{dy}{dt} + f(y,t) = 0 \qquad \text{...7.7}$$

where y and f are n-component vectors. Remember that there is no significance in the use of the letter t in the above equations. The variable is not necessarily time but could just as easily be space, as in (7.9), or some other physical quantity. Formally we can write the solution of (7.7) as

$$y(t) = y(t_0) - \int_{t_0}^{t} f(y(t'),t')dt' \qquad \text{...7.8}$$

by integrating both sides over the interval $t_0 \to t$. Although (7.8) is formally correct, in practice it is usually impossible to evaluate the integral on the right-hand-side as it presupposes the solution $y(t)$. We will have to employ an approximation. All differential equations require boundary conditions. Here we will consider cases in which all the boundary conditions are defined at a particular value of t (e.g. t=0). For higher order equations the boundary conditions may be defined at different values of t. The modes of a violin string at frequency ω obey the equation

$$\frac{d^2y}{dx^2} = -\frac{\omega^2}{c^2} y \qquad \text{...7.9}$$

with boundary conditions such that $y = 0$ at both ends of the string. We shall consider such problems later.

Euler Method

Consider an approximate solution of (7.8) over a small interval $\delta t = t_{n+1} - t_n$ by writing the integral as

$$\int_{t_n}^{t_{n+1}} f(y(t'),t')dt' \approx \delta t \, f(y(t_n),t_n). \qquad \text{...7.10}$$

to obtain

$$y(t_{n+1}) = y(t_n) - \delta t \, f(y(t_n),t_n).$$

or, in a more concise notation,

$$y_{n+1} = y_n - \delta t \, f(y_n,t_n) = y_n - \delta t f_n. \qquad \text{...7.11}$$

We can integrate over any larger interval by subdividing the range into sections of width δt and repeating (7.11) for each part. Equivalently we can consider that we have approximated the derivative with a forward difference

$$\left.\frac{dy}{dt}\right|_n \approx \frac{y_{n+1} - y_n}{\delta t}. \qquad ...7.12$$

We will also come across centred and backward differences,

$$\left.\frac{dy}{dt}\right|_n \approx \begin{cases} \dfrac{y_{n+1} - y_{n-1}}{2\delta t} & \text{centred} \\ \dfrac{y_n - y_{n-1}}{\delta t} & \text{backward} \end{cases} \qquad ...7.13$$

respectively. Here we have used a notation which is very common in computational physics, in which we calculate $y_n = y(t_n)$ at discrete values of t given by $t_n = n\delta t$, and $f_n = f(y_n, t_n)$. In what follows we will drop the vector notation y except when it is important for the discussion.

Order of Accuracy

How accurate is the Euler method? To quantify this we consider a Taylor expansion of $y(t)$ around t_n

$$y_{n+1} = y_n + \delta t \left.\frac{dy}{dt}\right|_n + \left.\frac{\delta t^2}{2}\frac{d^2y}{dt^2}\right|_n + \cdots \qquad ...7.14$$

and substitute this into (7.11)

$$y_n + \delta t \left.\frac{dy}{dt}\right|_n + \left.\frac{\delta t^2}{2}\frac{d^2y}{dt^2}\right|_n + \cdots \approx y_n - \delta t f(y_n, t_n) \qquad ...7.15$$

$$= y_n + \delta t \left.\frac{dy}{dt}\right|_n, \qquad ...7.16$$

where we have used (7.7) to obtain the final form. Hence, we see that the term in δt in the expansion has been correctly reproduced by the approximation, but that the higher order terms are wrong. We therefore describe the Euler method as 1st order accurate. An approximation to a quantity is *nth* order accurate if the term in δt^n in the Taylor expansion of the quantity is correctly reproduced. The order of accuracy of a method is the order of accuracy with which the unknown is approximated. Note that the term accuracy has a slightly different meaning in this context from that which you might use to describe the results of an experiment. Sometimes the term order of accuracy is used to avoid any ambiguity. The leading order deviation is called the truncation error. Thus, truncation error is the term in δt^2.

Stability

The Euler method is 1st order accurate. However there is another important consideration in analysing the method: stability. Let us suppose that at some time the actual numerical solution deviates from the true solution of the difference equation (7.11) (N.B. not the original differential equation (7.7)) by some small amount δy, due, for example, to the finite accuracy of the computer. Then adding this into (7.11) gives

$$y_{n+1} + \delta y_{n+1} = y_n + \delta y_n - \delta t\left[f(y_n, t_n) + \left.\frac{\partial f}{\partial y}\right|_n \delta y_n \right], \qquad ...7.17$$

where the term in brackets, [], is the Taylor expansion of $f(y,t)$ with respect to y. Subtracting (7.11) we obtain a linear equation for δy

$$\delta y_{n+1} = \left[1 - \delta t \left.\frac{\partial f}{\partial y}\right|_n \right] \delta y_n, \qquad ...7.18$$

which it is convenient to write in the form

$$\delta y_{n+1} = g \delta y_n. \qquad ...7.19$$

If g has a magnitude greater than one then δy_n will tend to grow with increasing n and may eventually dominate over the required solution. Hence the Euler method is stable only if $|g| \le 1$ or

$$-1 \le 1 - \delta t \frac{\partial f}{\partial y} \le +1. \qquad ...7.20$$

As δt is positive by definition the 2nd inequality implies that the derivative must also be positive. The 1st inequality leads to a restriction on δt, namely

$$\delta t \le 2 / \frac{\partial f}{\partial y}. \qquad ...7.21$$

When the derivative is complex more care is required in the calculation of $|g|$. In this case it is easier to look for solutions of the condition $|g|^2 \le 1$. For the oscillation equation (7.3) the condition becomes

$$1 + \delta t^2 \omega^2 \le 1 \qquad ...7.22$$

which is impossible to fulfil for real δt and ω. Comparing these result with our 3 types of differential equations (7.1) we find the following stability conditions

Decay	Growth	Oscillation
$\delta t \le 2/\alpha$	unstable	unstable

The Euler method is conditionally stable for the decay equation. A method is stable if a small deviation from the true solution does not tend to grow as the solution is iterated.

Growth Equation

Actually, our analysis doesn't make too much sense in the case of the growth equation as the true solution should grow anyway. A more sensible condition would be that the relative error in the solution does not grow. This can be achieved by substituting $y_n \in_n$ for δy_n above and looking for the condition that $\in_n$ does not grow. We will not treat this case further here but it is, in fact, very important in problems such as chaos, in which small changes in the initial conditions lead to solutions which diverge from one another.

Application to Non-Linear Differential Equations

The linear-differential equations in physics can often be solved analytically whereas most non-linear ones can only be solved numerically. It is important therefore to be able to apply the ideas developed here to such cases. Consider the simple example

$$\frac{dy}{dt} + \alpha y^2 = 0. \qquad \text{...7.23}$$

In this case $f(y,t) = \alpha y^2$ and $\partial f / \partial y = 2\alpha y$ which can be substituted into (7.21) to give the stability condition

$$\delta t \le 1/(\alpha y), \qquad \text{...7.24}$$

which depends on y, unlike the simpler cases. In writing a programme to solve such an equation it may therefore be necessary to monitor the value of the solution, y, and adjust δt as necessary to maintain stability.

Application to Vector Equations

A little more care is required when y and f are vectors. In this case δy is an arbitrary infinitesimal vector and the derivative $\partial f / \partial y$ is a matrix F with components

$$F_{ij} = \frac{\partial f_i}{\partial y_j} \qquad \text{...7.25}$$

in which f_i and y_j represent the components of f and y respectively. Hence (7.18) takes the form

$$\delta y_i^{(n+1)} = \delta y_i^{(n)} - \delta t \sum_j \left.\frac{\partial f_i}{\partial y_j}\right|_n \delta y_j^{(n)} \qquad \text{...7.26}$$

$$\delta y_{n+1} = [1 - \delta t F]\delta y_n = G \delta y_n. \qquad \text{...7.27}$$

This leads directly to the stability condition that all the eigenvalues of G must have modulus less than unity. In general any of the stability conditions derived in this course for scalar equations can be re-expressed in a form suitable for vector equations by applying it to all the eigenvalues of an appropriate matrix.

Leap-Frog Method

How can we improve on the Euler method? The most obvious way would be to replace the forward difference in (7.12) with a centred difference (7.13) to get the formula

$$y_{n+1} = y_{n-1} - 2\delta t\, f(y_n, t). \qquad ...7.28$$

If we expand both y_{n+1} and y_{n-1} as before (7.28) becomes

$$y_n + \delta t \left.\frac{dy}{dt}\right|_n + \frac{\delta t^2}{2}\left.\frac{d^2y}{dt^2}\right|_n + \frac{\delta t^3}{6}\left.\frac{d^3y}{dt^3}\right|_n + \cdots$$

$$= y_n - \delta t \left.\frac{dy}{dt}\right|_n + \frac{\delta t^2}{2}\left.\frac{d^2y}{dt^2}\right|_n - \frac{\delta t^3}{6}\left.\frac{d^3y}{dt^3}\right|_n + \cdots - 2\delta t f_n \qquad ...7.29$$

$$= y_n + \delta t \left.\frac{dy}{dt}\right|_n + \frac{\delta t^2}{2}\left.\frac{d^2y}{dt^2}\right|_n - \frac{\delta t^3}{6}\left.\frac{d^3y}{dt^3}\right|_n + \cdots \qquad ...7.30$$

from which all terms up to δt^2 cancel so that the method is clearly 2nd order accurate. Note in passing that using (7.28) 2 consecutive values of y_n are required in order to calculate the next one: y_n and y_{n-1} are required to calculate y_{n+1}. Hence 2 boundary conditions are required, even though (7.28) was derived from a 1st order differential equation. This so-called leap-frog method is more accurate than the Euler method, but is it stable? Repeating the same analysis as for the Euler method we again obtain a linear equation for δy_n

$$\delta y_{n+1} = \delta y_{n-1} - 2\delta t \left.\frac{\partial f}{\partial y}\right|_n \delta y_n. \qquad ...7.31$$

We analyse this equation by writing $\delta y_n = g\delta y_{n-1}$ and $\delta y_{n+1} = g^2 \delta y_{n-1}$ to obtain

$$g^2 = 1 - 2\delta t \left.\frac{\partial f}{\partial y}\right|_n g \qquad ...7.32$$

which has the solutions

$$g = \delta t \frac{\partial f}{\partial y} \pm \sqrt{\left(\delta t \frac{\partial f}{\partial y}\right)^2 + 1}. \qquad ...7.33$$

The product of the 2 solutions is equal to the constant in the quadratic equation, i.e. –1. Since the 2 solutions are different, one of them always has magnitude > 1. Since for a small random error it is impossible to guarantee that there will be no contribution with $|g| > 1$ this contribution will tend to dominate as the equation is iterated. Hence the method is unstable. There is an important exception to this instability: when the partial derivative is purely imaginary (but not when it has some general complex value), the quantity under the square root in (7.33) can be negative and both g's have modulus unity. Hence, for the case of oscillation (7.3) where $\partial f / \partial y = \pm i\omega$, the algorithm is just stable, as long as

$$\delta t \le 1/\omega. \qquad ...7.34$$

The stability properties of the leap-frog method are summarised below

Decay	Growth	Oscillation
unstable	unstable	$\delta t \le 1/\omega$

Again the growth equation should be analysed somewhat differently.

Runge-Kutta Method

So far we have found one method which is stable for the decay equation and another for the oscillatory equation. Can we combine the advantages of both? As a possible compromise consider the following two step algorithm (ignoring vectors)

$$y'_{n+1/2} = y_n - \frac{1}{2}\delta t\, f(y_n, t_n) \qquad ...7.35$$

$$y_{n+1} = y_n - \delta t\, f(y'_{n+1/2}, t_{n+1/2}). \qquad ...7.36$$

In practice the intermediate value $y'_{n+1/2}$ is discarded after each step. We see that this method consists of an Euler step followed by a Leap-Frog step. This is called the 2nd order Runge-Kutta or two-step method. It is in fact one of a hierarchy of related methods of different accuracies. The stability analysis for (7.4) is carried out in the same way as before. Here we simply quote the result

$$\delta y_{n+1} = \left[1 - \delta t\frac{\partial f}{\partial y} + \frac{1}{2}\left(\delta t\frac{\partial f}{\partial y}\right)^2\right]\delta y_n. \qquad ...7.37$$

In deriving this result it is necessary to assume that the derivatives, $\partial f / \partial y$ are independent of t. This is not usually a problem. From (7.37) we conclude the stability conditions

Decay	Growth	Oscillation
$\delta t \le 2/\alpha$	unstable	$1 + \frac{1}{4}(\delta t\omega)^4 \le 1$

Note that in the oscillatory case the method is strictly speaking unstable but the effect is so small that it can be ignored in most cases, as long as $\omega\delta t < 1$. This method is often used for damped oscillatory equations.

Predictor-Corrector Method

This method is very similar to and often confused with the Runge-Kutta method. We consider substituting the trapezoidal rule for the estimate of the integral in (7.8) to obtain the equation

$$y_{n+1} = y_n - \frac{1}{2}\delta t\left[f(y_{n+1},t_{n+1}) + f(y_n,t_n)\right]. \qquad ...7.38$$

Unfortunately the presence of y_{n+1} on the right hand side makes a direct solution of (7.38) impossible except for special cases. A possible compromise is the following method

$$y'_{n+1} = y_n - \delta t\, f(y_n,t_n) \qquad ...7.39$$

$$y_{n+1} = y_n - \frac{1}{2}\delta t\left[f(y'_{n+1},t_{n+1} + f(y_n,t_n)\right]. \qquad ...7.40$$

This method consists of a guess for y_{n+1} based on the Euler method (the Prediction) followed by a correction using the trapezoidal rule to solve the integral equation. The accuracy and stability properties are identical to those of the Runge-Kutta method.

Intrinsic Method

Returning to the possibility of solving the integral equation using the trapezoidal or trapezium rule, let us consider the case of a linear differential equation, such as our examples. For the decay equation we have

$$y_{n+1} = y_n - \frac{1}{2}\delta t\left[\alpha y_{n+1} + \alpha y_n\right] \qquad ...7.41$$

which can be rearranged into an explicit equation for y_{n+1} as a function of y_n

$$y_{n+1} = \left[\frac{1-\frac{1}{2}\delta t\cdot\alpha}{1+\frac{1}{2}\delta t\cdot\alpha}\right]y_n. \qquad ...7.42$$

This intrinsic method is 2nd order accurate as that is the accuracy of the trapezoidal rule for integration. What about the stability? Applying the same methodology as before we find that the crucial quantity, g, is the expression in square brackets, [], in (7.42) which is always <1 for the decay equation and has modulus unity in the oscillatory case (after substituting $\alpha \mapsto \pm i\omega$). Hence it is stable in both cases. Why is it not used instead of the other methods? Unfortunately only a small group of equations, such as our examples, can be rearranged in this way. For non-linear equations it may be impossible, and even for linear equations when y is a vector, there may be a formal solution which is not useful in practice. It is always possible to solve the resulting non-linear equation iteratively, using Newton-Raphson iteration, but this is usually not worthwhile in practice. In fact the intrinsic method is also stable for the growth equation when it is analysed as discussed earlier, so that the method is in fact stable for all 3 classes of equations.

Decay	Growth	Oscillation
stable	stable	stable

In these notes we have introduced some methods for solving ordinary differential equations. However, by far the most important lesson to be learned is that to be useful a method must be both accurate and stable. The latter condition is often the most difficult to fulfill and careful analysis of the equations to be solved may be necessary before choosing an appropriate method to use for finding a numerical solution. The stability conditions derived above tend to have the form $\delta t \le \omega^{-1}$ which may be interpreted as δt should be less than the characteristic period of oscillation. This conforms with common sense. In fact we can write down a more general common sense condition: δt should be small compared with the smallest time scale present in the problem. Finally, many realistic problems don't fall into the neat categories of (7.1). The simplest example is a damped harmonic oscillator. Often it is difficult to find an exact analytical solution for the stability condition. It pays in such cases to consider some extreme conditions, such as very weak damping or very strong damping, work out the conditions for these cases and simply choose the most stringent condition. In non-linear problems the cases when the unknown, y, is very large or very small may provide tractable solutions.

Classical Electrons in a Magnetic Field

The simplest source of ODEs in physics is classical mechanics, so we choose such a problem. The dynamics of a charge in a magnetic field is described by the equation

$$m\frac{dv}{dt} = qv \times B - \gamma v \qquad ...7.43$$

where m and q are the mass and charge of the particle respectively, B is the magnetic field and γ represents some sort of friction.

A Uniform Field

Before considering the more general problem we start with a particle in a spatially uniform field. This problem is analytically solvable and can be used to test the various methods before applying them to the general case.

Units

Note firstly that there are only 2 independent constants in the problem, qB/m and γ/m, and that these constants have the units of inverse time; in fact the former is the cyclotron frequency and the latter is a damping rate. In general in any programming problem it pays to think carefully about the units to be used in the programme. There are several reasons for this.

- If inappropriate units are used the programme may not work at all. An example of this would be the use of SI units to study the dynamics of galaxies or to study atomic physics. In the former R^3 might easily arise and be bigger than the largest number representable on the machine, whereas in the latter $\hbar^4$ may be smaller than the smallest number on the machine and be set automatically to zero with disastrous consequences.

- The problem often has its own natural units and it makes sense to work in these units. This has the consequence that most of the numbers in your programme will be of order unity rather than very large or very small.

In general you should look for the natural units of a problem and write your programme appropriately. Note that these will generally not be SI or cgs. In the problem we are considering here there are 2 natural time scales, m/qB and m/γ. If we decide to work in one of these, e.g. the cyclotron period m/qB, we can rewrite (7.43) in the simpler form

$$\frac{dv_x}{dt'} = +v_y - \left|\frac{\gamma}{qB}\right| v_x \qquad \text{...7.44}$$

$$\frac{dv_y}{dt'} = -v_x - \left|\frac{\gamma}{qB}\right| v_y \qquad \text{...7.45}$$

or perhaps

$$\frac{dv_x}{dt'} = -v_y - \left|\frac{\gamma}{qB}\right| v_x \qquad \text{...7.46}$$

$$\frac{dv_y}{dt'} = +v_x - \left|\frac{\gamma}{qB}\right| v_y \qquad \text{...7.47}$$

depending on the sign of qB/m. Here $t = t'|m/qB|$ and we have chosen our coordinate system such that the magnetic field, *B*, is in the *z*-direction. Note, in addition, that choosing the units appropriately has eliminated all but one of the constants from the problem. This cuts down on superfluous arithmetic in the programme.

The Analytical Solution

In order to understand the behaviour of the various methods for ODEs we need to know the analytical solution of the problem. 2 dimensional problems such as this one are often most easily solved by turning the 2D vector into a complex number. Thus by defining $\tilde{v} = v_x + iv_y$ we can rewrite (7.43) in the form

$$\frac{d\tilde{v}}{dt} = -i\tilde{v}\left(\frac{qB}{m}\right) - \tilde{v}\left(\frac{\gamma}{m}\right) \qquad \text{...7.48}$$

which can be easily solved using the integrating factor method to give

$$\tilde{v} = \tilde{v}_0 \exp\left[-i\left(\frac{qB}{m}\right)t - \left(\frac{\gamma}{m}\right)t\right]. \qquad \text{...7.49}$$

Finally we take real and imaginary parts to find the v_x and v_y components

$$v_x = +v_0 \cos\left[\left(\frac{qB}{m}\right)t + \phi_0\right] \exp\left[-\left(\frac{\gamma}{m}\right)t\right] \qquad \text{...7.50}$$

$$v_y = -v_0 \sin\left[\left(\frac{qB}{m}\right)t + \phi_0\right] \exp\left[-\left(\frac{\gamma}{m}\right)t\right] \qquad \text{...7.51}$$

Choosing an Algorithm

The problem as posed does not fit neatly into the categories defined in (7.1). By considering the accuracy and stability properties of the various algorithms described in these notes you have to decide which is the most appropriate algorithm to use for solving the problem. The computer time required by the algorithm as well as the ease with which it can be programmed may be legitimate considerations. It may not be possible to solve the the stability equations exactly in the most general case. Nevertheless you should be able to deduce enough information on which to base a decision. In your report you should discuss the merits and demerits of the various possible algorithms and explain the rationale behind your choice. You should also write a programme to implement your chosen algorithm and test it for various values of and in all the physically significant regimes.

Crossed Electric and Magnetic Fields

You are now in a position to apply your chosen algorithm to a more complicated problem. In addition to the uniform magnetic field, B, we now add an electric field in the x-direction, $E = (E_x, 0, 0)$.

$$\frac{dv_x}{dt} = +\left(\frac{qB}{m}\right)v_y - \left(\frac{\gamma}{m}\right)v_x + \left(\frac{qE}{m}\right) \qquad ...7.52$$

$$\frac{dv_y}{dt} = -\left(\frac{qB}{m}\right)v_x - \left(\frac{\gamma}{m}\right)v_y \qquad ...7.53$$

Oscillating Electric Field

Finally, if you have time, you might consider making the electric field explicitly time dependent

$$E_x = E_0 \cos \omega t \qquad ...7.54$$

and investigate the behaviour of the system as a function of frequency.

8

Integration of ODEs

Here, we shall discuss the standard numerical techniques used to integrate systems of ordinary differential equations (ODEs). We shall then employ these techniques to simulate the trajectories of various different types of baseball pitch.

By definition, an ordinary differential equation, or o.d.e., is a differential equation in which all dependent variables are functions of a *single* independent variable. Furthermore, an nth-order o.d.e. is such that, when it is reduced to its simplest form, the highest order derivative it contains is nth-order.

According to Newton's laws of motion, the motion of any collection of rigid objects can be reduced to a set of second-order o.d.e.s. in which time, t, is the common independent variable. For instance, the equations of motion of a set of n interacting point objects moving in 1-dimension might take the form:

$$\frac{d^2 x_j}{dt^2} = \frac{F_j(x_1 \ldots, x_n, t)}{m_j} \qquad \ldots 8.1$$

for $j = 1$ to n, where x_j is the position of the j th object, m_j is its mass, etc. Note that a set of n second-order o.d.e.s can always be rewritten as a set of $2n$ first-order o.d.e.s. Thus, the above equations of motion can be rewritten:

$$\frac{dx_j}{dt} \quad v_j, \qquad \ldots 8.2$$

$$\frac{dv_j}{dt} = \frac{F_j(x_1, \ldots, x_n, t)}{m_j}. \qquad \ldots 8.3$$

for $j=1$ to n. We conclude that a general knowledge of how to numerically solve a set of coupled first-order o.d.e.s would enable us to investigate the behaviour of a wide variety of interesting dynamical systems.

Euler's Method

Consider the general first-order o.d.e.,

$$y' = f(x,y), \qquad ...8.4$$

where $'$ denotes d/dx, subject to the general initial-value boundary condition

$$y(x_0) = y_0. \qquad ...8.5$$

Clearly, if we can find a method for numerically solving this problem, then we should have little difficulty generalising it to deal with a system of n simultaneous first-order o.d.e.s.

It is important to appreciate that the numerical solution to a differential equation is only an *approximation* to the actual solution. The actual solution, $y(x)$, to equation (8.4) is (presumably) a *continuous* function of a continuous variable, x. However, when we solve this equation numerically, the best that we can do is to evaluate approximations to the function $y(x)$ at a series of *discrete* grid-points, the x_n (say), where $n = 0,1,2,...$ and $x_0 < x_1 < x_2 \cdots$. For the moment, we shall restrict our discussion to *equally spaced* grid-points, where

$$x_n = x_0 + nh. \qquad ...8.6$$

Here, the quantity is referred to as the *step-length*. Let y_n be our approximation to $y(x)$ at the grid-point x_n. A numerical integration scheme is essentially a method which somehow employs the information contained in the original o.d.e., equation (8.4), to construct a series of rules interrelating the various y_n.

The simplest possible integration scheme was invented by the celebrated 18th century Swiss mathematician Leonhard Euler, and is, therefore, called *Euler's method*. Incidentally, it is interesting to note that virtually all of the standard methods used in numerical analysis were invented *before* the advent of electronic computers. In olden days, people actually performed numerical calculations *by hand*—and a very long and tedious process it must have been! Suppose that we have evaluated an approximation, y_n, to the solution, $y_{(x)}$, of equation (8.4) at the grid-point x_n. The approximate gradient of $y(x)$ at this point is, therefore, given by

$$y'_v = f(x_n, y_n). \qquad ...8.7$$

Let us approximate the curve $y(x)$ as a *straight-line* between the neighbouring grid-points x_n and x_{n+1}. It follows that

$$y_{n+1} = y_n + y'_n h, \qquad ...8.8$$

or

$$y_{n+1} = y_n + f(x_n, y_n)h. \qquad ...8.9$$

The above formula is the essence of Euler's method. It enables us to calculate all of the y_n, given the initial value, y_0, at the first grid-point, x_0. Euler's method is illustrated in figure 8.1.

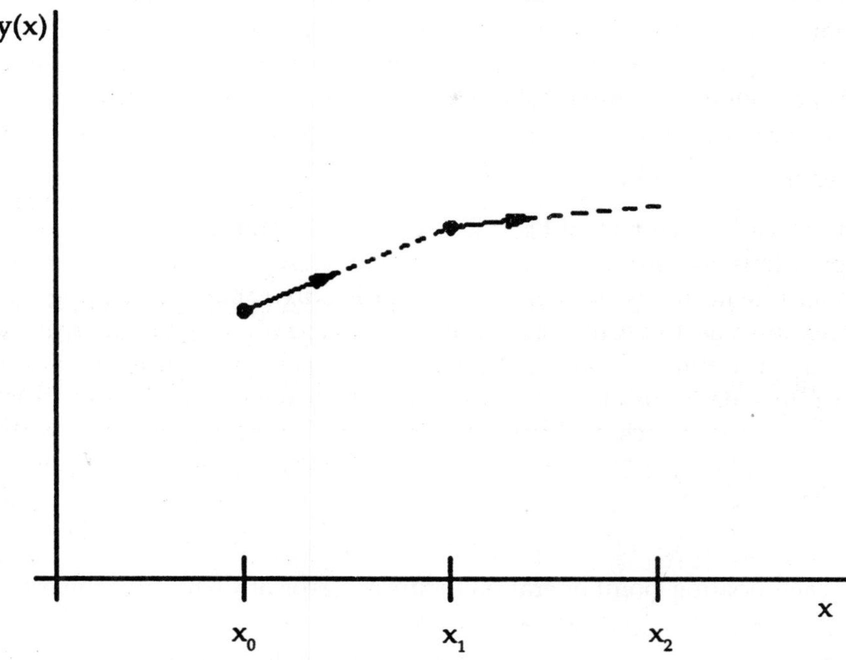

***Figure 8.1:** Illustration of Euler's method*

Numerical Errors

There are two major sources of error associated with a numerical integration scheme for o.d.e.s: namely, *truncation error* and *round-off error*. Truncation error arises in Euler's method because the curve $y(x)$ is *not* generally a straight-line between the neighbouring grid-points x_n and x_{n+1}, as assumed above. The error associated with this approximation can easily be assessed by Taylor expanding $y(x)$ about $x = x_n$:

$$y(x_n + h) = y(x_n) + hy'(x_n) + \frac{h^2}{2} y''(x_n) + \cdots$$
$$= y_n + h f(x_n, y_n) + \frac{h^2}{2} y''(x_n) + \cdots. \quad ...8.10$$

A comparison of equations (8.9) and (8.10) yields

$$y_{n+1} = y_n + h f(x_n, y_n) + O(h^2). \quad ...8.11$$

In other words, every time we take a step using Euler's method we incur a truncation error of $O(h^2)$, where is the step-length. Suppose that we use Euler's method to integrate our o.d.e. over an x-interval of order unity. This requires $O(h^{-1})$ steps. If each step incurs an error of $O(h^2)$, and the errors are simply cumulative (a fairly conservative assumption), then the net truncation error is $O(h)$. In other words, the error associated with integrating an o.d.e. over a finite interval using Euler's method is directly proportional to the step-length. Thus, if we want to keep the relative error in the integration below about 10^{-6} then we would need to take about one million steps per unit interval in x. Incidentally, Euler's method is termed a *first-order* integration method because the truncation error associated with integrating over a finite interval scales like h^1. More generally, an integration method is conventionally called th order if its truncation error *per step* is $O(h^{n+1})$.

Note that truncation error would be incurred even if computers performed floating-point arithmetic operations to infinite accuracy. Unfortunately, computers *do not* perform such operations to infinite accuracy. In fact, a computer is only capable of storing a floating-point number to a fixed number of decimal places. For every type of computer, there is a characteristic number, η, which is defined as the smallest number which when added to a number of order unity gives rise to a new number: *i.e.,* a number which when taken away from the original number yields a non-zero result. Every floating-point operation incurs a *round-off error* of $O(\eta)$ which arises from the finite accuracy to which floating-point numbers are stored by the computer. Suppose that we use Euler's method to integrate our o.d.e. over an -interval of order unity. This entails $O(h^{-1})$ integration steps, and, therefore, $O(h^{-1})$ floating-point operations. If each floating-point operation incurs an error of $O(\eta)$, and the errors are simply cumulative, then the net round-off error is $O(\eta/h)$.

The total error, ϵ, associated with integrating our o.d.e. over an x-interval of order unity is (approximately) the sum of the truncation and round-off errors. Thus, for Euler's method,

$$\epsilon \sim \frac{\eta}{h} + h. \qquad ...8.12$$

Clearly, at large step-lengths the error is dominated by truncation error, whereas round-off error dominates at small step-lengths. The net error attains its minimum value, $\epsilon_0 \sim \eta^{1/2}$, when $h = h_0 \sim \eta^{1/2}$. There is clearly no point in making the step-length, h, any smaller than h_0, since this increases the number of floating-point operations but does not lead to an increase in the overall accuracy. It is also clear that the ultimate accuracy of Euler's method (or any other integration method) is determined by the accuracy, η, to which floating-point numbers are stored on the computer performing the calculation.

The value of η depends on how many bytes the computer hardware uses to store floating-point numbers. For IBM-PC clones, the appropriate value for *double precision* floating point numbers is $\eta = 2.22 \times 10^{-16}$ (this value is specified in the system header file float.h). It follows that the minimum practical step-length for Euler's method on such a computer is $h_0 \sim 10^{-8}$, yielding a minimum relative integration error of $\epsilon_0 \sim 10^{-8}$. This level of accuracy is perfectly adequate for most scientific calculations. Note, however, that the corresponding η value for *single precision* floating-point numbers is only $\eta = 1.19 \times 10^{-7}$, yielding a minimum practical step-length and a minimum relative error for Euler's method of $h_0 \sim 3\times10^{-4}$ and $\epsilon_0 \sim 3\times10^{-4}$, respectively. This level of accuracy is generally *not* adequate for scientific calculations, which explains why such calculations are invariably performed using double, rather than single, precision floating-point numbers on IBM-PC clones (and most other types of computer).

Numerical Instabilities

Consider the following example. Suppose that our o.d.e. is

$$y' = -\alpha y, \qquad ...8.13$$

where $\alpha > 0$, subject to the boundary condition

$$y(0) = 1. \qquad ...8.14$$

Of course, we can solve this problem analytically to give

$$y(x) = \exp(-\alpha x). \qquad ...8.15$$

Note that the solution is a *monotonically decreasing* function of x. We can also solve this problem numerically using Euler's method. Appropriate grid-points are

$$x_n = nh, \qquad ...8.16$$

where $n = 0,1,2,\cdots$. Euler's method yields

$$y_{n+1} = (1-\alpha h)y_n. \qquad ...8.17$$

Note one curious fact. If $h > 2/\alpha$ then $|y_{n+1}| > |y_n|$. In other words, if the step-length is made too large then the numerical solution becomes an oscillatory function of x of *monotonically increasing* amplitude: i.e., the numerical solution *diverges* from the actual solution. This type of catastrophic failure of a numerical integration scheme is called a *numerical instability*. All simple integration schemes become unstable if the step-length is made sufficiently large.

Runge-Kutta Methods

There are two main reasons why Euler's method is *not* generally used in scientific computing. Firstly, the truncation error per step associated with this method is far larger than those associated with other, more advanced, methods (for a given value of h). Secondly, Euler's method is too prone to numerical instabilities. The methods most commonly employed by scientists to integrate o.d.e.s were first developed by the German mathematicians C.D.T. Runge and M.W. Kutta in the latter half of the nineteenth century.The basic reasoning behind so-called *Runge-Kutta* methods is outlined in the following.

The main reason that Euler's method has such a large truncation error per step is that in evolving the solution from x_n to x_{n+1} the method only evaluates derivatives at the beginning of the interval: i.e., at x_n. The method is, therefore, very *asymmetric* with respect to the beginning and the end of the interval. We can construct a more symmetric integration method by making an Euler-like trial step to the midpoint of the interval, and then using the values of both x and y at the midpoint to make the real step across the interval. To be more exact,

$$k_1 = hf(x_n, y_n), \quad ...8.18$$

$$k_2 = hf(x_n + h/2, y_n + k_1/2), \quad ...8.19$$

$$y_{n+1} = y_n + k_2 + O(h^3). \quad ...8.20$$

As indicated in the error term, this symmetrisation cancels out the first-order error, making the method *second-order*. In fact, the above method is generally known as a *second-order Runge-Kutta* method. Euler's method can be thought of as a first-order Runge-Kutta method.

Of course, there is no need to stop at a second-order method. By using two trial steps per interval, it is possible to cancel out both the first and second-order error terms, and, thereby, construct a third-order Runge-Kutta method. Likewise, three trial steps per interval yield a fourth-order method, and so on.

The general expression for the total error, ϵ, associated with integrating our o.d.e. over an x-interval of order unity using an nth-order Runge-Kutta method is approximately

$$\epsilon \sim \frac{\eta}{h} + h^n. \quad ...8.21$$

Here, the first term corresponds to round-off error, whereas the second term represents truncation error. The minimum practical step-length, h_0, and the minimum error, ϵ_0, take the values

$$h_0 \sim \eta^{1/(n+1)}, \quad ...8.22$$

$$\epsilon_0 \sim \eta^{n/(n+1)}, \quad ...8.23$$

respectively. In Tab. 1, these values are tabulated against $n$using $\eta = 2.22 \times 10^{-16}$ (the value appropriate to double precision arithmetic on IBM-PC clones). It can be seen that h_0 increases and ϵ_0 decreases as gets larger. However, the relative change in these quantities becomes progressively less dramatic as $n$increases.

Table 8.1: *The minimum practical step-length, h_0, and minimum error, ϵ_0, for an th-order Runge-Kutta method integrating over a finite interval using double precision arithmetic on an IBM-PC clone.*

n	h_0	ϵ_0
1	1.5×10^{-8}	1.5×10^{-8}
2	6.1×10^{-6}	3.7×10^{-11}
3	1.2×10^{-4}	1.8×10^{-12}
4	7.4×10^{-4}	3.0×10^{-13}
5	2.4×10^{-3}	9.0×10^{-14}

In the majority of cases, the limiting factor when numerically integrating an o.d.e. is not round-off error, but rather the computational effort involved in calculating the function $f(x,y)$. Note that, in general, an nth-order Runge-Kutta method requires evaluations of this function per step. It can easily be appreciated that as n is increased a point is quickly reached beyond which any benefits associated with the increased accuracy of a higher order method are more than offset by the computational "cost" involved in the necessary additional evaluation $f(x,y)$ of per step. Although there is no hard and fast general rule, in most problems encountered in computational physics this point corresponds to $n = 4$. In other words, in most situations of interest a *fourth-order Runge Kutta* integration method represents an appropriate compromise between the competing requirements of a low truncation error per step and a low computational cost per step.

The standard *fourth-order Runge-Kutta* method takes the form:

$$k_1 = h\,f(x_n, y_n), \quad \text{...8.24}$$

$$k_2 = h\,f(x_n + h/2, y_n + k_1/2), \quad \text{...8.25}$$

$$k_3 = h\,f(x_n + h/2, y_n + k_2/2), \quad \text{...8.26}$$

$$k_4 = h\,f(x_n + h, y_n + k_3), \quad \text{...8.27}$$

$$y_{n+1} = y_n + \frac{k_1}{6} + \frac{k_2}{3} + \frac{k_3}{3} + \frac{k_4}{6} + O(h^5). \quad \text{...8.28}$$

This is the method which we shall use, throughout this course, to integrate first-order o.d.e.s. The generalisation of this method to deal with systems of coupled first-order o.d.e.s is (hopefully) fairly obvious.

An Example Fixed-step RK4 Routine

Listed below is an example fixed-step, fourth-order Runge-Kutta (RK4) integration routine which utilises the Blitz++ library.

```
// rk4_fixed.cpp
/*
  Function to advance set of coupled first-order o.d.e.s by single step
   using fixed step-length fourth-order Runge-Kutta scheme
       x              ... independent variable
       y              ... array of dependent variables
       h              ... fixed step-length
   Requires right-hand side routine
     void rhs_eval (double x, Array<double,1> y, Array<double,1>& dydx)
   which evaluates derivatives of y (w.r.t. x) in array dydx
*/
#include <blitz/array.h>
using namespace blitz;
```

```
void rk4_fixed      (double& x, Array<double,1>& y,
                     void(*rhs_eval)(double,Array<double,1>, Array<double,1>&),
                     double h)
{
  // Array y assumed to be of extent n, where n is no. of coupled
  // equations
  int n = y.extent(0);
  // Declare local arrays
  Array<double,1> k1(n), k2(n), k3(n), k4(n), f(n), dydx(n);
  // Zeroth intermediate step
  (*rhs_eval) (x, y, dydx);
  for (int j = 0; j < n; j++)
    {
      k1(j) = h * dydx(j);
      f(j) = y(j) + k1(j) / 2.;
    }
  // First intermediate step
  (*rhs_eval) (x + h / 2., f, dydx);
  for (int j = 0; j < n; j++)
    {
      k2(j) = h * dydx(j);
      f(j) = y(j) + k2(j) / 2.;
    }
  // Second intermediate step
  (*rhs_eval) (x + h / 2., f, dydx);
  for (int j = 0; j < n; j++)
    {
      k3(j) = h * dydx(j);
      f(j) = y(j) + k3(j);
    }
  // Third intermediate step
  (*rhs_eval) (x + h, f, dydx);
  for (int j = 0; j < n; j++)
    {
      k4(j) = h * dydx(j);
    }
  // Actual step
  for (int j = 0; j < n; j++)
    {
     y(j) += k1(j) / 6. + k2(j) / 3. + k3(j) / 3. + k4(j) / 6.;
    }
  x += h;
  return;
}
```

Adaptive Integration Methods

Consider the following system of o.d.e.s:

$$\frac{dx}{dt} = v, \qquad ...8.29$$

$$\frac{dv}{dt} = \frac{v}{t} - 4kt^2 x, \qquad ...8.30$$

subject to the boundary conditions $x = \sqrt{k}v^2$ and $v = 2\sqrt{k}v$ at $x = v$, where $0 < v \ll 1$. This system can be solved analytically to give

$$x = \sin\left(\sqrt{k}t^2\right). \qquad ...8.31$$

One peculiarity of the above solution is that its variation scale-length *decreases rapidly* as t increases.

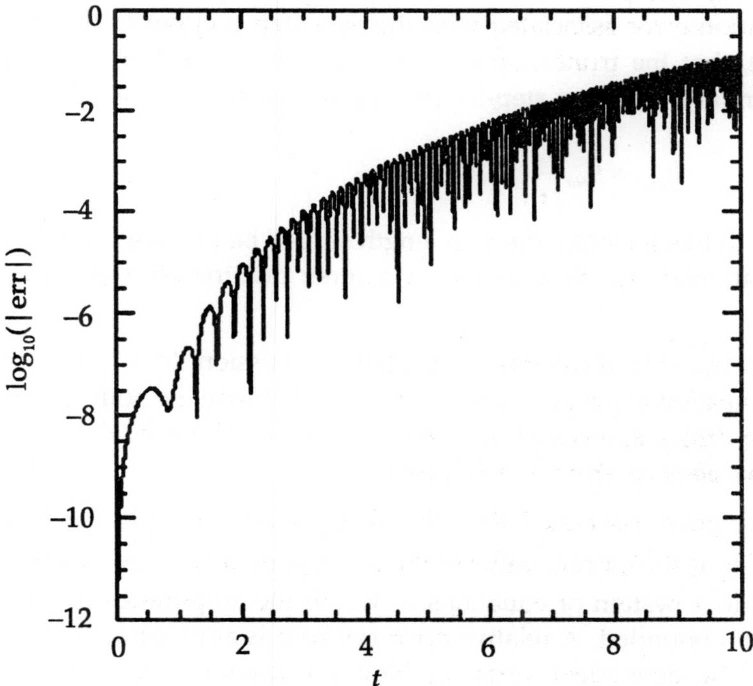

Figure 8.2: *Global integration error associated with a fixed step-length ($h = 0.01$), fourth-order Runge-Kutta method, plotted against the independent variable, t, for a system of o.d.e.s in which the variation scale-length decreases rapidly with increasing t. Double precision calculation*

Let us compare the above solution with that obtained numerically using a fourth-order Runge-Kutta method. Figure 8.2 shows the integration error associated with such a method (calculated by integrating the above system, with $k = 10$, from $t = 10^{-3}$, to $t = t$, and then taking the difference between the numerical and analytic solutions) with a fixed step-length of $h = 0.01$, plotted against the independent variable, t. It can be seen that, although the error starts off small, it rises rapidly as the variation scale-length of the solution decreases (*i.e.*, as increases), and quickly becomes unacceptably large. Of course, we could reduce the error by simply reducing the step-length, *h*. However, this is a very inefficient solution. The step-length only needs to be reduced at large *t*. There is no need to reduce it, at all, at small *t*. Clearly, the ideal solution to this problem would be an integration method in which the step-length is *varied* so as to maintain a relatively constant truncation error per step. Such an *adaptive integration method* would take large steps when variation scale-length of the solution was large, and *vice versa*.

Let us investigate how we could convert our fixed step-length, fourth-order Runge-Kutta method into a corresponding adaptive method. First of all, we need an estimate of the truncation error at each step. Suppose that the current step-length is h. We can estimate the truncation error, ϵ, associated with the current step by taking the difference between the solutions obtained by stepping by $h/2$ twice and by once (starting from the same point, in both cases). Let ϵ_0 be the desired truncation error per step. How do we adjust so as to ensure that the truncation error associated with the next step is closer to this value? Observe, from equation (8.28), that the truncation error per step in a fourth-order scheme scales like h^5. It follows, therefore, that our step-length adjustment formula should take the form

$$h_{new} = h_{old} \left| \frac{\epsilon_0}{\epsilon} \right|^{1/5} . \quad ...8.32$$

According to this formula, the step-length should be increased if the truncation error per step is too small, and *vice versa,* in such a manner that the error per step remains relatively constant at ϵ_0.

There are a number of caveats to the above discussion. In a system of n coupled o.d.e.s, the overall truncation error per step, ϵ, should, of course, be some appropriately weighted average of the errors associated with each equation. There is also a question of whether ϵ should be an *absolute error* or a *relative error*.

The relative error associated with the *i*th equation is simply the absolute error divided by $|y_i|$, where y_i is the current value of the th dependent variable. An absolute error estimate is appropriate to a system of equations in which the amplitudes of the various dependent variables remain bounded. A relative error estimate is appropriate to a system in which the amplitudes of the dependent variables blow-up at some point, but the variables always remain the same sign.

Finally, a *mixed error* estimate—usually the minimum of the absolute and relative errors—is appropriate to a system in which the amplitudes of the dependent variables blow-up at some point, but the signs of the variables oscillate. It is usually a good idea to place some limits on the allowed variation of the step-length from step to step: *e.g.*, by preventing the step-length from increasing or decreasing by more than some factor $S > 1$ per step. This prevents *h* from oscillating unduly about its optimum value. Obviously, if *h* becomes absurdly small then the integration method has failed, and should abort with an appropriate error message. Finally, a limit should be placed on how large h can become—unfortunately, adaptive methods have a tendency to become a little over optimistic when integration is easy.

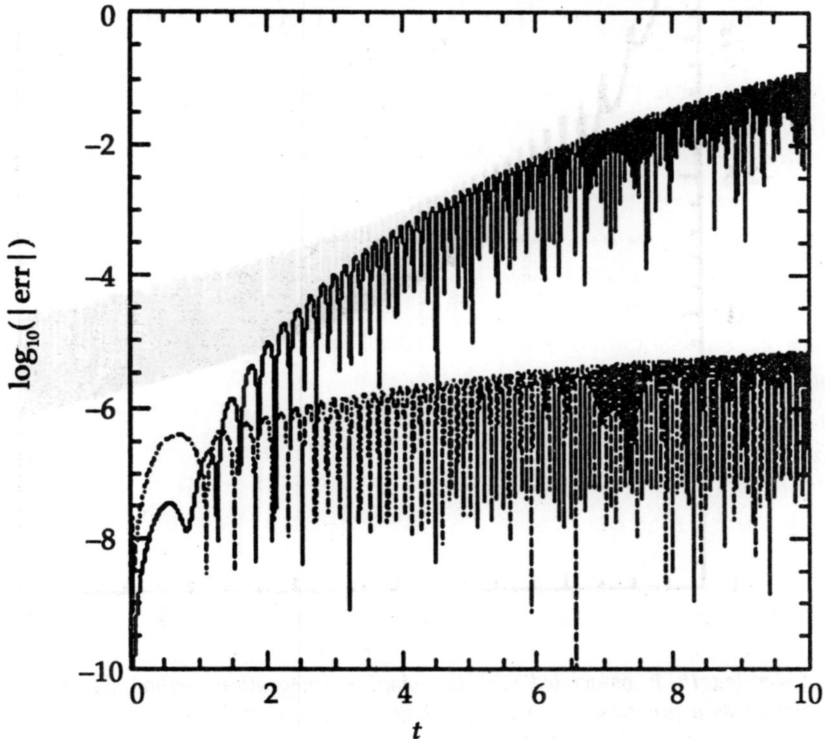

Figure 8.3: *Global integration errors associated with a fixed step-length* $(h = 0.01)$, *fourth-order Runge-Kutta method (solid curve) and a corresponding adaptive method* ($\epsilon_0 = 10^{-8}$) *(dotted curve), plotted against the independent variable, t, for a system of o.d.e.s in which the variation scale-length decreases rapidly with increasing t. Double precision calculation*

Figure 8.3 shows the integration errors associated with a fixed step-length, fourth-order Runge-Kutta method and a corresponding adaptive method—constructed along the lines discussed above—as functions of the independent variable, t.

The errors are calculated by integrating the current system, with $k = 10$, from $t = 10^{-3}$ to $t = t$, and then taking the difference between the numerical and analytic solutions. The fixed step-length associated with the former method is $h = 0.01$. The desired truncation error per step associated with the latter is $\epsilon_0 = 10^{-8}$.

It can be seen that the performance of the adaptive method is far superior to that of the fixed step-length method, since the former method maintains a relatively constant integration error as the variation scale-length of the solution decreases (*i.e.*, as t increases). Figure 8.4 illustrates how this is achieved. This figure shows the step-length, h, associated with the adaptive method as a function of t. It can be seen that the adaptive method maintains a relatively constant truncation error per step by decreasing h as t increases.

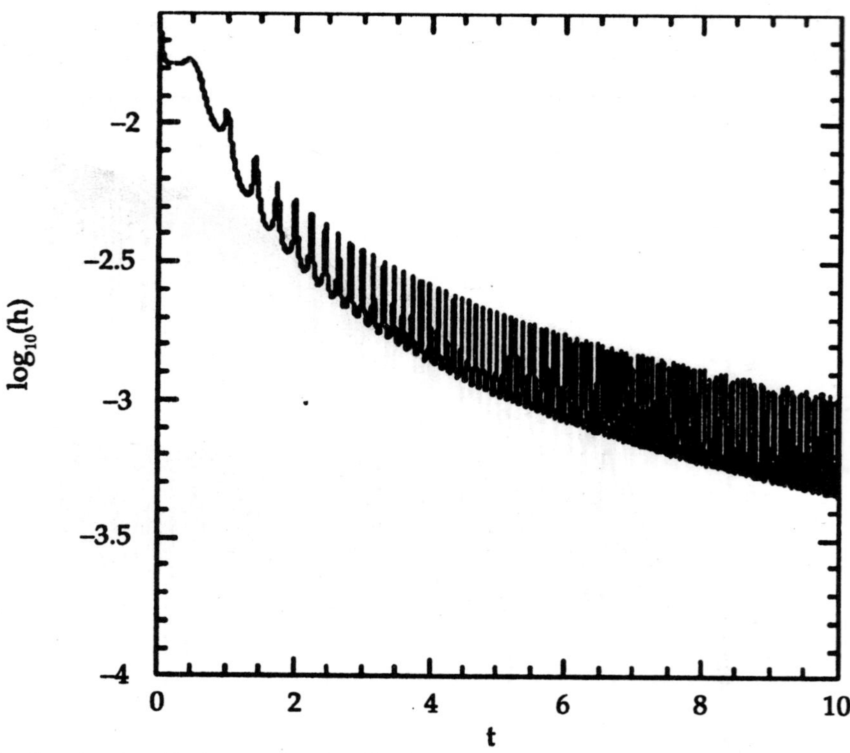

Figure 8.4: The step-length, h, associated with the adaptive integration method shown in the previous figure, plotted as a function of the independent variable, t. Double precision calculation

An Example Adaptive-step RK4 Routine

Listed below is an example adaptive-step RK4 routine which makes use of the previously listed fixed-step routine. Note that the routine recalculates all steps whose truncation error exceeds the desired value acc.

```
// rk4_adaptive.cpp
/*
  Function to advance set of coupled first-order o.d.e.s by single step
   using adaptive fourth-order Runge-Kutta scheme
      x           ... independent variable
      y           ... array of dependent variables
      h           ... step-length
      t_err       ... actual truncation error per step
      acc         ... desired truncation error per step
      S           ... step-length cannot change by more than this factor from
                      step to step
```

```
      rept         ... number of step recalculations
      maxrept      ... maximum allowable number of step recalculations
      h_min        ... minimum allowable step-length
      h_max        ... maximum allowable step-length
      flag         ... controls manner in which truncation error is calculated
   Requires right-hand side routine
     void rhs_eval (double x, Array<double,1> y, Array<double,1>& dydx)
    which evaluates derivatives of y (w.r.t. x) in array dydx.
  Function advances equations by single step whilst attempting to maintain
   constant truncation error per step of acc:
     flag = 0    ... error is absolute
     flag = 1    ... error is relative
     flag = 2    ... error is mixed
   If step-length falls below h_min then routine aborts
*/
#include <stdio.h>
#include <stdlib.h>
#include <math.h>
#include <blitz/array.h>
using namespace blitz;
void rk4_fixed (double&, Array<double,1>&,
               void (*)(double, Array<double,1>, Array<double,1>&),
              double);
void rk4_adaptive (double& x, Array<double,1>& y,
                           void (*rhs_eval)(double, Array<double,1>,
Array<double,1>&),
                         double& h, double& t_err, double acc,
                         double S, int& rept, int maxrept,
                         double h_min, double h_max, int flag)
{
   // Array y assumed to be of extent n,  where n is no. of coupled
   // equations
   int n = y.extent(0);
   // Declare local arrays
```

```
Array<double,1> y0(n), y1(n);
// Declare repetition counter
static int count = 0;
// Save initial data
double x0 = x;
y0 = y;
// Take full step
rk4_fixed (x, y, rhs_eval, h);
// Save data
y1 = y;
// Restore initial data
x = x0;
y = y0;
// Take two half-steps
rk4_fixed (x, y, rhs_eval, h/2.);
rk4_fixed (x, y, rhs_eval, h/2.);
// Calculate truncation error
t_err = 0.;
double err, err1, err2;
if (flag == 0)
  {
    // Use absolute truncation error
    for (int i = 0; i < n; i++)
      {
        err = fabs (y(i) - y1(i));
        t_err = (err > t_err) ? err : t_err;
      }
  }
else if (flag == 1)
  {
    // Use relative truncation error
    for (int i = 0; i < n; i++)
      {
```

```
                err = fabs ((y(i) - y1(i)) / y(i));
                t_err = (err > t_err) ? err : t_err;
              }
         }
      else
         {
            // Use mixed truncation error
            for (int i = 0; i < n; i++)
              {
                err1 = fabs ((y(i) - y1(i)) / y(i));
                err2 = fabs (y(i) - y1(i));
                err = (err1 < err2) ? err1 : err2;
                t_err = (err > t_err) ? err : t_err;
              }
         }
      // Prevent small truncation error from rounding to zero
      if (t_err == 0.) t_err = 1.e-15;
      // Calculate new step-length
      double h_est = h * pow (fabs (acc / t_err), 0.2);
      // Prevent step-length from changing by more than factor S
      if (h_est / h > S)
         h *= S;
      else if (h_est / h < 1. / S)
         h /= S;
      else
         h = h_est;
      // Prevent step-length from exceeding h_max
      h = (fabs(h) > h_max) ? h_max * h / fabs(h) : h;
      // Abort if step-length falls below h_min
      if (fabs(h) < h_min)
         {
            printf ("Error - |h| < hmin\n");
            exit (1);
```

```
    }
  // If truncation error acceptable take step
  if ((t_err <= acc) || (count >= maxrept))
    {
      rept = count;
      count = 0;
    }
  // If truncation error unacceptable repeat step
  else
    {
      count++;
      x = x0;
      y = y0;
      rk4_adaptive (x, y, rhs_eval, h, t_err, acc,
                    S, rept, maxrept, h_min, h_max, flag);
    }

  return;
}
```

Advanced Integration Methods

Of course, Runge-Kutta methods are not the last word in integrating o.d.e.s. Far from it! Runge-Kutta methods are sometimes referred to as *single-step* methods, since they evolve the solution from x_n to x_{n+1} without needing to know the solutions at x_{n-1}, x_{n-2}, etc. There is a broad class of more sophisticated integration methods, known as *multi-step* methods, which utilise the previously calculated solutions at x_{n-1}, x_{n-2}, etc. in order to evolve the solution from x_n to x_{n+1}. Examples of these methods are the various Adams methods and the various Predictor-Corrector methods.

The main advantages of Runge-Kutta methods are that they are easy to implement, they are very stable, and they are "self-starting" (*i.e.*, unlike muti-step methods, we do not have to treat the first few steps taken by a single-step integration method as special cases). The primary disadvantages of Runge-Kutta methods are that they require significantly more computer time than multi-step methods of comparable accuracy, and they do not easily yield good *global* estimates of the truncation error. However, for the straightforward dynamical systems under investigation in this course, the advantage of the relative simplicity and ease of use of Runge-Kutta methods far outweighs the disadvantage of their relatively high computational cost.

The Physics of Baseball Pitching

Baseball is the oldest professional sport in the US. It is a game of great subtlety (like cricket!) which has fascinated fans for over a hundred years. It has also fascinated physicists—partly, because many physicists are avid baseball fans, but, partly, also, because there are clearly delineated physics principles at work in this game. Indeed, more books and papers have been written on the physics of baseball than on any other sport.

A baseball is formed by winding yarn around a small sphere of cork. The ball is then covered with two interlocking pieces of white cowhide, which are tightly stitched together. The mass and circumference of a regulation baseball are 5oz and 9in (*i.e.*, about 150g and 23 cm), respectively. In the major leagues, the ball is pitched a distance of 60 feet 6 inches (*i.e.*, 18.44 m), towards the hitter, at speeds which typically lie in the range 60 to 100 mph (*i.e.*, about 30 to 45 m/s). As is well-known to baseball fans, there are a surprising variety of different pitches. "Sliders" deviate sideways through the air. "Curveballs" deviate sideways, but also dip unusually rapidly. Conversely, "fastballs" dip unusually slowly. Finally, the mysterious "knuckleball" can weave from side to side as it moves towards the hitter. How is all this bizarre behaviour possible? Let us investigate.

Air Drag

A baseball in flight is subject to three distinct forces. The first is *gravity*, which causes the ball to accelerate vertically downwards at . The second is *air drag*, which impedes the ball's motion through the air. The third is the *Magnus force*, which permits the ball to curve laterally. Let us discuss the latter two forces in more detail.

As is well-known, the drag force acting on an object which moves *very slowly* through a viscous fluid is directly proportional to the speed of that object with respect to the fluid. For example, a sphere of radius, moving with speed v through a fluid whose coefficient of viscosity is η, experiences a drag force given by Stokes' law:

$$f_D = 6\pi\eta r v. \qquad \text{...8.33}$$

As students who have attempted to reproduce Millikan's oil drop experiment will recall, this force is the dominant drag force acting on a microscopic oil drop falling through air. However, for most *macroscopic* projectiles moving through air, the above force is dwarfed by a second drag force which is proportional to v^2.

The origin of this second force is fairly easy to understand. At velocities sufficiently low for Stokes' law to be valid, air is able to flow smoothly around a passing projectile. However, at higher velocities, the projectile's motion is too rapid for this to occur. Instead, the projectile effectively knocks the air out of its way. The total mass of air which the projectile comes into contact with per second is $\rho v A$, where ρ is v the air density, A the projectile speed, and the projectile's cross-sectional area. Suppose, as seems reasonable, that the projectile imparts to this air mass a speed v' which is directly proportional to v. The rate of momentum gain of the air, which is equal to the drag force acting on the projectile, is approximately

$$f_D = \frac{1}{2} C_D(v)\rho A v^2, \qquad \text{...8.34}$$

where the *drag coefficient*, $C_D(v)$, is a dimensionless quantity.

The drag force acting on a projectile, passing through air, always points in the *opposite* direction to the projectile's instantaneous direction of motion. Thus, the vector drag force takes the form

$$f_D = -\frac{1}{2} C_D(v) \rho A v \mathbf{v}. \qquad ...8.35$$

When a projectile moves through air it leaves a *turbulent* wake. The value of the drag coefficient, C_D, is closely related to the properties of this wake. Turbulence in fluids is conventionally characterised in terms of a dimensionless quantity known as a *Reynolds number*:

$$R_e = \frac{\rho v d}{\eta}. \qquad ...8.36$$

Here, d is the typical length-scale (*e.g.*, the diameter) of the projectile. For sufficiently small Reynolds numbers, the air flow immediately above the surface of the projectile remains smooth, despite the presence of the turbulent wake, and C_D takes an approximately constant value which depends on the projectile's shape. However, above a certain critical value of R_e—which corresponds to 2×10^5 for a *smooth* projectile—the air flow immediately above the surface of the projectile becomes turbulent, and the drag coefficient drops: i.e., a projectile slips more easily through the air when the surrounding flow is completely turbulent. In this high Reynolds number regime, the drag coefficient generally falls rapidly by a factor of between 3 and 10, as R_e is increased, and then settles down to a new, roughly constant value. Note that the critical Reynolds number is significantly less than 2×10^5 for projectiles with rough surfaces. Paradoxically, a rough projectile generally experiences less air drag, at high velocities, than a smooth one. The typical dependence of the drag coefficient on the Reynolds number is illustrated schematically in figure 8.5.

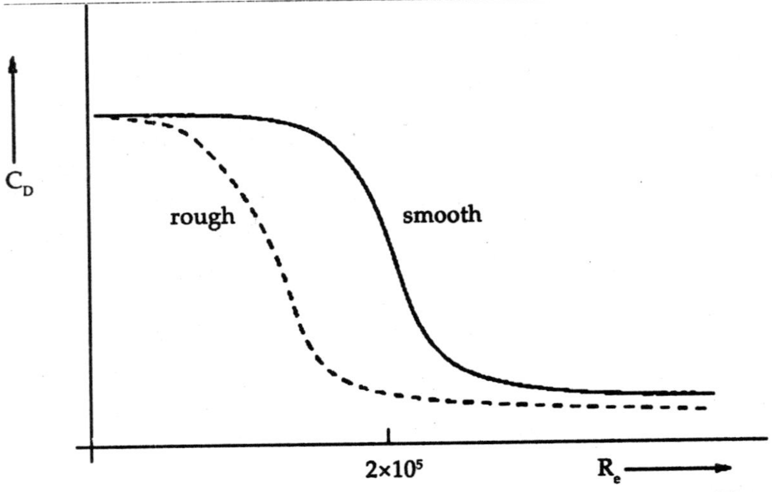

Figure 8.5: *Typical dependence of the drag coefficient,* C_D*, on the Reynolds number,* R_e

Wind tunnel measurements reveal that the drag coefficient is a strong function of speed for baseballs, as illustrated in figure 8.6. At low speeds, the drag coefficient is approximately constant. However, as the speed increases, C_D drops substantially, and is more than a factor of 2 smaller at high speeds.

This behaviour is similar to that described above. The sudden drop in the drag coefficient is triggered by a transition from laminar to turbulent flow in the air layer immediately above the ball's surface. The critical speed (to be more exact, the critical Reynolds number) at which this transition occurs depends on the properties of the surface.

Thus, for a completely smooth baseball the transition occurs at speeds well beyond the capabilities of the fastest pitchers. Conversely, the transition takes place at comparatively low speeds if the surface of the ball is rough.

In fact, the raised stitches on the otherwise smooth regulation baseball cause the transition to occur at an intermediate speed which is easily within the range of major league pitchers. Note that the magnitude of the drag force is substantial—it actually exceeds the force due to gravity for ball speeds above about 95mph.

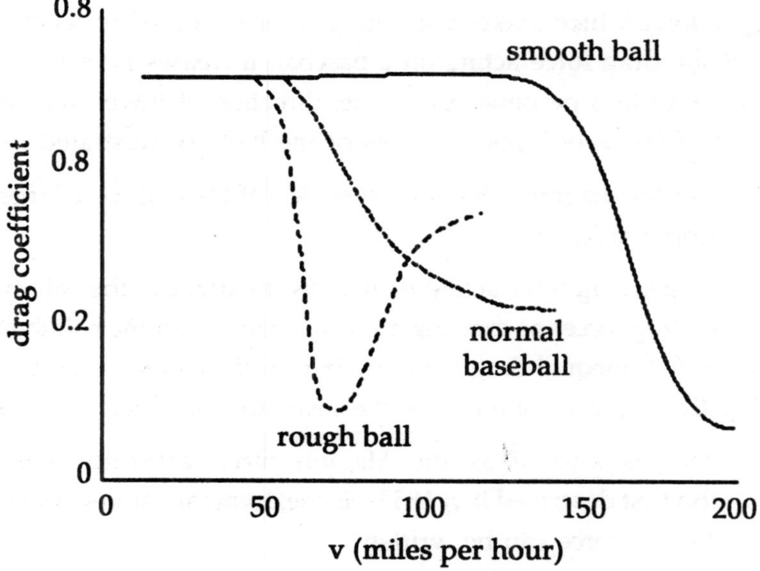

Figure 8.6: Variation of the drag coefficient, C_D, with speed, v, for normal, rough, and smooth baseballs.

The above discussion leads to a number of interesting observations. First, the raised stitches on a baseball have an important influence on its aerodynamic properties. Without them, the air drag acting on the ball at high speeds would increase substantially.

Indeed, it seems unlikely that major league pitchers could throw 95mph fastballs if baseballs were completely smooth. On the other hand, a scuffed-up baseball experiences even less air drag than a regulation ball. Presumably, such a ball can be thrown faster—which explains why balls are so regularly renewed in major league games.

Giordano has developed the following useful formula which quantifies the drag force acting on a baseball:

$$\frac{f_D}{m} = -F(v)v\mathrm{v}, \qquad ...8.37$$

where

$$F(v) = 0.0039 + \frac{0.0058}{1 + \exp[(v - v_d)/\Delta]}. \qquad ...8.38$$

Here, $v_d = 35m/s$ and $\Delta = 5m/s$.

The Magnus Force

We have not yet explained how a baseball is able to *curve* through the air. Physicists in the last century could not account for this effect, and actually tried to dismiss it as an "optical illusion." It turns out that this strange phenomenon is associated with the fact that the balls thrown in major league games tend to *spin* fairly rapidly—typically, at 1500rpm.

The origin of the force which makes a spinning baseball curve can readily be appreciated once we recall that the drag force acting on a baseball increases with increasing speed.For a ball spinning about an axis perpendicular to its direction of travel, the speed of the ball, relative to the air, is *different* on opposite sides of the ball, as illustrated in figure 8.7.

It can be seen, from the diagram, that the lower side of the ball has a larger speed relative to the air than the upper side.

This results in a larger drag force acting on the lower surface of the ball than on the upper surface. If we think of drag forces as exerting a sort of pressure on the ball then we can readily appreciate that when the unequal drag forces acting on the ball's upper and lower surfaces are added together there is a component of the resultant force acting *upwards*.

This force—which is known as the *Magnus force*, after the German physicist Heinrich Magnus, who first described it in 1853—is the dominant spin-dependent force acting on baseballs. The Magnus force can be written

$$f_M = S(v)\omega \times \mathrm{v}, \qquad ...8.39$$

where is the angular velocity vector of the ball. According to Adair and Giordano, it is a fairly good approximation to take

$$B = \frac{S}{m} = 4.1 \times 10^{-4} \qquad ...8.40$$

for baseballs. Note that B is a dimensionless quantity. The magnitude of the Magnus force is about one third of the force due to gravity for typical curveballs.

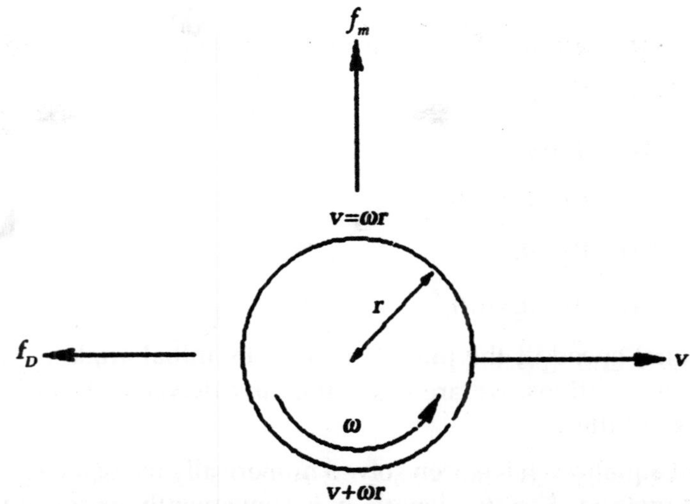

***Figure 8.7:** Origin of the Magnus force for a ball of radius r, moving with speed v, and spinning with angular velocity ω about an axis perpendicular to its direction of motion*

Simulations of Baseball Pitches

Let us adopt a set of coordinates such that x measures displacement from the pitcher to the hitter, y measures horizontal displacement (a displacement in the $+y$ direction corresponds to a displacement to the hitter's right-hand side), and measures vertical displacement (a displacement in the +z direction corresponds to an upward displacement). Using these coordinates, the equations of motion of a baseball can be written as the following set of coupled first-order o.d.e.s:

$$\frac{dx}{dt} = v_x, \qquad \text{...8.41}$$

$$\frac{dy}{dt} = v_y, \qquad \text{...8.42}$$

$$\frac{dz}{dt} = v_z, \qquad \text{...8.43}$$

$$\frac{dv_x}{dt} = -F(v)vv_x + B\omega\left(v_z \sin\phi - v_y \cos\phi\right), \qquad \text{...8.44}$$

$$\frac{dv_y}{dt} = -F(v)vv_y + B\omega v_x \cos\phi, \qquad \text{...8.45}$$

$$\frac{dv_z}{dt} = -g - F(v)vv_z - B\omega v_x \sin\phi. \qquad \text{...8.46}$$

Here, the ball's angular velocity vector has been written $\omega = \omega(0, \sin\phi, \cos\phi)$. Appropriate boundary conditions at $t = 0$ are:

$$x(t=0)=0, \quad \text{...8.47}$$

$$y(t=0)=0, \quad \text{...8.48}$$

$$z(t=0)=0, \quad \text{...8.49}$$

$$v_x(t=0)=v_0\cos\theta \quad \text{...8.50}$$

$$v_y(t=0)=0, \quad \text{...8.51}$$

$$v_z(t=0)=v_0\sin\theta, \quad \text{...8.52}$$

where v_0 is the initial speed of the pitch, and θ is its initial angle of elevation. Note that, in writing the above equations, we are neglecting any decrease in the ball's rate of spin as it moves towards the hitter.

The above set of equations have been solved numerically using a fixed step-length, fourth-order Runge-Kutta method. The step-length, h, is conveniently expressed as a fraction of the pitch's estimated time-of-flight, $T=l/v_0$, where $l=18.44m$ is the horizontal distance between the pitcher and the hitter. Thus, $1/h$ is approximately the number of steps used to integrate the trajectory.

It is helpful, at this stage, to describe the conventional classification of baseball pitches in terms of the direction of the ball's axis of spin. Figure 8.8 shows the direction of rotation of various different pitches thrown by a right-handed pitcher, as seen by the hitter. Obviously, the directions are reversed for the corresponding pitches thrown by a left-handed pitcher. Note, from equation (8.39), that the arrows in figure 8.8 show both the direction of the ball's spin and the direction of the Magnus force.

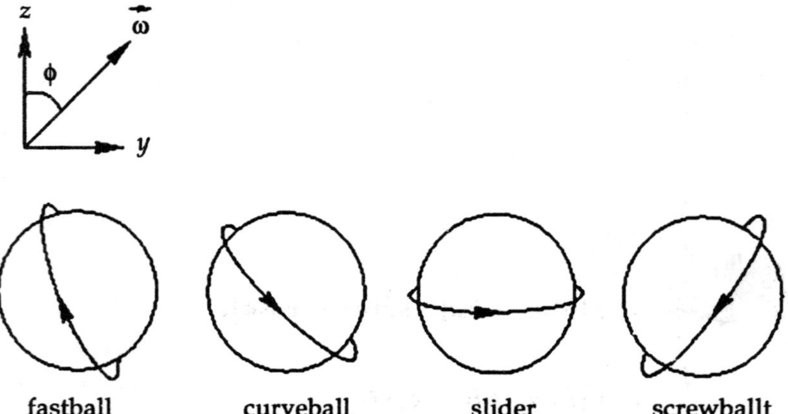

Figure 8.8: Rotation direction, as seen by hitter, for various pitches thrown by a right-handed pitcher. The arrow shows the direction of rotation, which is also the direction of the Magnus force.

Figure 8.9 shows the numerical trajectory of a "slider" delivered by a right-handed pitcher. The ball is thrown such that its axis of rotation points vertically upwards, which is the direction of spin generated by the natural clockwise rotation of a right-handed pitcher's wrist. The associated Magnus force causes the ball to curve sideways, *away* from a right-handed hitter (*i.e.*, in the $+y$-direction). As can be seen, from the figure, the sideways displacement for an 85 mph pitch spinning at 1800 rpm is over 1 foot. Of course, a slider delivered by a left-handed pitcher would curve *towards* a right-handed hitter.

It turns out that pitches which curve towards a hitter are far harder to hit than pitches which curve away.

Thus, a right-handed hitter is at a distinct disadvantage when facing a left-handed pitcher, and *vice versa*. A lot of strategy in baseball games is associated with teams trying to match the handedness of hitters and pitchers so as to gain an advantage over their opponents.

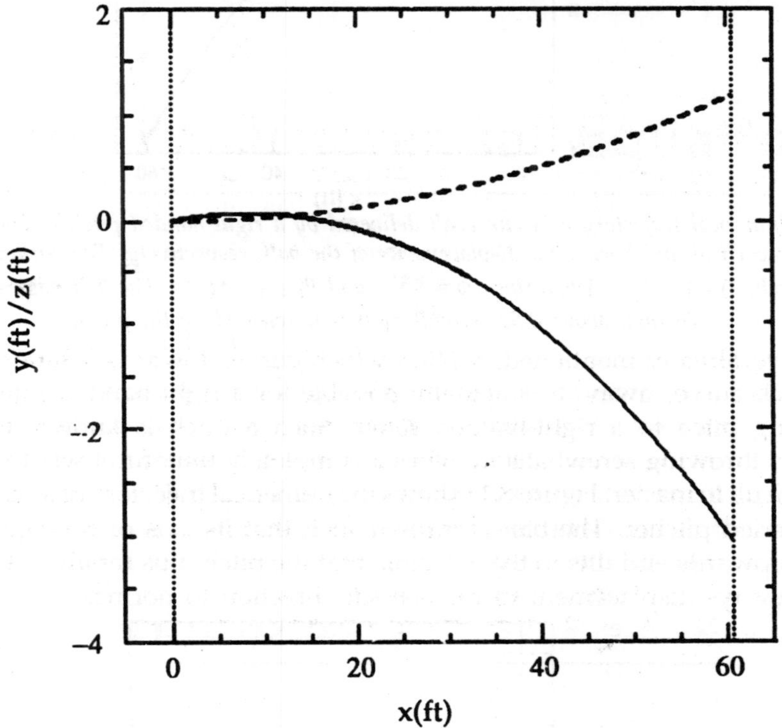

Figure 8.9: *Numerical trajectory of a slider delivered by a right-handed pitcher. The solid and dashed curves show the vertical and horizontal displacements of the ball, respectively. The parameters for this pitch are* $v_0 = 85$ *mph,* $\theta = 1°, \omega = 1800$ *rpm,* $\phi = 0°$ *, and* $h = 1 \times 10^{-4}$ *. The ball passes over the plate at 76 mph about 0.52 seconds after it is released by the pitcher.*

Figure 8.9 shows the numerical trajectory of a "curveball" delivered by a right-handed pitcher. The ball is thrown such that its axis of rotation, as seen by the hitter, points upwards, but also *tilts* to the right. The hand action associated with throwing a curveball is actually somewhat more natural than that associated with a slider.

The Magnus force acting on a curveball causes the ball to deviate both sideways and *downwards*. Thus, although a curveball generally does not move laterally as far as a slider, it dips unusually rapidly—which makes it difficult to hit. The anomalously large dip of a typical curveball is apparent from a comparison of figure 8.9 and 8.10.

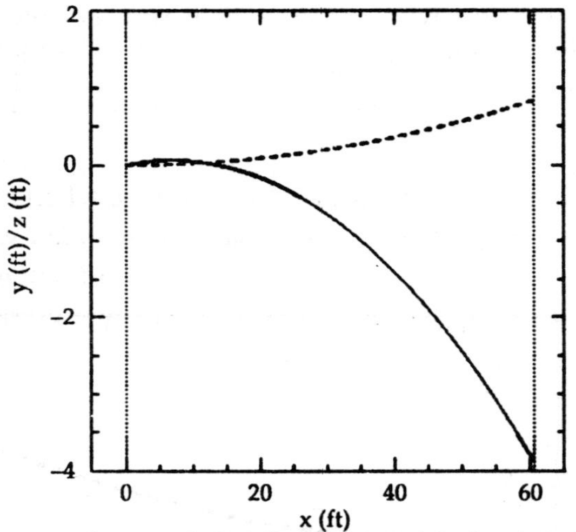

Figure 8.10: Numerical trajectory of a curveball delivered by a right-handed pitcher. The solid and dashed curves show the vertical and horizontal displacements of the ball, respectively. The parameters for this pitch are $v_0 = 85$ *mph,* $\theta - 1^\circ$, $\omega = 1800$ *rpm,* $\phi = 45^\circ$, *and* $h = 1 \times 10^{-4}$. *The ball passes over the plate at 76 mph about 0.52 seconds after it is released by the pitcher.*

As we have already mentioned, a pitch which curves towards a hitter is harder to hit than one which curves away. It is actually possible for a right-handed pitcher to throw an inward curving pitch to a right-handed hitter. Such a pitch is known as a "screwball." Unfortunately, throwing screwballs involves a completely unnatural wrist rotation which is extremely difficult to master. Figure 8.11 shows the numerical trajectory of a screwball delivered by a right-handed pitcher. The ball is thrown such that its axis of rotation, as seen by the hitter, points upwards and tilts to the *left*. Note that the pitch dips rapidly—like a curveball—but has a sideways displacement in the *opposite* direction to normal.

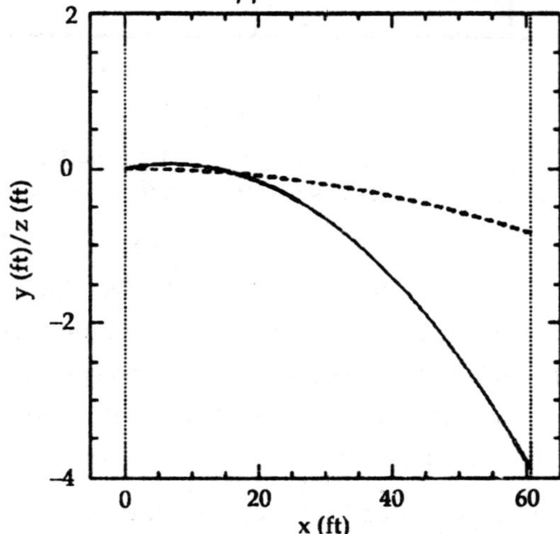

Figure 8.11: Numerical trajectory of a screwball delivered by a right-handed pitcher. The solid and dashed curves show the vertical and horizontal displacements of the ball, respectively. The parameters for this pitch are $v_0 = 85$ *mph,* $\theta = 1^\circ$, $\omega = 1800$ *rpm,* $\phi = 135^\circ$, *and* $h = 1 \times 10^{-4}$. *The ball passes over the plate at 76 mph about 0.52 seconds after it is released by the pitcher.*

Probably the most effective pitch in baseball is the so-called "fastball." As the name suggests, a fastball is thrown extremely rapidly—typically, at 95 mph. The natural hand action associated with throwing this type of pitch tends to impart a significant amount of *backspin* to the ball.

Thus, the Magnus force acting on a fastball has a large upwards component, slowing the ball's rate of fall. Figure 8.12 shows the numerical trajectory of a fastball delivered by a right-handed pitcher. The ball is thrown such that its axis of rotation, as seen by the hitter, points *downwards* and tilts to the left.

Note that the pitch falls an unusually small distance. This is partly because the ball takes less time than normal to reach the hitter, but partly also because the Magnus force has a substantial upward component.

This type of pitch is often called a "rising fastball," because hitters often claim that the ball rises as it moves towards them. Actually, this is an optical illusion created by the ball's smaller than expected rate of fall.

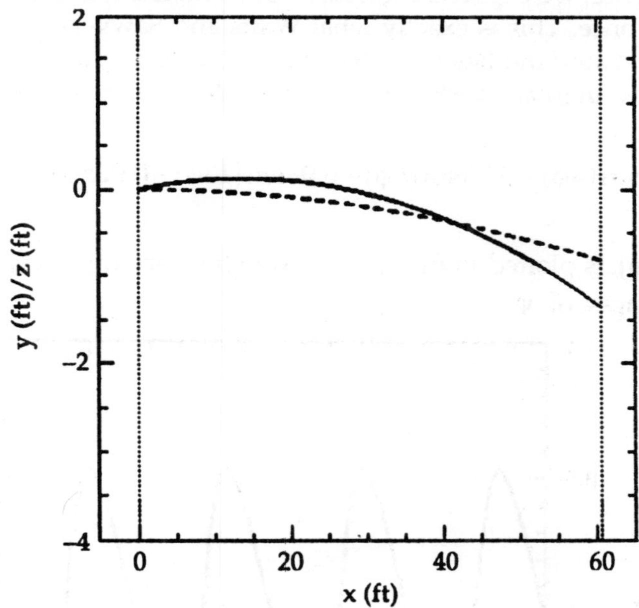

Figure 8.12: *Numerical trajectory of a fastball delivered by a right-handed pitcher. The solid and dashed curves show the vertical and horizontal displacements of the ball, respectively. The parameters for this pitch are* $v_0 = 95$ *mph,* $\theta = 1°$, $\omega = 1800$ *rpm,* $\phi = 225°$, *and* $h = 1\times10^{-4}$. *The ball passes over the plate at 86 mph about 0.46 seconds after it is released by the pitcher.*

The Knuckleball

Probably the most entertaining pitch in baseball is the so-called "knuckleball." Unlike the other types of pitch we have encountered, knuckleballs are low speed pitches (typically, 65mph) in which the ball is purposely thrown with as little spin as possible. It turns out that knuckleballs are hard to hit because they have unstable trajectories which shift from side to side in a highly unpredictable manner. How is this possible?

Suppose that a moving ball is not spinning at all, and is orientated such that one of its stitches is exposed on one side, whereas the other side is smooth. It follows, from figure 8.6, that the drag force on the smooth side of the ball is greater than that on the stitch side.

Hence, the ball experiences a lateral force in the direction of the exposed stitch. Suppose, now, that the ball is rotating slowly as it moves towards the hitter. As the ball moves forward its orientation changes, and the exposed stitch shifts from side to side, giving rise to a lateral force which also shifts from side to side.

Of course, if the ball is rotating sufficiently rapidly then the oscillations in the lateral force average out. However, for a slowly rotating ball these oscillations can profoundly affect the ball's trajectory.

Watts and Sawyer have performed wind tunnel measurements of the lateral force acting on a baseball as a function of its angular orientation. Note that the stitches on a baseball pass any given point *four* times for each complete revolution of the ball about an axis passing through its centre. Hence, we expect the lateral force to exhibit four maxima and four minima as the ball is rotated once. This is exactly what Watts and Sawyer observed. For the case of a 65mph knuckleball, Giordano has extracted the following useful expression for the lateral force, as a function of angular orientation, φ, from Watts and Sawyer's data:

$$\frac{f_y}{mg} = G(\varphi) = 0.5[\sin(4\varphi) - 0.25\sin(8\varphi) + 0.08\sin(12\varphi) - 0.025\sin(16\varphi)]. \qquad \text{...8.53}$$

The function $G(\varphi)$ is plotted in figure 8.13. Note the very rapid changes in this function in certain narrow ranges of φ.

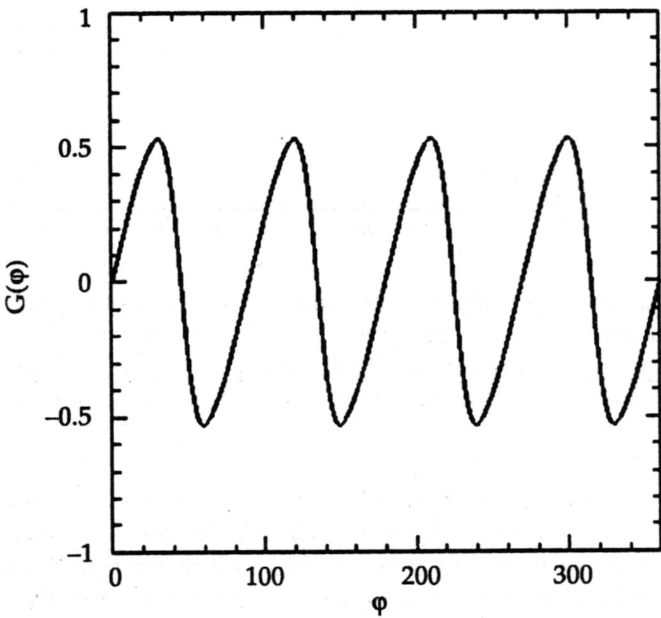

Figure 8.13: *The lateral force function,* $G(\varphi)$, *for a 65mph knuckleball.*

Using the above expression, the equations of motion of a knuckleball can be written as the following set of coupled first-order o.d.e.s:

$$\frac{dx}{dt} = v_x, \quad ...8.54$$

$$\frac{dy}{dt} = v_y, \quad ...8.55$$

$$\frac{dz}{dt} = v_z, \quad ...8.56$$

$$\frac{dv_x}{dt} = -F(v)v v_x, \quad ...8.57$$

$$\frac{dv_y}{dt} = -F(v)v v_y + gG(\varphi), \quad ...8.58$$

$$\frac{dv_z}{dt} = -g - F(v)v v_z. \quad ...8.59$$

$$\frac{d\varphi}{dt} = \omega. \quad ...8.60$$

Note that we have added an equation of motion for the angular orientation, φ, of the ball, which is assumed to rotate at a fixed angular velocity, ω, about a vertical axis.

Here, we are neglecting the Magnus force, which is expected to be negligibly small for a slowly spinning pitch. Appropriate boundary conditions at $t = 0$ are:

$$x(t = 0) = 0, \quad ...8.61$$

$$y(t = 0) = 0, \quad ...8.62$$

$$z(t = 0) = 0, \quad ...8.63$$

$$v_x(t = 0) = v_0 \cos\theta \quad ...8.64$$

$$v_y(t = 0) = 0, \quad ...8.65$$

$$v_z(t = 0) = v_0 \sin\theta, \quad ...8.66$$

$$\varphi(t = 0) = \varphi_0, \quad ...8.67$$

where v_0 is the initial speed of the pitch, θ is the pitch's initial angle of elevation, and φ_0 is the ball's initial angular orientation.

The above set of equations have been solved numerically using a fixed step-length, fourth-order Runge-Kutta method. The step-length, h, is conveniently expressed as a fraction of the pitch's estimated time-of-flight,

$$T = l / v_0,$$

where

$$l = 18.44m$$

is the horizontal distance between the pitcher and the hitter.

Thus, 1/h is approximately the number of steps used to integrate the trajectory.

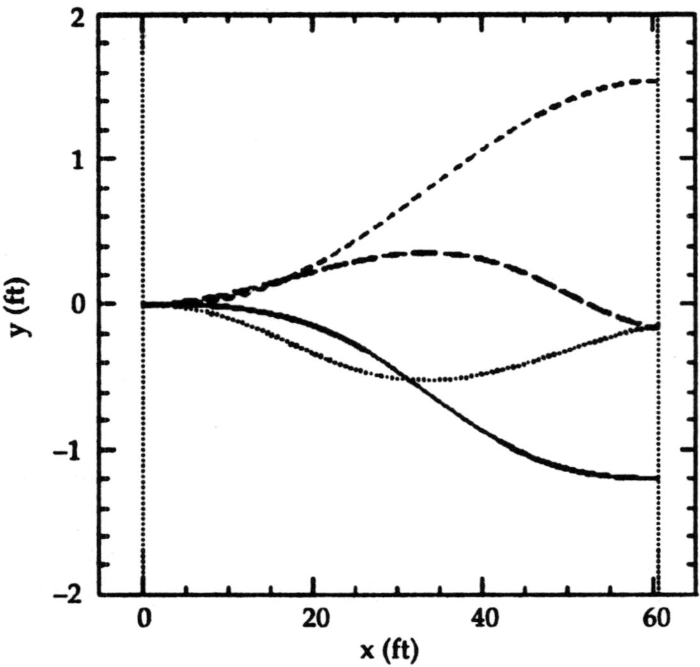

Figure 8.14: Numerical trajectories of knuckleballs with the same angular velocity but different initial orientations. The solid, dotted, long-dashed, and short-dashed curves show the horizontal displacement of the ball for $\varphi_0 = 0°$, $22.5°$, $45°$, *and* $67.5°$, *respectively. The other parameters for this pitch are* $v_0 = 65$ *mph,* $\theta = 4°$, $\omega = 20$ *rpm, and* $h = 1 \times 10^{-4}$. *The ball passes over the plate at 56 mph about 0.69 seconds after it is released by the pitcher. The ball rotates about 83° whilst it is in the air.*

Figure 8.14 shows the lateral displacement of a set of four knuckleballs thrown with the same rate of spin, $\omega = 20$ rpm, but starting with different angular orientations. Note the striking difference between the various trajectories shown in this figure.

Clearly, even a fairly small change in the initial orientation of the ball translates into a large change in its subsequent trajectory through the air. For this reason, knuckleballs are extremely unpredictable—neither the pitcher, the hitter, nor the catcher can be really sure where one of these pitches is going to end up.

Needless to say, baseball teams always put their best catcher behind the plate when a knuckleball pitcher is on the mound!

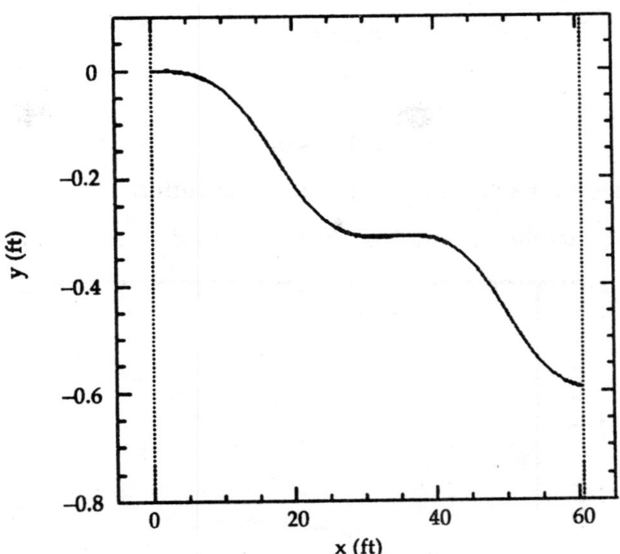

Figure 8.15: *Numerical trajectory of a knuckleball. The parameters for this pitch are* $v_0 = 65$ *mph,* $\theta = 4°$, $\varphi_0 = 0^0$, $\omega = 40$ *rpm, and* $h = 1 \times 10^{-4}$. *The ball passes over the plate at 56 mph about 0.69 seconds after it is released by the pitcher. The ball rotates about 166° whilst it is in the air.*

Figure 8.15 shows the lateral displacement of a knuckleball thrown with a somewhat higher rate of spin than those shown previously: i.e., $\omega = 40$ rpm. Note the curious way in which the ball "dances" through the air. Given that the difference between a good hit and a bad hit can correspond to a shift in the strike point on the bat by as little as 1/4 of an inch, it must be quite a challenge to hit such a pitch well!

9

Partial Differential Equations

Partial differential equations (PDE) are a type of differential equation, i.e., a relation involving an unknown function (or functions) of several independent variables and their partial derivatives with respect to those variables. Partial differential equations are used to formulate, and thus aid the solution of, problems involving functions of several variables; such as the propagation of sound or heat, electrostatics, electrodynamics, fluid flow, and elasticity. Seemingly distinct physical phenomena may have identical mathematical formulations, and thus be governed by the same underlying dynamic. They find their generalisation in Stochastic partial differential equations. Just as ordinary differential equations are often modelled by Dynamical systems, partial differential equations are often modelled by multidimensional systems.

Types of Equations

The PDE's which occur in physics are mostly second order. For linear equations in 2 dimensions there is a simple classification in terms of the general equation

$$a\frac{\partial^2 V}{\partial x^2} + 2b\frac{\partial^2 V}{\partial x \partial y} + c\frac{\partial^2 V}{\partial y^2} + d\frac{\partial V}{\partial x} + e\frac{\partial V}{\partial y} + fV + g = 0 \quad \text{...9.1}$$

as shown in the following table

Condition	*Type*	*Example*
$b^2 < ac$	Elliptic	Laplace's equation (9.2)
$b^2 > ac$	Hyperbolic	Wave equation (9.3)
$b^2 = ac$	Parabolic	Diffusion/Schrodinger equation (9.4)

These are listed in their simplest form as follows (with the substitution $y \mapsto t$ where appropriate)

Laplace's equation $$\frac{\partial^2 V}{\partial x^2} + \frac{\partial^2 V}{\partial y^2} = 0 \qquad ...9.2$$

Wave equation $$\frac{\partial^2 V}{\partial x^2} - \frac{1}{c^2}\frac{\partial^2 V}{\partial t^2} = 0 \qquad ...9.3$$

Diffusion equation $$D\frac{\partial^2 V}{\partial x^2} - \frac{\partial V}{\partial t} = 0 \qquad ...9.4$$

We shall consider each of these cases separately as different methods are required for each.

Elliptic Equations — Laplace's Equation

We shall deal with elliptic equations later when we come to consider matrix methods. For the moment it suffices to note that, apart from the formal distinction, there is a very practical distinction to be made between elliptic equations on the one hand and hyperbolic and parabolic on the other hand. Generally speaking elliptic equations have boundary conditions which are specified around a closed boundary. Usually all the derivatives are with respect to spatial variables, such as in Laplace's or Poisson's Equation. Hyperbolic and Parabolic equations, by contrast, have at least one open boundary. The boundary conditions for at least one variable, usually time, are specified at one end and the system is integrated indefinitely. Thus, the wave equation and the diffusion equation contain a time variable and there is a set of initial conditions at a particular time. These properties are, of course, related to the fact that an ellipse is a closed object, whereas hyperbolae and parabolae are open.

Hyperbolic Equations — Wave Equations

The classical example of a hyperbolic equation is the wave equation

$$\frac{\partial^2 y}{\partial t^2} - c^2\frac{\partial^2 y}{\partial x^2} = 0. \qquad ...9.5$$

The wave equation can be rewritten in the form

$$\left(\frac{\partial}{\partial t} + c\frac{\partial}{\partial x}\right)\left(\frac{\partial}{\partial t} - c\frac{\partial}{\partial x}\right)y = 0 \qquad ...9.6$$

or as a system of 2 equations

$$\frac{\partial z}{\partial t} + c\frac{\partial z}{\partial x} = 0 \qquad ...9.7$$

$$\frac{\partial y}{\partial t} - c\frac{\partial y}{\partial x} = z. \qquad ...9.8$$

Following Advective equation is in fact the conservation of mass equation which is an incompressible fluid

$$\frac{\partial u}{\partial t}+c\frac{\partial u}{\partial x}=0. \quad ...9.9$$

Note also that the boundary conditions will usually be specified in the form

$$u(x,t)=u_0(x) \text{at } t=0. \quad ...9.10$$

which gives the value of $u(x,t)$ for all x at a particular time.

A Simple Algorithm

As a first attempt to solve (9.9) we consider using centred differences for the space derivative and Euler's method for the time part

$$u_j^{n+1}=u_j^n-\frac{c\delta t}{2\delta x}(u_{j+1}^n-u_{j-1}^n) \quad ...9.11$$

where the subscripts j represent space steps and the superscripts n time steps. By analogy with the discussion of the Euler and Leap-Frog methods we can see immediately that this method is 1st order accurate in t and 2nd order in x. We note firstly that

$$u_j^{n+1}-u_j^n\approx\delta t\left.\frac{\partial u}{\partial t}\right|_j^n+\frac{1}{2}\delta t^2\left.\frac{\partial^2 u}{\partial t^2}\right|_j^n+\cdots \quad ...9.12$$

whereas

$$u_{j+1}^n-u_{j-1}^n\approx 2\delta x\left.\frac{\partial u}{\partial x}\right|_j^n+\frac{1}{3}\delta x^3\left.\frac{\partial^3 u}{\partial x^3}\right|_j^n+\cdots \quad ...9.13$$

and substitute these forms into (9.11) to obtain

$$\delta t\left.\frac{\partial u}{\partial t}\right|_j^n+\frac{1}{2}\delta t^2\left.\frac{\partial^2 u}{\partial t^2}\right|_j^n+\ldots\approx\frac{c\delta t}{2\delta x}\left(2\delta x\left.\frac{\partial u}{\partial x}\right|_j^n+\frac{1}{3}\delta x^3\left.\frac{\partial^3 u}{\partial x^3}\right|_j^n+\ldots\right. \quad ...9.14$$

so that, when the original differential equation (9.9) is subtracted, we are left with a truncation error which is 2nd order in the time but 3rd order in the spatial part. The stability is a property of the time, t, integration rather than the space, x. We analyse this by considering a plane wave solution for the x-dependence by substituting $u_j^n=\upsilon^n\exp(ikx_j)$ to obtain

$$\upsilon^{n+1}e^{ikx_j}=\upsilon^n e^{ikx_j}-\frac{c\delta t}{2\delta x}\upsilon^n(e^{ikx_{j+1}}-e^{ikx_{j-1}}) \quad ...9.15$$

or, after dividing out the common exponential factor,

$$\upsilon^{n+1}=\left[1-i\frac{c\delta t}{\delta x}\sin(k\delta x)\right]\upsilon^n. \quad ...9.16$$

Since the wave and advection equations express a conservation law the solution should neither grow nor decay as a function of time. If we substitute $\upsilon^n\mapsto\upsilon^n+\delta\upsilon^n$ and subtract (9.16) we obtain an equation for $\delta\upsilon^n$

$$\delta \upsilon^{n+1} = \left[1 - i\frac{c\delta t}{\delta x}\sin(k\delta x)\right]\delta \upsilon^n \qquad ...9.17$$

which is identical to (9.16). Hence the stability condition is simply given by the quantity in square brackets. We call this the amplification factor and write it as

$$g = 1 - i\alpha \qquad ...9.18$$

where

$$\alpha = \frac{c\delta t}{\delta x}\sin(k\delta x). \qquad ...9.19$$

As g is complex the stability condition becomes

$$|g|^2 = 1 + \alpha^2 \leq 1 \text{ for all } k. \qquad ...9.20$$

The condition must be fulfilled for all wave vectors k; otherwise a component with a particular k will tend to grow at the expense of the others. This is known as the von Neumann stability condition. Unfortunately it is never fulfilled for the simple method applied to the advection equation: i.e. the method is unstable for the advection equation.

An Improved Algorithm — the Lax Method

We see from (9.20) that our simple method is in fact unstable for the advection equation, for all finite values of δx and δt. How might we improve on this? Let us consider a minor (?) modification of (9.11)

$$u_j^{n+1} = \frac{1}{2}(u_{j+1}^n + u_{j-1}^n) - \frac{c\delta t}{2\delta x}(u_{j+1}^n - u_{j-1}^n). \qquad ...9.21$$

in which the term in u_j^n has been replaced by an average over its 2 neighbours. When we apply the same (von Neumann) analysis to this algorithm we find

$$\upsilon^{n+1} = \left[\cos(k\delta x) - i\frac{c\delta t}{\delta x}\sin(k\delta x)\right]\upsilon^n \qquad ...9.22$$

so that

$$|g|^2 = \cos^2(k\delta x) + \left(\frac{c\delta t}{\delta x}\right)^2 \sin^2(k\delta x) \qquad ...9.23$$

$$= 1 - \sin^2(k\delta x)\left\{1 - \left(\frac{c\delta t}{\delta x}\right)^2\right\} \qquad ...9.24$$

which is stable for all k as long as

$$\frac{\delta x}{\delta t} \geq c \qquad ...9.25$$

which is an example of the Courant-Friedrichs-Lewy condition applicable to hyperbolic equations. There is a simple physical explanation of this condition: if we start with the initial condition such that $u(x,t=0)=0$ everywhere except at one point on the spatial grid, then a point *m* steps away on the grid will remain zero until at least *m* time steps later. If, however, the equation is supposed to describe a physical phenomenon which travels faster than that then something must go wrong. This is equivalent to the condition that the time step, δt, must be smaller than the time taken for the wave to travel the distance of the spatial step, δx; or that the speed of propagation of information on the grid, $\delta x/\delta t$, must be greater than any other speed in the problem.

Non-Linear Equations

When applying the von Neumann analysis to non-linear equations the substitution $y_i \mapsto y_i + \delta y_i$ should be performed before setting $\delta y_i = \delta y \exp(ikx_i)$. y_i should then be treated as a constant, independent of *x*(or *i*). The resulting stability condition will then depend on *y*, just as in the case of ODE's.

Other Methods for Hyperbolic Equations

There is a large number of algorithms applicable to hyperbolic equations in general. As before, their suitability for solving a particular problem must be based on an analysis of their accuracy and stability for any wavelength. Here we simply name a few of them:

- The Lelevier method.
- The one-step and two-step Lax-Wendroff methods.
- The Leap-Frog method.
- The Quasi-Second Order method.

Eulerian and Lagrangian Methods

In the methods discussed so far the differential equations have been discretised by defining the value of the unknown at fixed points in space. Such methods are known as Eulerian methods. We have confined ourselves to a regular grid of points, but sometimes it is advantageous to choose some other grid. An obvious example is when there is some symmetry in the problem, such as cylindrical or spherical: often it is appropriate to base our choice of grid on cylindrical or spherical coordinates rather than the Cartesian ones used so far. Suppose, however, we are dealing with a problem in electromagnetism or optics, in which the dielectric constant varies in space. Then it might be appropriate to choose a grid in which the points are more closely spaced in the regions of higher dielectric constant. In that way we could take account of the fact that the wavelengths expected in the solution will be shorter. Such an approach is known as an adaptive grid. In fact it is not necessary for the grid to be fixed in space. In fluid mechanics, for example, it is often better to define a volume of space containing a fixed mass of fluid and to let the boundaries of these cells move in response to the dynamics of the fluid. The differential equation is transformed into a form in which the variables are the positions of the boundaries of the cells rather than the quantity of fluid in each cell. A simple example of such a Lagrangian method is described in the Lagrangian Fluid project.

Parabolic Equations — Diffusion

Parabolic equations such as the diffusion and time-dependent Schrödinger equations are similar to the hyperbolic case in that the boundary is open and we consider integration with respect to time, but, as we shall see, they present some extra problems.

A Simple Method

We consider the diffusion equation and apply the same simple method we tried for the hyperbolic case.

$$\frac{\partial u}{dt} - \kappa \frac{\partial^2 u}{\partial x^2} = 0 \qquad ...9.26$$

and discretise it using the Euler method for the time derivative and the simplest centred 2nd order derivative to obtain

$$u_j^{n+1} = u_j^n + \frac{\kappa \delta t}{\delta x^2}(u_{j+1}^n - 2u_j^n + u_{j-1}^n). \qquad ...9.27$$

Applying the von Neumann analysis to this system by considering a single Fourier mode in x space, we obtain

$$u^{n+1} = v^n \left[1 - \frac{4\kappa\delta t}{\delta x^2}\sin^2\left(\frac{\kappa \delta x}{2}\right)\right] \qquad ...9.28$$

so that the condition that the method is stable for all k gives

$$\delta t \le \frac{1}{2}\frac{\delta x^2}{\kappa}. \qquad ...9.29$$

Although the method is in fact conditionally stable the condition (9.29) hides an uncomfortable property: namely, that if we want to improve accuracy and allow for smaller wavelengths by halving δx we must divide δt by 4. Hence, the number of space steps is doubled and the number of time steps is quadrupled: the time required is multiplied by 8. Note that this is different from the sorts of conditions we have encountered up to now, in that it doesn't depend on any real physical time scale of the problem.

The Dufort-Frankel Method

We consider here one of many alternative algorithms which have been designed to overcome the stability problems of the simple algorithm. The Dufort-Frankel method is a trick which exploits the unconditional stability of the intrinsic method for simple differential equations. We modify (9.27) to read

$$u_j^{n+1} = u_j^{n-1} + \frac{2\kappa\delta t}{\delta x^2}\{u_{j+1}^n - (u_j^{n+1} + u_j^{n-1}) + u_{j-1}^n\}. \qquad ...9.30$$

which can be solved explicitly for u_j^{n+1} at each mesh point

$$u_j^{n+1} = \left(\frac{1-\alpha}{1+\alpha}\right)u_j^{n-1} + \left(\frac{\alpha}{1+\alpha}\right)(u_{j+1}^n + u_{j-1}^n) \qquad ...9.31$$

where

$$\alpha = 2\frac{\kappa\delta t}{\delta x^2}. \quad ...9.32$$

When the usual von Neumann analysis is applied to this method it is found to be unconditionally stable. Note however that this does not imply that δx and δt can be made indefinitely large; common sense tells us that they must be small compared to any real physical time or length scales in the problem. We must still worry about the accuracy of the method. Another difficulty this method shares with the Leap-Frog method is that it requires boundary conditions at 2 times rather than one, even though the original diffusion equation is only 1st order in time.

Conservative Methods

Most of the differential equations we encounter in physics embody some sort of conservation law. It is important therefore to try to ensure that the method chosen to solve the equation obeys the underlying conservation law exactly and not just approximately. This principle can also have the side effect of ensuring stability. The simplest approach to deriving a method with such properties is to go back to the original derivation of the differential equation.

The Equation of Continuity

The archetypal example of a differential equation which implies a conservation law is the equation of continuity, which, in its differential form says that

$$\frac{\partial\rho}{\partial t} + \nabla\cdot j = 0 \quad ...9.33$$

where ρ represents a density and j a current density. The density and current density could be of mass, charge, energy or something else which is conserved. Here we shall use the words charge and current for convenience. We consider here the 1*D* form for simplicity. The equation is derived by considering space as divided into sections of length δx. The change in the total charge in a section is equal to the total current coming in (going out) through its ends

$$\frac{\partial}{\partial t}\int_{x_i}^{x_{i+1}} \rho\, dx = -\int j\cdot dS_i, \quad ...9.34$$

It is therefore useful to re-express the differential equation in terms of the total charge in the section and the total current coming in through each face, so that we obtain a discrete equation of the form

$$\frac{\partial}{\partial t}Q_j = I_{j-1,j} - I_{j,j+1} \quad ...9.35$$

where Q_j represents the total charge in part j and $I_{j,j+1}$ is the current through the boundary between parts j and $j+1$. This takes care of the spatial part. What about the time derivative? We can express the physics thus: *The change in the charge in a cube is equal to the total charge which enters through its faces.*

The Diffusion Equation

In many cases the current through a face is proportional to the difference in density (or total charge) between neighbouring cubes

$$I_{j,j+1} = -\beta(Q_{j+1} - Q_j). \qquad ...9.36$$

Substituting this into the equation of continuity leads directly to the diffusion equation in discrete form

$$Q_j^{n+1} = Q_j^n + \delta t\beta(Q_{j+1}^n - 2Q_j^n + Q_{j-1}^n) \qquad ...9.37$$

which is of course our simple method of solution. To check whether this algorithm obeys the conservation law we sum over all j, as $\Sigma_j Q_j$ should be conserved. Note that it helps to consider the whole process as taking place on a circle as this avoids problems associated with currents across the boundaries. In this case (e.g.) $\Sigma_j Q_{j+1} = \Sigma_j Q_j$ and it is easy to see that the conservation law is obeyed for (9.37).

Maxwell's Equations

Let us now try to apply the same ideas to Maxwell's equations. In free space we have

$$\nabla \times E = -\mu_0 \frac{\partial H}{\partial t} \qquad ...9.38$$

$$\nabla \times H = -\varepsilon_0 \frac{\partial E}{\partial t}. \qquad ...9.39$$

In order to reverse the derivation of these equations we consider space to be divided into cubes as before. We integrate over a face of the cube and apply Stokes' theorem

$$\oint_S \nabla \times E \cdot dS = \int E \cdot dl \qquad ...9.40$$

$$= -\frac{\partial}{\partial t}\oint_S \mu_0 H \cdot dS \qquad ...9.41$$

and the integral of the electric field, E, round the edges of the face is equal to minus the rate of change of the magnetic flux through the face, i.e. Faraday's law. Here we can associate the electric field, E with the edges and the magnetic field with the face of the cube. In the case of the diffusion equation we had to think in terms of the total charge in a cube instead of the density. Now we replace the electric field with the integral of the field along a line and the magnetic field with the flux through a face.

Dispersion

Let us return to the Lax method for hyperbolic equations. The solution of the differential equation has the form

$$u(x,t) = u_0 e^{i(\omega t - kx)} \qquad ...9.42$$

where $\omega = ck$. Let us substitute (9.42) into the Lax algorithm

$$ve^{i(\omega t^{n+1}-kx_j)} = \frac{1}{2}ve^{i(\omega t^{n}-kx_j)}\left\{(e^{+ik\delta x}+e^{-ik\delta x})-\frac{c\delta t}{\delta t}(e^{+ik\delta x}-e^{-ik\delta x})\right\}. \quad ...9.43$$

By cancelling the common factors we now obtain

$$e^{i\omega\delta t} = \cos(k\delta x) - i\frac{c\delta t}{\delta x}\sin(k\delta x). \quad ...9.44$$

From this we can derive a dispersion relation, ω vs k, for the discrete equation and compare the result with $\omega = ck$ for the original differential equation. Since, in general, ω could be complex we write it as $\omega = \Omega + i\gamma$ and compare real and imaginary parts on both sides of the equation to obtain

$$\cos\Omega\delta t e^{-\gamma\delta t} = \cos k\delta x \quad ...9.45$$

$$\sin\Omega\delta t e^{-\gamma\delta t} = -\frac{c\delta t}{\delta x}\sin k\delta x. \quad ...9.46$$

Taking the ratio of these or the sum of their squares respectively leads to the equations

$$\tan(\Omega\delta t) = \frac{c\delta t}{\delta x}\tan(k\delta x) \quad ...9.47$$

$$e^{-2\gamma\delta t} = \cos^2(k\delta x) + \left(\frac{c\delta t}{\delta x}\right)^2 \sin^2(k\delta x). \quad ...9.48$$

The first of these equations tells us that in general the phase velocity is not c, although for long wavelengths, $k\delta x \ll 1$ and $\Omega\delta x \ll 1$, we recover the correct dispersion relationship. This is similar to the situation when we compare lattice vibrations with classical elastic waves: the long wavelength sound waves are OK but the shorter wavelengths deviate. The second equation (9.7) describes the damping of the modes. Again for small $k\delta x$ and $\gamma\delta t$, $\gamma = 0$, but (e.g.) short wavelengths, $\lambda = 4\delta x$, are strongly damped. This may be a desirable property as short wavelength oscillations may be spurious. After all we should have chosen δx to be small compared with any expected features. Nevertheless with this particular algorithm $\lambda = 2\delta x$ is not damped. Other algorithms, such as Lax-Wendroff, have been designed specifically to damp anything with a $k\delta x > \frac{1}{4}$.

Lagrangian Fluid Code

Small amplitude sound waves travel at speed $c = \sqrt{\gamma P/\rho}$ without changing shape (so that a sine wave remains a sine wave). However, when the amplitude is large, the sound speed differs between the peaks and the troughs of the wave and the wave changes shape as it propagates. This project is to simulate this effect in one-dimension. A difference method is used in which the system is split up into cells and the cell boundaries are allowed to move with the local fluid velocity (a so-called Lagrangian method). You should not require any library routines to complete this project.

The Difference Equations

The difference equations are,

$$u^{(m+\frac{1}{2})}_{n+\frac{1}{2}} = u^{(m-\frac{1}{2})}_{n+\frac{1}{2}} + (P^{(m)}_{n} - P^{(m)}_{n+1})\delta t/\Delta m, \quad \text{...9.49}$$

$$x^{(m+1)}_{n+\frac{1}{2}} = x^{(m)}_{n+\frac{1}{2}} + u^{(m+\frac{1}{2})}_{n+\frac{1}{2}}\delta t, \quad \text{...9.50}$$

where,

$$P^{(m)}_{n} = \text{constant} \times (\rho^{(m)}_{n})^{\gamma} \quad \text{...9.51}$$

and,

$$\rho^{(m)}_{n} = \Delta m/(x^{(m)}_{n+\frac{1}{2}} - x^{(m)}_{n-\frac{1}{2}}), \quad \text{...9.52}$$

$x^{(m)}_{n+\frac{1}{2}}$ and $u^{(m)}_{n+\frac{1}{2}}$ are the positions and velocities of the cell boundaries. $\rho^{(m)}_{n}$ and $P^{(m)}_{n}$ are the densities and pressures in the cells. Δm is the mass in each cell. It is useful for the purpose of programming to redefine things to get rid of the various half integer indices. Hence we can write the equations as

$$u'^{(m+1)}_{n} = u'^{(m)}_{n} + (P^{(m)}_{n} - P^{(m)}_{n+1})\delta t/\Delta m \quad \text{...9.53}$$

$$x'^{(m+1)}_{n} = x'^{(m)}_{n} + u'^{(m+1)}_{n}\delta t \quad \text{...9.54}$$

$$\rho^{(m)}_{n} = \Delta m/(x'^{(m)}_{n} - x'^{(m)}_{n-1}), \quad \text{...9.55}$$

which maps more easily onto the arrays in a programme.

Boundary Conditions

Use periodic boundary conditions to allow simulation of an infinite wave train. Note, though, the presence of P_{n+1} and x_{n-1}. If n runs from 1 to N then P_{N+1} and x_0 will be needed. This is best done by creating ``ghost'' cells at the end of the arrays. In C(++) this can be done by a declaration such as

.........double x[N+1], u[N+1], P[N+2];

and, after each loop in which P or x are updated, by setting $P_{N+1} = P_1$ and $x_0 = x_N - L$, where is the equilibrium length of the grid. Note that this requires you to leave a couple of empty array elements, but for large N this is not significant.

Initial Conditions

Sensible initial conditions might be,

$$u'^{(0)}_{n} = Mc\sin(2\pi n/N + \pi c\delta t/L) \quad \text{...9.56}$$

$$x'^{(0)}_{n} = nL/N + (ML/2\pi)\cos(2\pi n/N), \quad \text{...9.57}$$

where M, the amplitude of the sound wave, is the ratio of u to the sound speed (i.e. the Mach number). Be careful to choose sensible values for the parameters: e.g. the values of x_n should rise monotonically with n, otherwise some cells will have negative fluid densities. The essential physics of the problem is independent of the absolute values of the equilibrium pressure and density so you can set $L = 1, \Delta m = (1/N)$ and $P = \rho^{\gamma}$. In addition you can assume that $\gamma = C_P / C_V = \frac{5}{3}$. Note that the scheme is leap-frog but not of the dangerous variety: alternate steps solve for u and x.

Start with a relatively small value for $M(\ll 1)$ and show that the wave maintains its shape and moves at the correct velocity. Then increase M to find out what h.s. The stability analysis for a non-linear equation like this is difficult. Try halving the distance between grid points. How does the behaviour of the system change? Do the effects you observe describe real physics or are they numerical artefacts? One common numerical problem only manifests itself for large M. Try running for a few steps at $M = 1$, look at the values for x and try to work out what has gone wrong. Think about how to prevent this problem

Consider the operation:

$$y_n \mapsto \frac{1}{4}(y_{n+1} + 2y_n + y_{n-1}) \qquad \text{...9.58}$$

Solitons

The Korteweg de Vries equation,

$$\frac{\partial^3 y}{\partial x^3} + y\frac{\partial y}{\partial x} + \frac{\partial y}{\partial t} = 0, \qquad \text{...9.59}$$

is one of a class of non-linear equations which have so-called soliton solutions. In this case a solution can be written in the form,

$$y = 12\alpha^2 \operatorname{sech}^2[\alpha(x - 4\alpha^2 t)], \qquad \text{...9.60}$$

which has the form of a pulse which moves unchanged through the system. Ever since the phenomenon was first noted (on a canal in Scotland) it has been recognised in a wide range of different physical situations. The "bore" which occurs on certain rivers, notably the Severn, is one such. The dynamics of phase boundaries in various systems, such as domain boundaries in a ferromagnet, and some meteorological phenomena can also be described in terms of solitons.

Discretisation

The simplest discretisation of (9.59), based on the Euler method, gives the equation

$$y_j^{(n+1)} = y_j^{(n)} - \left[\frac{1}{4}\frac{\delta t}{\delta x}\left\{\left(y_{j+1}^{(n)}\right)^2 - \left(y_{j-1}^{(n)}\right)^2\right\} + \frac{1}{2}\frac{\delta t}{\delta x^3}\left\{y_{j+2}^{(n)} - 2y_{j+1}^{(n)} + 2y_{j-1}^{(n)} - y_{j-2}^{(n)}\right\}\right]. \qquad \text{...9.61}$$

You should check that this is indeed a sensible discretisation of (9.59) but that it is unstable. Note that when analysing the non-linear term in (9.61) you should make the substitution

$y \mapsto y + \delta y$ and retain the linear terms in δy. Thereafter you should treat y as a constant, independent of x and t, and apply the von Neumann method to δy. If you find the full stability analysis difficult you might consider the 2 limits of large and small y. In the former case the 3rd derivative is negligible and (9.59) reduces to a non-linear advection equation, whereas in the latter the non-linear term is negligible and the equation is similar to the diffusion equation but with a 3rd derivative. In any case you will require to choose δt so that the equation is stable in both limits. You are free to choose any method you wish to solve the equation but you will find the Runge-Kutta or Predictor-Corrector methods most reliable. Hence treating y_j as a long vector y and the terms on the right-hand-side of (9.61) as a vector function $f(y)$ the R-K method can be written concisely as

$$y'^{(n+1/2)} = y^{(n)} - \frac{1}{2}\delta t\, f(y^{(n)}) \qquad \text{...9.62}$$

$$y^{(n+1)} = y^{(n)} - \delta t\, f(y'^{(n+1/2)}), \qquad \text{...9.63}$$

where $\delta t\, f(y)$ is the quantity in square brackets [] in (9.61). Check that this method is stable, at least in the 2 limiting cases. Bear in mind that the Runge-Kutta method is usable for oscillatory equations in spite of the small instability as long as the term $(\omega\delta t)^4$ is small. By studying the analytical solution (9.60) you should be able to choose a sensible value for δx in terms of and from the stability conditions you can deduce an appropriate δt. Again, by looking at (9.60) you should be able to decide on a sensible size for the total system. You should use periodic boundary conditions, so that your solitons can run around your system several times if necessary. The easiest way to do this is by using ``ghost'' elements at each end of your arrays. Suppose your arrays should run from y_1 to y_N. Then you can add a couple of extra elements to each end: $y_{-1}, y_0, y_{N+1}, y_{N+2}$. After each step you can then assign these values as

$$y_{-1} = y_{N-1}, \quad y_0 = y_N, \quad y_{N+1} = y_1, \quad y_{N+2} = y_2 \qquad \text{...9.66}$$

so that the derivative terms in (9.61) can be calculated without having to take any special measures.

10

Wave Equation

Wave is a disturbance that travels through space and time, usually by transference of energy. Waves are described by a wave function that can take on many forms depending on the type of wave. A mechanical wave is a wave that propagates through a medium due to restoring forces produced upon its deformation. For example, sound waves propagate via air molecules bumping into their neighbours. This transfers some energy into these neighbours, which will cause a chain reaction of collisions between the molecules. When air molecules collide with their neighbours, they also bounce away from them (restoring force). This keeps the molecules from actually travelling with the wave. Waves travel and transfer energy from one point to another, often with no permanent displacement of the particles of the medium—that is, with little or no associated mass transport. They consist instead of oscillations or vibrations around almost fixed locations. Imagine a cork on rippling water, it would bob up and down staying in about the same place while the wave itself moves outward. When we say that a wave carries energy but not mass, we are referring to the fact that even as a wave travels outward from the centre (carrying energy of motion), the medium itself does not flow with it.

The wave equation is an important second-order linear partial differential equation of waves, such as sound waves, light waves and water waves. It arises in fields such as acoustics, electromagnetics, and fluid dynamics. Historically, the problem of a vibrating string such as that of a musical instrument was studied by Jean le Rond d'Alembert, Leonhard Euler, Daniel Bernoulli, and Joseph-Louis Lagrange. The wave equation, which in one dimension takes the form

$$\frac{\partial^2 \xi}{\partial t^2} = c^2 \frac{\partial^2 \xi}{\partial x^2}, \qquad \text{...10.1}$$

occurs so frequently in physics that it is not necessary to enumerate examples. Here, ξ is usually some sort of displacement or perturbation, whereas is the (constant) wave speed. The wave equation possesses the formal solution

$$\xi(x,t) = F(x-ct) + G(x+ct), \qquad ...10.2$$

where F and G are arbitrary functions. The above solution represents arbitrarily shaped wave pulses propagating with speed c in the $+x$ and $-x$ directions, respectively, without changing shape.

The wave equation, which is second-order in space and time, can be written as two coupled first-order equations by defining the new variables $v = \partial\xi/\partial t$ and $\theta = -c\partial\xi/\partial x$. Expressing equation (10.1) in terms of these new variables, we obtain

$$\frac{\partial v}{\partial t} + c\frac{\partial \theta}{\partial x} = 0, \qquad ...10.3$$

$$\frac{\partial \theta}{\partial t} + c\frac{\partial v}{\partial x} = 0. \qquad ...10.4$$

Note that when solving the wave equation numerically it is generally preferable to write it as a set of coupled first-order equations, as shown above.

1-d Advection Equation

The wave equation is closely related to the so-called *advection equation*, which in one dimension takes the form

$$\frac{\partial u}{\partial t} = -v\frac{\partial u}{\partial x}. \qquad ...10.5$$

This equation describes the passive advection of some scalar field $u(x,t)$ carried along by a flow of constant speed . Since the advection equation is somewhat simpler than the wave equation, we shall discuss it first. The advection equation possesses the formal solution

$$u(x,t) = F(x - vt), \qquad ...10.6$$

where F is an arbitrary function. This solution describes an arbitrarily shaped pulse which is swept along by the flow, at constant speed v, without changing shape.

We seek the solution of equation (10.5) in the region $x_l \le x \le x_h$, subject to the simple Dirichlet boundary conditions $u(x_l) = u(x_h) = 0$. As usual, we discretise in time on the uniform grid $t_n = t_0 + n\delta t$, for $n = 0,1,2,\cdots$. Furthermore, in the x-direction, we discretise on the uniform grid $x_i = x_l + i\delta x$, for $i = 0, N+1$, where $\delta x = (x_h - x_l)/(N+1)$. Adopting an explicit temporal differencing scheme, and a centred spatial differencing scheme, equation (10.5) yields

$$\frac{u_i^{n+1} - u_i^n}{\delta t} = -v\frac{u_{i+1}^n - u_{i-1}^n}{2\delta x}, \qquad ...10.7$$

where $u_i^n \equiv u(x_i, t_n)$. The above equation can be rewritten

$$u_i^{n+1} = u_i^n - \frac{C}{2}\left(u_{i+1}^n - u_{i-1}^n\right), \qquad ...10.8$$

where $C = v\delta t/\delta x$.

Let us perform a von Neumann stability analysis of the above differencing scheme. Writing $u(x,t) = \hat{u}(t)\exp(ikx)$, we obtain $u_i^{n+1} = A\hat{u}_i^n$, where

$$A = 1 - iC\sin(k\delta x). \qquad ...10.9$$

Note that

$$|A|^2 = 1 + C^2\sin^2(k\delta x) > 1. \qquad ...10.10$$

Thus, the magnitude of the amplification factor is greater than unity for all . This implies, unfortunately, that the simple differencing scheme (10.8) is *unconditionally unstable.*

Lax Scheme

The instability in the differencing scheme (10.8) can be fixed by replacing u_i^n on the right-hand side by the *spatial average* of taken over the neighbouring grid points. Thus, we obtain

$$u_i^{n+1} = \frac{1}{2}\left(u_{i+1}^n + u_{i-1}^n\right) - \frac{C}{2}\left(u_{i+1}^n - u_{i-1}^n\right), \qquad ...10.11$$

which is known as the *Lax* scheme. A von Neumann stability analysis of the Lax scheme yields the following expression for the amplification factor:

$$A = \cos(k\delta x) - \mathrm{i}C\sin(k\delta x). \qquad ...10.12$$

Now

$$|A|^2 = 1 - \left(1 - C^2\right)\sin^2(k\delta x). \qquad ...10.13$$

It follows that the Lax scheme is unconditionally stable (*i.e.*, $|A| < 1$ for all k), provided that $C < 1$. From the definition of C, the inequality $C < 1$ can also be written

$$\delta t < \frac{\delta x}{v}. \qquad ...10.14$$

This is the famous Courant-Friedrichs-Lewy (or CFL) stability criterion. In fact, *all* stable *explicit* differencing schemes for solving the advection equation are subject to the CFL constraint, which determines the maximum allowable time-step.

Listed below is a routine which solves the 1-d advection equation via the Lax method.

```
// Lax1D.cpp
// Function to evolve advection equation in 1-d:
//   du / dt + v du / dx = 0   for  xl <= x <= xh
//   u = 0   at x=xl and x=xh
// Array u assumed to be of extent N+2.
// Now, ith element of array corresponds to
//   x_i = xl + i * dx       i=0,N+1
// Here, dx = (xh - xl) / (N+1) is grid spacing.
// Function evolves u by single time-step.
// C = v dt / dx, where dt is time-step.
// Uses Lax scheme.
```

```
#include <blitz/array.h>
using namespace blitz;
void Lax1D (Array<double,1>& u, double C)
{
  // Set N. Declare local array.
  int N = u.extent(0) - 2;
  Array<double,1> u0(N+2);
  // Evolve u
  u0 = u;
  for (int i = 1; i <= N; i++)
    u(i) = 0.5 * (u0(i+1) + u0(i-1)) - 0.5 * C * (u0(i+1) - u0(i-1));
  // Set boundary conditions
  u(0) = 0.;
  u(N+1) = 0.;
}
```

Figure 10.1 shows an example calculation which uses the above routine to advect a Gaussian pulse. The initial condition is

$$u(x,0) = \exp\left[-100(x-0.5)^2\right], \qquad \text{...10.15}$$

and the calculation is performed with $v=1$, $\delta t = 2.49\times10^{-3}$, and $N = 200$. Furthermore, $x_l = vt$ and $x_h = 1+vt$. Note that $C = 0.5$ for these parameters. It can be seen that the pulse is advected at the correct speed: i.e., the pulse appears approximately stationary when plotted versus $x - vt$. Unfortunately, the pulse does not remain the same shape (as it should). Instead, the pulse becomes gradually lower and wider as it propagates, and eventually diffuses away entirely.

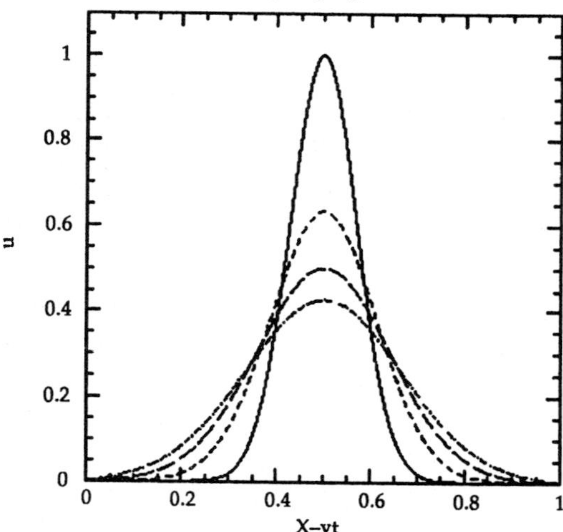

Figure 10.1: Advection of a 1-d Gaussian pulse. Numerical calculation performed using $v=1$, $\delta t = 2.49\times10^{-3}$, *and* $N = 200$. *The solid curve shows the initial condition at* $t = 0$, *the short-dashed curve the numerical solution at* $t = 1$, *the long-dashed curve the numerical solution at* $t = 2$, *and the dot-dashed curve the numerical solution at* $t = 3$

It is clear, from the above calculation, that the Lax scheme introduces a spurious dispersion effect into the advection problem. We can understand the origin of this effect by attempting a Fourier solution, $u(x,t) = \hat{u}_k(t)\exp(ikx)$, of equation (10.5). We easily obtain

$$u_k(t) = \hat{u}_k(0)\exp(-ikvt). \qquad \text{...10.16}$$

Note that $|\hat{u}_k|$ is *constant* in time for all values of *k*. In other words, the *amplitudes* of the Fourier harmonics of a true solution of the advection equation remain constant in time—it is the *phases* of the harmonics which evolve. Let us now examine equation (10.13). It can be seen that, provided the CFL condition $C < 1$ is satisfied, the magnitude of the amplification factor, $|A|$, is *less than unity* for all Fourier harmonics. In other words, the Lax differencing scheme causes the Fourier harmonics to *decay* in time. It is this unphysical attenuation of the Fourier harmonics which gives rise to the strong dispersion effect illustrated in figure 10.1.

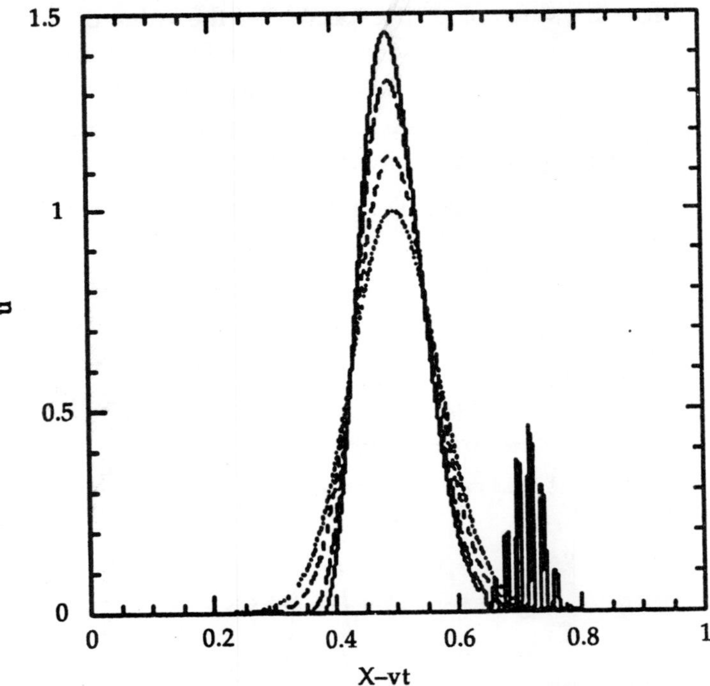

Figure 10.2: *Advection of a 1-d Gaussian pulse. Numerical calculation performed using* $v = 1$, $\delta t = 9.95\times10^{-3}$, *and* $N = 200$. *The dotted curve shows the initial condition at* $t = 0.00$, *the short-dashed curve the numerical solution at* $t = 0.15$, *the long-dashed curve the numerical solution at* $t = 0.30$, *and the solid curve the numerical solution at* $t = 0.37$

Figure 10.2 shows a calculation made using the Lax scheme in which the CFL condition is violated. This calculation is identical to the one discussed previously, except that the time-step has been increased to $\delta t = 9.95\times10^{-3}$, yielding a CFL parameter, $C = 2.0$, which exceeds unity. It can be seen that the pulse grows in amplitude, and eventually starts to break up due to a short-wavelength instability.

Crank-Nicholson Scheme

The Crank-Nicholson implicit scheme for solving the diffusion equation can be adapted to solve the advection equation. Thus, taking the average of the right-hand side of equation (10.5) between the beginning and end of the time-step, we obtain the differencing scheme written below:

$$u_i^{n+1}+\frac{C}{4}\left(u_{i+1}^{n+1}-u_{i-1}^{n+1}\right)=u_i^n-\frac{C}{4}\left(u_{i+1}^n-u_{i-1}^n\right). \quad ...10.17$$

A von Neumann stability analysis of the above scheme yields the following expression for the amplification factor:

$$A=\frac{1-i(C/2)\sin(k\delta x)}{1+i(C/2)\sin(k\delta x)}. \quad ...10.18$$

Note that $|A|=1$ for all values of k, irrespective of the value of C. This implies that the Crank-Nicholson implicit scheme is *not* subject to the CFL constraint, $C<1$ (since there is no value k of for which $|A|>1$). Moreover, there is no spurious decay in the Fourier harmonics of the solution (since $|A|=1$). Hence, unlike the Lax scheme, we would not expect the Crank-Nicholson scheme to introduce strong numerical dispersion into the advection problem.

Listed below is a routine which solves the 1-d advection equation via the Crank-Nicholson method.

```
// Advect1D.cpp
// Function to evolve advection equation in 1-d:
//   du / dt + v du / dx = 0   for  xl <= x <= xh
//   u = 0   at x=xl and x=xh
// Array u assumed to be of extent N+2.
// Now, ith element of array corresponds to
//   x_i = xl + i * dx       i=0,N+1
// Here, dx = (xh - xl) / (N+1) is grid spacing.
// Function evolves u by single time-step.
// C = v dt / dx, where dt is time-step.
// Uses Crank-Nicholson scheme.
#include <blitz/array.h>
using namespace blitz;
void Tridiagonal (Array<double,1> a, Array<double,1> b, Array<double,1> c,
                  Array<double,1> w, Array<double,1>& u);
void Advect1D (Array<double,1>& u, double C)
{
   // Find N. Declare local arrays.
   int N = u.extent(0) - 2;
   Array<double,1> a(N+2), b(N+2), c(N+2), w(N+2);
```

```
    // Initialise tridiagonal matrix
    for (int i = 2; i <= N;     i++) a(i)  = - 0.25  * C;
    for (int i = 1; i <= N; i++) b(i)  = 1.;
    for (int i = 1; i <= N-1; i++) c(i)  = + 0.25  * C;
    // Initialise right-hand side vector
    for (int i = 1; i <= N; i++)
      w(i)  = u(i)  - 0.25  * C  *  (u(i+1)  - u(i-1));
    // Invert tridiagonal matrix equation
    Tridiagonal (a, b, c, w, u);
    // Calculate i=0 and i=N+1 values
    u(0)  = 0.;
    u(N+1)  = 0.;
}
```

Figure 10.1 shows an example calculation which uses the above routine to advect a Gaussian pulse. The initial condition is as specified in equation (10.15), and the calculation is performed with $v = 1$, $\delta t = 9.95 \times 10^{-3}$, and $N = 200$. Note that $C = 2.0$ for these parameters. It can be seen that the pulse propagates at (almost) the correct speed, and maintains approximately the same shape. Clearly, the performance of the Crank-Nicholson scheme is vastly superior to that of the Lax scheme.

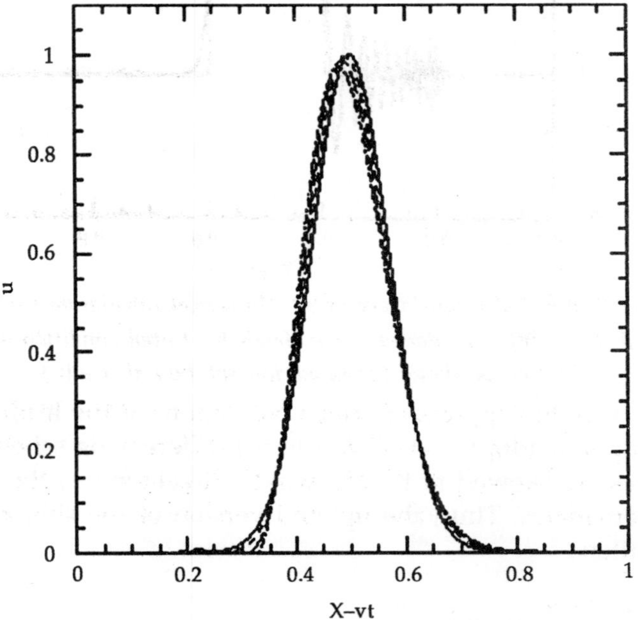

***Figure 10.3:** Advection of a 1-d Gaussian pulse. Numerical calculation performed using $v = 1$, $\delta t = 9.95 \times 10^{-3}$, and $N = 200$. The solid curve shows the initial condition at $t = 0$, the short-dashed curve the numerical solution at $t = 1$, the long-dashed curve the numerical solution at $t = 2$, and the dot-dashed curve the numerical solution at $t = 3$*

Upwind Differencing

We might be forgiven for concluding that the Crank-Nicholson scheme represents an efficient and accurate general purpose numerical method for solving the advection equation. This is indeed the case, provided we restrict ourselves to fairly *smooth* waveforms. Unfortunately, the Crank-Nicholson scheme does a very poor job at advecting waveforms with sharp leading or trailing edges. This is illustrated in figure 10.4, which shows a calculation in which the Crank-Nicholson scheme is used to advect a square wave-pulse. It can be seen that spurious oscillations are generated at both the leading and trailing edges of the wave-form. It turns out that all *central difference* schemes for solving the advection equation suffer from a similar problem.

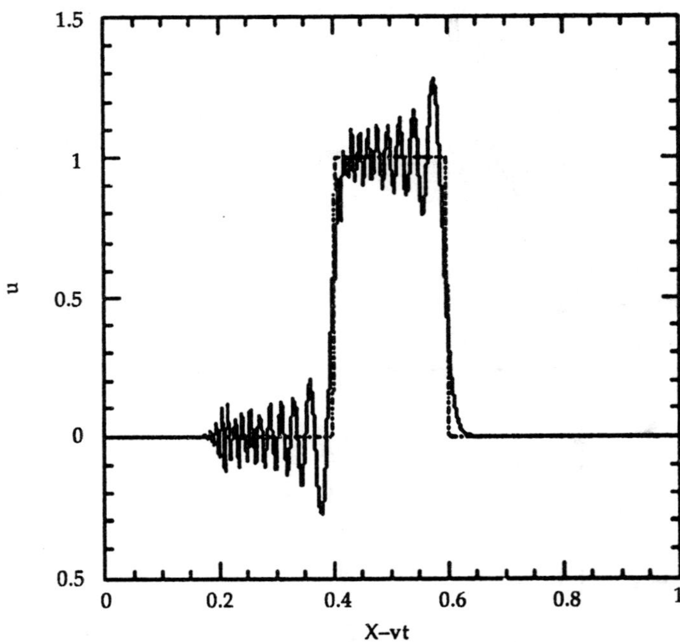

Figure 10.4: *Advection of a 1-d square wave-pulse. Numerical calculation performed using* $v = 1$, $\delta t = 2.49 \times 10^{-3}$, *and* $N = 200$. *The dotted curve shows the initial condition at* $t = 0.0$, *whereas the solid curve shows the numerical solution at* $t = 0.1$

The only known way to suppress spurious oscillations at the leading and trailing edges of a sharp wave-form is to adopt a so-called *upwind* differencing scheme. In such a scheme, the spatial differences are skewed in the "upwind" direction: i.e., the direction from which the advecting flow emanates. Thus, the upwind version of the simple explicit differencing scheme (10.7) is written

$$\frac{u_i^{n+1} - u_i^n}{\delta t} = -v \frac{u_i^n - u_{i-1}^n}{\delta x}, \quad \text{...10.19}$$

or

$$u_i^{n+1} = u_i^n - C\left(u_i^n - u_{i-1}^n\right), \quad \text{...10.20}$$

Note that this scheme is only *first-order* in space, whereas every other scheme we have discussed has been *second-order*. A von Neumann stability analysis of the above scheme yields

$$A = 1 - C\left[1 - \cos(k\delta x)\right] - iC\sin(k\delta x). \qquad ...10.21$$

Note that

$$|A|^2 = 1 - 2C(1-C)\left[1 - \cos(k\delta x)\right]. \qquad ...10.22$$

It follows that $|A| < 1$ for all k provided that $C < 1$. Thus, the upwind differencing scheme is stable provided that the CFL condition is satisfied. figure 10.5 shows a calculation in which the above scheme is used to advect a square wave-pulse. There are now no spurious oscillations generated at the sharp edges of the waveform. On the other hand, the waveform shows evidence of dispersion. Indeed, the upwind differencing scheme suffers from the same type of spurious dispersion problem as the Lax scheme. Unfortunately, there is no known differencing scheme which is both non-dispersive and capable of dealing well with sharp wave-fronts. In fact, sophisticated codes which solve the advection (or wave) equation generally employ an upwind scheme in regions close to sharp wave-fronts, or shocks, and a more accurate non-dispersive scheme elsewhere.

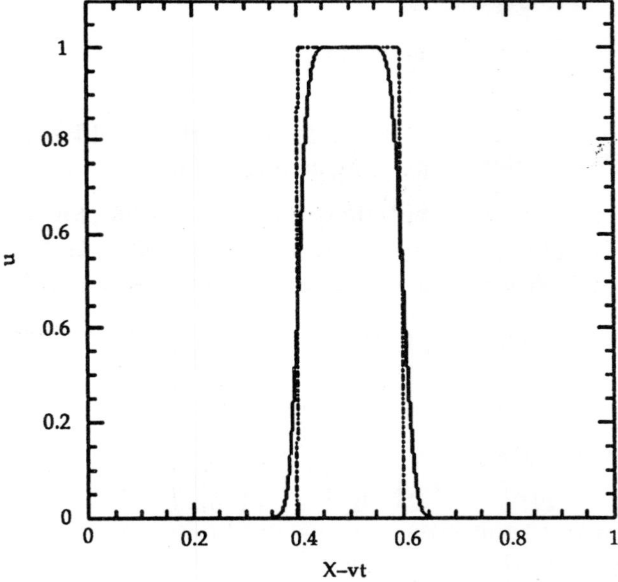

Figure 10.5: *Advection of a 1-d square wave-pulse. Numerical calculation performed using* $v = 1$, $\delta t = 2.49 \times 10^{-3}$, *and* $N = 200$. *The dotted curve shows the initial condition at* $t = 0.0$, *whereas the solid curve shows the numerical solution at* $t = 0.1$

Incidentally, it is easily demonstrated that the downwind differencing scheme,

$$\frac{u_i^{n+1} - u_i^n}{\delta t} = -v\frac{u_{i+1}^n - u_i^n}{\delta x}, \qquad ...10.23$$

is unconditionally unstable.

1-d Wave Equation

Consider a plane polarised electromagnetic wave propagating *in vacuo* along the -axis. Suppose that the electric and magnetic fields take the form $E=[E_x(z,t)0,0]$, and $B=[0,B_y(z,t),0]$. Now, according to Maxwell's equations,

$$\frac{\partial E_x}{\partial t}+c\frac{\partial H_y}{\partial z}=0, \qquad ...10.24$$

$$\frac{\partial H_y}{\partial t}+c\frac{\partial E_x}{\partial z}=0, \qquad ...10.25$$

where $H_y=cB_y$, and is the velocity of light. Note that the above equations take the form of two coupled advection equations. Let us find the numerical solution of these equations in some region $0\le z\le L$ which is bounded by *perfectly conducting walls* at $z=0$ and $z=L$. Now, both the *tangential* electric field and the *normal* magnetic field must be zero at a perfect conductor. It follows that

$$E_x(0,t)=E_x(L,t)=0. \qquad ...10.26$$

Moreover, equation (10.24) yields

$$\frac{\partial H_y(0,t)}{\partial z}=\frac{\partial H_y(L,t)}{\partial z}=0. \qquad ...10.27$$

We expect equation (10.24) and (10.25) to support wavelike solutions which bounce backwards and forwards between the conducting walls.

As before, we discretise in time on the uniform grid $t_n=t_0+n\delta t$, for $n=0,1,2,\cdots$. Furthermore, in the z-direction, we discretise on the uniform grid $z_i=i\delta z$, for $i=0,I$, where $\delta z=L/I$. Adopting a Crank-Nicholson temporal differencing scheme equation (10.24)-(10.25) yield

$$\frac{(E_x)_i^{n+1}-(E_x)_i^n}{\delta t}+\frac{c}{2}\left(\frac{\partial H_y}{\partial z}\right)_i^{n+1}+\frac{c}{2}\left(\frac{\partial H_y}{\partial z}\right)_i^{n}=0, \qquad ...10.28$$

$$\frac{(H_y)_i^{n+1}-(H_y)_i^n}{\delta t}+\frac{c}{2}\left(\frac{\partial E_x}{\partial z}\right)_i^{n+1}+\frac{c}{2}\left(\frac{\partial E_x}{\partial z}\right)_i^{n}=0, \qquad ...10.29$$

where $(E_x)_i^n\equiv E_x(z_i,t_n)$, etc.

Adopting a Fourier approach, we write

$$(E_x)_i^n=\sum_{j=0,I}\hat{E}_j^n\sin(ij\pi/I), \qquad ...10.30$$

$$(H_y)_i^n=\sum_{j=0,I}\hat{H}_j^n\cos(ij\pi/I), \qquad ...10.31$$

which automatically satisfies the boundary conditions (10.26) and (10.27). Equations (10.28) and (10.29) yield

$$\hat{E}_i^{n+1} - \hat{E}_i^n - iD\left(\hat{H}_i^{n+1} + \hat{H}_i^n\right) = 0, \qquad ...10.32$$

$$\hat{H}_i^{n+1} - \hat{H}_i^n + iD\left(\hat{E}_i^{n+1} + \hat{E}_i^n\right) = 0, \qquad ...10.33$$

where $D = \pi c \delta t/(2L)$. It follows that

$$\hat{E}_i^{n+1} = +\frac{2iD}{1+i^2D^2}\hat{H}_i^n + \frac{1-i^2D^2}{1+i^2D^2}\hat{E}_i^n, \qquad ...10.34$$

$$\hat{H}_i^{n+1} = -\frac{2iD}{1+i^2D^2}\hat{E}_i^n + \frac{1-i^2D^2}{1+i^2D^2}\hat{H}_i^n, \qquad ...10.35$$

for $i = 0, I$.

Let us determine the eigenvalues λ of the above linear system, assuming that $\left(\hat{E}_i^{n+1}, \hat{H}_i^{n+1}\right) = \lambda\left(\hat{E}_i^n, \hat{H}_i^n\right)$. After a little analysis, we obtain

$$\lambda^2 - 2\frac{1-i^2D^2}{1+i^2D^2}\lambda + 1 = 0. \qquad ...10.36$$

For all values of *i*, the two roots of the above quadratic are complex, but have modulus unity: i.e., $|\lambda| = 1$. This implies that our differencing scheme is both numerically stable (if $|\lambda| > 1$ then the scheme would be unstable, since the associated eigenfunction would be amplified at each time-step) and non-dispersive (if $|\lambda| < 1$ then all Fourier harmonics would eventually decay away, and, hence, so would the solution). Note that the above calculation is equivalent to von Neumann stability analysis.

The routine listed below solves the 1-d wave equation using the Crank-Nicholson scheme discussed above. The routine first Fourier transforms E_x and H_y, takes a time-step using equation (10.34) and (10.35), and then reconstructs E_x and H_y via an inverse Fourier transform.

```
// Wave1D.cpp
// Function to evolve 1-d wave equation:
//   d E_x / dt + c d H_y / dz = 0
//   d H_y / dt + c d E_x / dz = 0
// in region 0 < z < L.
// Boundary conditions:
//   E_x(0,t) = E_x(L,t) = 0
//   d H_y(0,t) / dz =  d H_y(L,t) / dz = 0
// Arrays Ex, Hy assumed to be of extent I+1.
// Now, ith elements of arrays correspond to
//   z_i = i * L / J          i=0,I
// Also, D = pi c dt / (2 L), where dt is time-step.
// Uses Crank-Nicholson scheme.
```

```
#include <blitz/array.h>
using namespace blitz;
void fft_forward_cos (Array<double,1> f, Array<double,1>& F);
void fft_backward_cos (Array<double,1> F, Array<double,1>& f);
void fft_forward_sin (Array<double,1> f, Array<double,1>& F);
void fft_backward_sin (Array<double,1> F, Array<double,1>& f);
void Wave1D (Array<double,1>& Ex, Array<double,1>& Hy, double D)
{
  // Find I. Declare local arrays
  int I = Ex.extent(0) - 1;
  Array<double,1> EE(I+1), HH(I+1), EE0(I+1), HH0(I+1);
  // Fourier transform Ex and Hy
  fft_forward_sin (Ex, EE0);
  fft_forward_cos (Hy, HH0);
  // Evolve EE and HH
  for (int i = 0; i <= I; i++)
    {
      double x = double (i) * D;
      double fp = 1. + x*x;
      double fm = 1. - x*x;
      EE(i) = 2. * x * HH0(i) + fm * EE0(i);
      EE(i) /= fp;
      HH(i) = - 2. * x * EE0(i) + fm * HH0(i);
      HH(i) /= fp;
    }
  // Reconstruct Ex and Hy via inverse Fourier transform
  fft_backward_sin (EE, Ex);
  fft_backward_cos (HH, Hy);
}
```

Equation 10.1 shows an example calculation which uses the above routine to propagate a Gaussian wave-pulse. The initial condition is

$$E_x(z,0) = H_y(z,0) = \exp\left[-100(z-0.5)^2\right], \qquad \text{...10.37}$$

and the calculation is performed with $L=1$, $c=1$, $\delta t = 5\times10^{-3}$, and $N=100$. It can be seen that the pulse moves to the right, reflects off the conducting wall at $z=1$, and then moves to the left. Note, that $E_x = +H_y$ when the wave is far from the conducting walls and moving to the right, whereas $E_x = -H_y$ when the wave is far from the conducting walls and moving to the left.

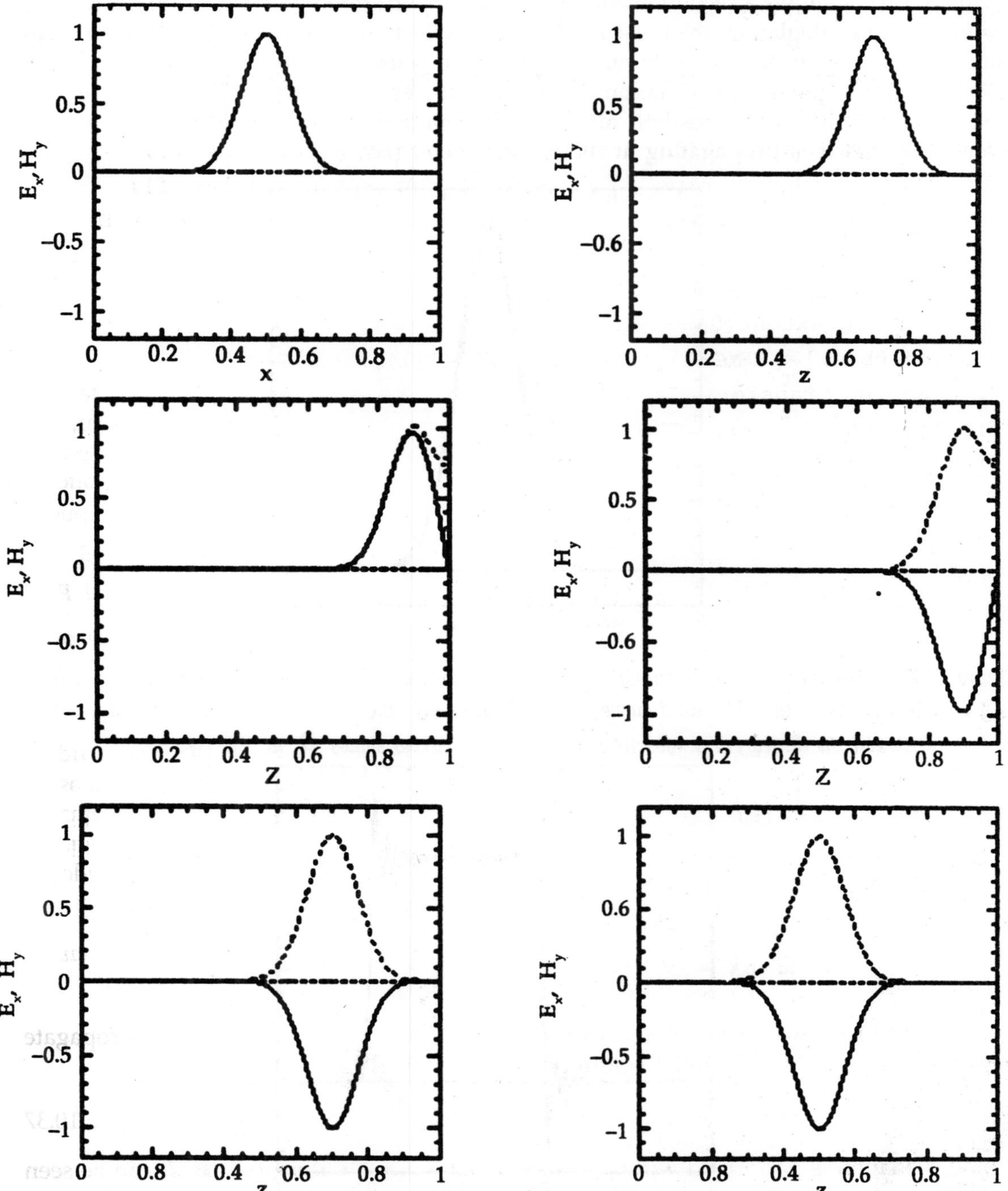

Figure 10.6: *Propagation of a 1-d Gaussian wave-pulse. Numerical calculation performed using* $c = 1$, $\delta t = 5 \times 10^{-3}$, *and* $N = 100$. *The solid curves show* E_x, *whereas the dashed curves show* H_y. *The top-left, top-right, middle-left, middle-right, bottom-left, and bottom-right panels show the solution at* $t = 0.0$, *0.2, 0.4, 0.6, 0.8, and 1.0, respectively*

Equation 10.24 shows a second example calculation performed with the same parameters as the first. In this calculation, the pulse is allowed to reflect off the conducting walls *ten* times before returning to its initial position. Note that the pulse amplitude and shape remain constant to a very good approximation during this process. The fact that the pulse returns almost exactly to its initial position after ten time periods have elapsed (*i.e.*, at $t = 10$) demonstrates that it is propagating at the correct speed (*i.e.*, $c = 1$).

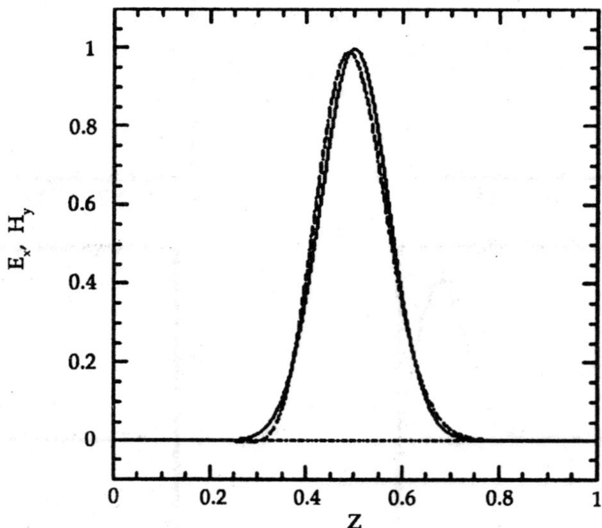

Figure 10.7: Propagation of a 1-d Gaussian wave-pulse. Numerical calculation performed using $c = 1$, $\delta t = 5 \times 10^{-3}$, *and* $N = 100$. *The solid curve shows* E_x *at* $t = 0$, *the dashed curve shows* E_x *at* $t = 10$, *and the dotted curve (obscured by the dashed curve) shows* B_y *at* $t = 10$

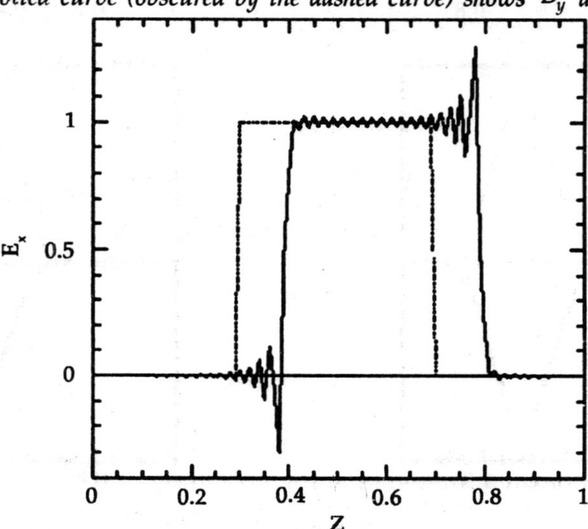

Figure 10.8: Propagation of a 1-d square wave-pulse. Numerical calculation performed using $c = 1$, $\delta t = 5 \times 10^{-3}$, *and* $N = 100$. *The dotted curve shows* E_x *at* $t = 0.0$, *and the solid curve shows* E_x *at* $t = 0.1$

Figure 10.8 shows a third example calculation which uses the above listed routine to propagate a square wave-pulse. Note that the routine does a very poor job, since spurious oscillations are generated at the sharp leading and trailing edges of the waveform. Such oscillations can only be suppressed by adopting an *upwind* differencing scheme, which in this case means that the spatial differences must be skewed in the direction from which the wave is propagating. Unfortunately, simple explicit upwind schemes are subject to the CFL constraint,

$$\delta t < \frac{\delta z}{c}, \quad \text{...10.38}$$

and also tend to be highly dispersive.

2-d Resonant Cavity

Figure 10.9 shows a 2-d resonant cavity consisting of a hollow, rectangular, perfectly conducting channel of dimensions $L_x \times L_y$. Suppose that the walls of the channel are aligned along the x- and y-axes. We shall excite this cavity in a rather artificial manner by imposing a z-directed alternating current pattern of frequency f, which has the same spatial structure as the mode in which we are interested. Let us calculate the electric and magnetic field patterns excited within the cavity by such a current pattern.

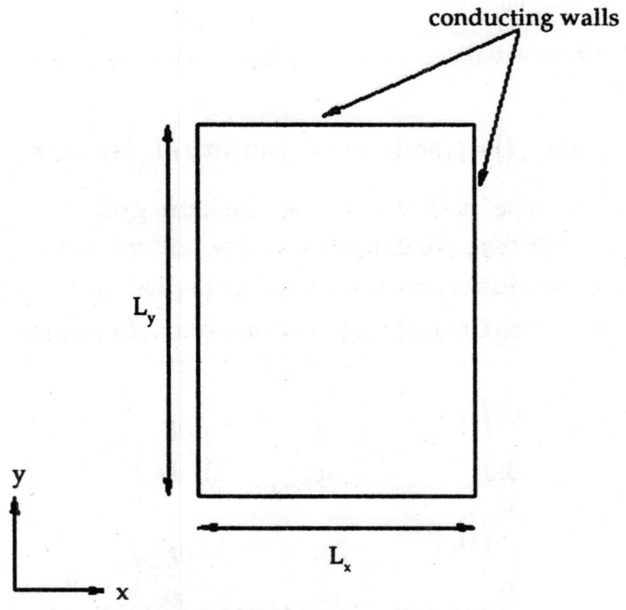

Figure 10.9: *A 2-d resonant cavity*

The electric and magnetic fields within the cavity can be written $E = [0, 0, E_z(x, y, t)]$, and $B = [B_x(x, y, t), B_y(x, y, t), 0]$, respectively. It follows from Maxwell's equations that

$$\frac{\partial H_x}{\partial t} + c\frac{\partial E_z}{\partial y} = 0, \quad \text{...10.39}$$

$$\frac{\partial H_y}{\partial t}+c\frac{\partial E_z}{\partial x}=0, \qquad ...10.40$$

$$\frac{\partial E_z}{\partial t}+c\frac{\partial H_y}{\partial x}+c\frac{\partial H_x}{\partial y}=J_z, \qquad ...10.41$$

where c is the velocity of light, $H_x = cB_x$, $H_y = -cB_y$, and $J_z = -\mu_0 c^2 j_z$. Note that the above system of equations takes the form of three coupled advection equations with a source term. The boundary conditions are that the *tangential* electric field and the *normal* magnetic field must be zero at the conducting walls. It follows that

$$E_z = 0 \qquad ...10.42$$

at all the walls (which are located at $x = 0, L_x$ and $y = 0, L_y$),

$$H_x = \frac{\partial H_y}{\partial x} = 0 \qquad ...10.43$$

at $x = 0, L_x$, and

$$H_y = \frac{\partial H_x}{\partial_y} = 0 \qquad ...10.44$$

at $y = 0, L_y$. Finally, the normalised current pattern associated with the (*m, n*) mode takes the form

$$J_z(x,y,t) = J_0 \sin(m\pi x / L_x)\sin(n\pi y / L_y)\sin(2\pi f t). \qquad ...10.45$$

As usual, we discretise in time on the uniform grid $t_n = t_0 + n\delta t$, for $n = 0,1,2,\cdots$. Furthermore, in the *x*-direction, we discretise on the uniform grid $x_i = i\delta x$, for $i = 0, I$, where $\delta x = L_x / I$. Finally, in the *y*-direction, we discretise on the uniform grid $y_j = j\delta y$, for $j = 0, J$, where $\delta y = L_y / J$. Adopting a Crank-Nicholson temporal differencing scheme, equation (10.39)-(10.41) yield

$$\frac{(H_x)_{i,j}^{n+1}-(H_x)_{i,j}^{n}}{\delta t}+\frac{c}{2}\left(\frac{\partial E_z}{\partial y}\right)_{i,j}^{n+1}+\frac{c}{2}\left(\frac{\partial E_z}{\partial y}\right)_{i,j}^{n}=0, \qquad ...10.46$$

$$\frac{(H_y)_{i,j}^{n+1}-(H_y)_{i,j}^{n}}{\delta t}+\frac{c}{2}\left(\frac{\partial E_z}{\partial x}\right)_{i,j}^{n+1}+\frac{c}{2}\left(\frac{\partial E_z}{\partial x}\right)_{i,j}^{n}=0, \qquad ...10.47$$

$$\frac{(E_z)_{i,j}^{n+1}-(E_z)_{i,j}^{n}}{\delta t}+\frac{c}{2}\left(\frac{\partial H_y}{\partial x}\right)_{i,j}^{n+1}+\frac{c}{2}\left(\frac{\partial H_y}{\partial x}\right)_{i,j}^{n}$$

$$+\frac{c}{2}\left(\frac{\partial H_x}{\partial y}\right)_{i,j}^{n+1}+\frac{c}{2}\left(\frac{\partial H_x}{\partial y}\right)_{i,j}^{n}=(J_z)_{i,j}^{n}, \qquad ...10.48$$

where $(H_x)^n_{i,j} \equiv H_x(x_i, y_j, t_n)$, etc.

Adopting a Fourier approach, we write

$$(E_z)^n_{i,j} = \sum_{i'=0,I}^{j'=0,j} E^n_{i',j'} \sin(ii'\pi/I)\sin(jj'\pi/J), \qquad ...10.49$$

$$(H_x)^n_{i,j} = \sum_{i'=0.I}^{j'=0,J} \hat{X}^n_{i',j'} \sin(ii'\pi/I)\sin(jj'\pi/J), \qquad ...10.50$$

$$(H_y)^n_{i,j} = \sum_{i'=0.I}^{j'=0,J} \hat{Y}^n_{i',j'} \cos(ii'\pi/I)\sin(jj'\pi/J), \qquad ...10.51$$

$$(J_z)^n_{i,j} = \sum_{i'=0.I}^{j'=0,J} \hat{J}^n_{i',j'} \sin(ii'\pi/I)\sin(jj'\pi/J), \qquad ...10.52$$

which automatically satisfies the boundary conditions (10.42)-(10.44). Equations (10.46)-(10.48) yield

$$\hat{X}^{n+1}_{i,j} - \hat{X}^n_{i,j} + jD_y\left(\hat{E}^{n+1}_{i,j} + \hat{E}^n_{i,j}\right) = 0, \qquad ...10.53$$

$$\hat{Y}^{n+1}_{i,j} - \hat{Y}^n_{i,j} + iD_x\left(\hat{E}^{n+1}_{i,j} + \hat{E}^n_{i,j}\right) = 0, \qquad ...10.54$$

$$\hat{E}^{n+1}_{i,j} - \hat{E}^n_{i,j} + iD_x\left(\hat{Y}^{n+1}_{i,j} + \hat{Y}^n_{i,j}\right) - jD_j\left(\hat{X}^{n+1}_{i,j} + \hat{X}^n_{i,j}\right) = \delta t\, \hat{J}^n_{i,j}, \qquad ...10.55$$

for $i = 0, I$ and $j = 0, J$, where $D_x = \pi c\delta t/(2L_x)$ and $D_y = \pi c\delta t/(2L_y)$. It follows that

$$\hat{E}^{n+1}_{i,j} = \frac{\left(1 - i^2D_x^2 - j^2D_y^2\right)\hat{E}^n_{i,j} + 2jD_y\hat{X}^n_{i,j} + 2iD_x\hat{Y}^n_{i,j} + \delta t\, \hat{J}^n_{i,j}}{1 + i^2D_x^2 + j^2D_y^2} \qquad ...10.56$$

$$\hat{X}^{n+1}_{i,j} = \hat{X}^n_{i,j} - jD_y\left(\hat{E}^{n+1}_{i,j} + \hat{E}^n_{i,j}\right), \qquad ...10.57$$

$$\hat{Y}^{n+1}_{i,j} = \hat{Y}^n_{i,j} - iD_x\left(\hat{E}^{n+1}_{i,j} + \hat{E}^n_{i,j}\right). \qquad ...10.58$$

The routine listed below solves the 2-d wave equation in a resonant cavity using the Crank-Nicholson scheme discussed above. The routine first Fourier transforms H_x, H_y, E_z, and J_z in both the x- and y-directions, takes a time-step using equation (10.56)-(10.58), and then reconstructs H_x, H_y, and E_z via an double inverse Fourier transform.

```
// Wave2D.cpp
// Function to evolve 2-d wave equation:
//   d H_x / dt + c d E_z / dy = 0
//   d H_y / dt + c d E_z / dx = 0
//   d E_z / dt + c d H_y / dx + c d H_x / dy = J_z
// in region 0 < x < L_x and 0 < y < L_y
```

```
// Boundary conditions:
//   E_z(0, y) = E_z(L_x, y) = E_z(x, 0) = E_z(x, L_y) = 0
//   H_x(0, y) = H_x(L_x, y) = d H_y(0, y) / dx =  d H_y(L_x, y) / dx = 0
//   H_y(x, 0) = H_y(x, L_y) = d H_x(x, 0) / dy =  d H_x(x, L_y) /
dy = 0
// Matrices Hx, Hy, Ez, Jz assumed to be of extent I+1, J+1.
// Now, (i,j)th elements of matrices correspond to
//   x_i = i * dx      i=0,I
//   y_j = j * dy      j=0,J
// Here, dx = L_x / I is grid spacing in x-direction,
// and dy = L_y / J is grid spacing in x-direction.
// Now, Dx = pi c dt / (2 L_x) and Dy = pi c dt / (2 L_y),
// where dt is time-step.
// Uses Crank-Nicholson scheme.
#include <blitz/array.h>
using namespace blitz;
void fft_forward_cos (Array<double,1> f, Array<double,1>& F);
void fft_backward_cos (Array<double,1> F, Array<double,1>& f);
void fft_forward_sin (Array<double,1> f, Array<double,1>& F);
void fft_backward_sin (Array<double,1> F, Array<double,1>& f);
void Wave2D (Array<double,2>& Hx, Array<double,2>& Hy, Array<double,2>& Ez,
             Array<double,2> Jz, double Dx, double Dy, double dt)
{
  // Find I and J. Declare local arrays
  int I = Hx.extent(0) - 1;
  int J = Hx.extent(1) - 1;
  Array<double,2> X(I+1, J+1), XX(I+1, J+1), XXX(I+1, J+1);
  Array<double,2> Y(I+1, J+1), YY(I+1, J+1), YYY(I+1, J+1);
  Array<double,2> E(I+1, J+1), EE(I+1, J+1), EEE(I+1, J+1);
  Array<double,2> K(I+1, J+1), KK(I+1, J+1);
  // Fourier transform solution in x-direction
  for (int j = 0; j <= J; j++)
    {
      Array<double,1> In(I+1), Out(I+1);
      // Fourier transform Hx
      for (int i = 0; i <= I; i++) In(i) = Hx(i, j);
```

```
            fft_forward_sin (In, Out);
            for (int i = 0; i <= I; i++) X(i, j) = Out(i);
            // Fourier transform Hy
            for (int i = 0; i <= I; i++) In(i) = Hy(i, j);
            fft_forward_cos (In, Out);
            for (int i = 0; i <= I; i++) Y(i, j) = Out(i);
            // Fourier transform Ez
            for (int i = 0; i <= I; i++) In(i) = Ez(i, j);
            fft_forward_sin (In, Out);
            for (int i = 0; i <= I; i++) E(i, j) = Out(i);
            // Fourier transform Jz
            for (int i = 0; i <= I; i++) In(i) = dt * Jz(i, j);
            fft_forward_sin (In, Out);
            for (int i = 0; i <= I; i++) K(i, j) = Out(i);
        }
    // Fourier transform solution in y-direction
    for (int i = 0; i <= I; i++)
        {
            Array<double,1> In(J+1), Out(J+1);

            // Fourier transform Hx
            for (int j = 0; j <= J; j++) In(j) = X(i, j);
            fft_forward_cos (In, Out);
            for (int j = 0; j <= J; j++) XX(i, j) = Out(j);
            // Fourier transform Hy
            for (int j = 0; j <= J; j++) In(j) = Y(i, j);
            fft_forward_sin (In, Out);
            for (int j = 0; j <= J; j++) YY(i, j) = Out(j);
            // Fourier transform Ez
            for (int j = 0; j <= J; j++) In(j) = E(i, j);
            fft_forward_sin (In, Out);
            for (int j = 0; j <= J; j++) EE(i, j) = Out(j);
            // Fourier transform Jz
            for (int j = 0; j <= J; j++) In(j) = K(i, j);
            fft_forward_sin (In, Out);
            for (int j = 0; j <= J; j++) KK(i, j) = Out(j);
        }
```

```
// Evolve XX, YY, and EE
for (int i = 0; i <= I; i++)
   for (int j = 0; j <= J; j++)
      {
         double x = double (i) * Dx;
         double y = double (j) * Dy;
         double fp = 1. + x*x + y*y;
         double fm = 1. - x*x - y*y;
         EEE(i, j) = fm * EE(i, j) + 2. * y * XX(i, j) +
            2. * x * YY(i,j) + KK(i, j);
         EEE(i, j) /= fp;
         XXX(i, j) = XX(i, j) - y * (EEE(i, j) + EE(i, j));
         YYY(i, j) = YY(i, j) - x * (EEE(i, j) + EE(i, j));
      }
// Reconstruct solution via inverse Fourier transform in y-direction
for (int i = 0; i <= I; i++)
   {
      Array<double,1> In(J+1), Out(J+1);
      // Reconstruct Hx
      for (int j = 0; j <= J; j++) In(j) = XXX(i, j);
      fft_backward_cos (In, Out);
      for (int j = 0; j <= J; j++) X(i, j) = Out(j);
      // Reconstruct Hy
      for (int j = 0; j <= J; j++) In(j) = YYY(i, j);
      fft_backward_sin (In, Out);
      for (int j = 0; j <= J; j++) Y(i, j) = Out(j);
      // Reconstruct Ez
      for (int j = 0; j <= J; j++) In(j) = EEE(i, j);
      fft_backward_sin (In, Out);
      for (int j = 0; j <= J; j++) E(i, j) = Out(j);
   }
// Reconstruct solution via inverse Fourier transform in x-direction
for (int j = 0; j <= J; j++)
   {
      Array<double,1> In(I+1), Out(I+1);
      // Reconstruct Hx
```

```
        for (int i = 0; i <= I; i++) In(i) = X(i, j);
        fft_backward_sin (In, Out);
        for (int i = 0; i <= I; i++) Hx(i, j) = Out(i);
        // Reconstruct Hy
        for (int i = 0; i <= I; i++) In(i) = Y(i, j);
        fft_backward_cos (In, Out);
        for (int i = 0; i <= I; i++) Hy(i, j) = Out(i);
        // Reconstruct Ez
        for (int i = 0; i <= I; i++) In(i) = E(i, j);
        fft_backward_sin (In, Out);
        for (int i = 0; i <= I; i++) Ez(i, j) = Out(i);
    }
}
```

The numerical calculations discussed below were performed using the above routine. The electromagnetic fields H_x, H_y, and E_z were all initialised to zero everywhere at $t = 0$. Figure 10.10 shows the maximum amplitude of E_z versus the frequency, *f*, for an $m = 1/n = 1$ driving current distribution. It can be seen that there is a clear resonance at $f \simeq 0.7$.

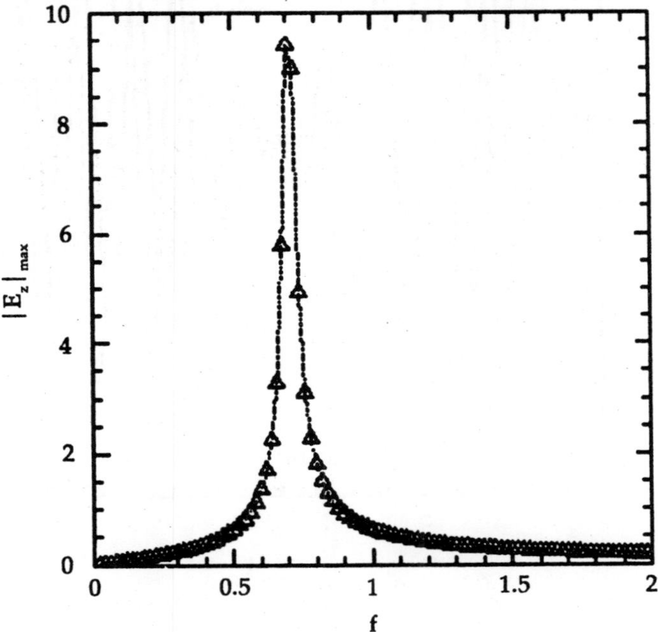

Figure 10.10: *Electromagnetic waves in a 2-d resonant cavity. The maximum amplitude of* $E_z(x = 0.5, y = 0.5)$ *between* $t = 0$ *and* $t = 20$ *versus the driving frequency, f. Numerical calculation performed using* $m = 1, n = 1$, $L_x = 1, L_y = 1$, $c = 1, I = J = 32$, *and* $\delta t = 10^{-2}$

Figures 10.11 and 10.12 illustrate the typical time variation of E_z, H_x, and H_y for a non-resonant and a resonant case, respectively. For the non-resonant case, the traces take the form of interference patterns between the directly driven response, which oscillates at the driving frequency f, and the transient response, which oscillates at the natural frequency f_0 of the cavity. Note that the transients never decay, since there is no dissipation in the present problem. Incidentally, it is easily demonstrated that

$$f_0 = \frac{c}{2}\sqrt{\frac{1}{(nL_x)^2} + \frac{1}{(mL_y)^2}}. \quad ...10.59$$

Hence, it follows that $f_0 = 1/\sqrt{2} = 0.7071$, for $n = m = L_x = L_y = c = 1$ which corresponds very well to the resonant frequency found in figure 10.10. For the resonant case, the traces take the form of waves of ever increasing amplitude which oscillate at the natural frequency f_0.

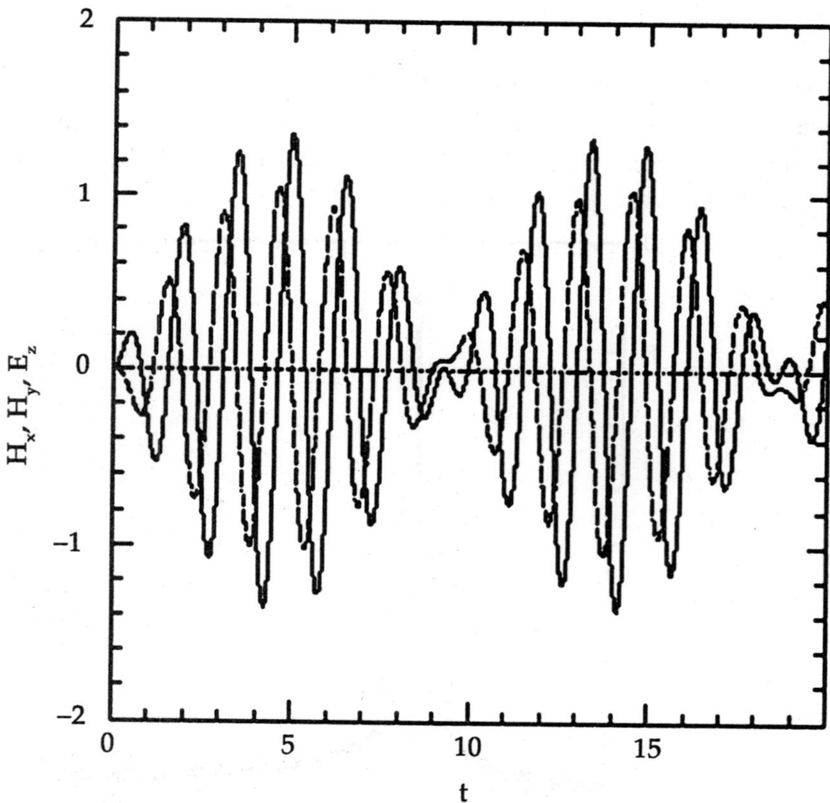

Figure 10.11: *Electromagnetic waves in a 2-d resonant cavity. Time traces of* $E_z(x = 0.5, y = 0.5)$ *(solid curve),* $H_x(x = 0.5, y = 0.0)$ *(dashed curve), and* $H_y(x = 0.0, y = 0.5)$ *(dotted curve—obscured by dashed curve). Numerical calculation performed using* $m = 1$, $n = 1$, $L_x = 1$, $L_y = 1$, $c = 1$, $I = J = 32$, $\delta t = 10^{-2}$, *and* $f = 0.6$

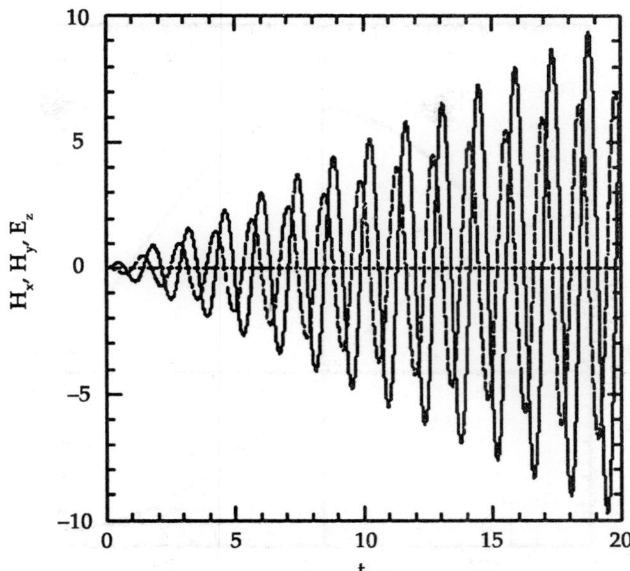

Figure 10.12: *Electromagnetic waves in a 2-d resonant cavity. Time traces of* $E_z(x=0.5, y=0.5)$ *(solid curve),* $H_x(x=0.5, y=0.0)$ *(dashed curve), and* $H_y(x=0.0, y=0.5)$ *(dotted curve—obscured by dashed curve). Numerical calculation performed using* $m=1$, $n=1$, $L_x=1$, $L_y=1$, $c=1$, $I=J=32$, $\delta t=10^{-2}$, *and* f=0.7071

Finally, figure 10.12 and 10.13 illustrate the spatial variation of the electromagnetic fields driven within the cavity when $m=1$ and $n=1$.

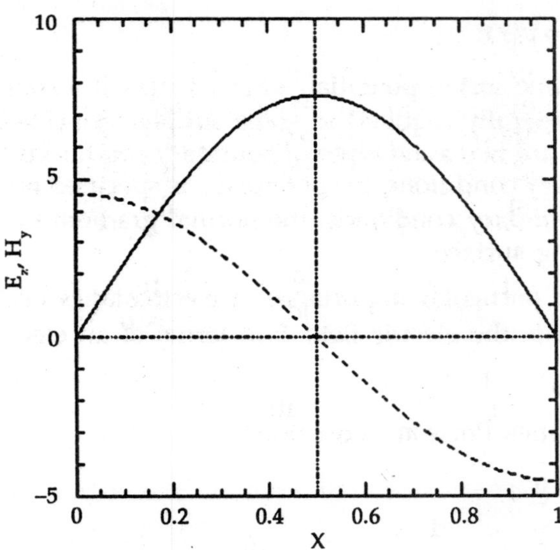

Figure 10.13: *Electromagnetic waves in a 2-d resonant cavity. Spatial variation of* E_z *(solid curve) and* H_y *(dashed curve) in the x-direction at* $t=20$ *and* $y=0.5$. *Numerical calculation performed using* $m=1$, $n=1$, $L_x=1$, $L_y=1$, $c=1$, $I=J=32$, $\delta t=10^{-2}$, *and* $f=0.7071$

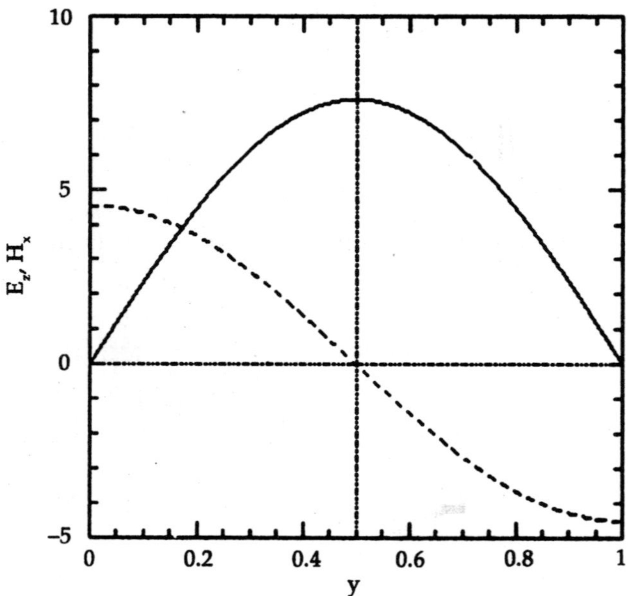

Figure 10.14: Electromagnetic waves in a 2-d resonant cavity. Spatial variation of E_z (solid curve) and H_x (dashed curve) in the y-direction at $t = 20$ and $x = 0.5$. Numerical calculation performed using $m = 1$, $n = 1$, $L_x = 1$, $L_y = 1$, $c = 1$, $I = J = 32$, $\delta t = 10^{-2}$, and $f = 0.7071$

Poisson's Equation

We shall discuss some simple numerical techniques for solving Poisson's equation:

$$\nabla^2 u(r) = v(r). \qquad ...10.60$$

Here, $u(r)$ is usually some sort of potential, whereas $v(r)$ is a source term. The solution to the above equation is generally required in some simply-connected volume V bounded by a closed surface S. There are two main types of boundary conditions to Poisson's equation. In so-called *Dirichlet* boundary conditions, the potential u is specified on the bounding surface S. In so-called *Neumann* boundary conditions, the normal gradient of the potential $\nabla u \cdot dS$ is specified on the bounding surface.

Poisson's equation is of particular importance in electrostatics and Newtonian gravity. In electrostatics, we can write the electric field E in terms of an electric potential ϕ:

$$E = -\nabla\phi. \qquad ...10.61$$

The potential itself satisfies Poisson's equation:

$$\nabla^2\phi = -\frac{\rho}{\epsilon_0}, \qquad ...10.62$$

where $\rho(r)$ is the charge density, and ϵ_0 the permittivity of free-space. In Newtonian gravity, we can write the force f exerted on a unit test mass in terms of a gravitational potential ϕ:

$$f = -\nabla\phi. \qquad ...10.63$$

The potential satisfies Poisson's equation:

$$\nabla^2\phi = 4\pi^2 G\rho, \qquad ...10.64$$

where $\rho(r)$ is the mass density, and G the universal gravitational constant.

1-d Problem with Dirichlet Boundary Conditions

As a simple test case, let us consider the solution of Poisson's equation in one dimension. Suppose that

$$\frac{d^2u(x)}{dx^2} = v(x), \qquad ...10.65$$

for $x_l \le x \le x_h$, subject to the Dirichlet boundary conditions $u(x_l) = u_l$ and $u(x_h) = u_h$.

As a first step, we divide the domain $x_l \le x \le x_h$ into equal segments whose vertices are located at the grid-points

$$x_i = x_l + \frac{i(x_h - x_l)}{N+1}, \qquad ...10.66$$

for $i = 1, N$. The boundaries, x_l and x_h, correspond to $i = 0$ and $i = N+1$, respectively.

Next, we discretise the differential term d^2u/dx^2 on the grid-points. The most straightforward discretisation is

$$\frac{d^2u(x_i)}{dx^2} = \frac{u_{i-1} - 2u_i + u_i + 1}{(\delta x)^2} + O(\delta x)^2. \qquad ...10.67$$

Here, $\delta x = (x_h - x_l)/(N+1)$, and $u_i \equiv u(x_i)$. This type of discretisation is termed a *second-order, central difference* scheme. It is "second-order" because the truncation error is $O(\delta x)^2$, as can easily be demonstrated via Taylor expansion. Of course, an nth order scheme would have a truncation error which is $O(\delta x)^n$. It is a "central difference" scheme because it is symmetric about the central grid-point, x_i. Our discretised version of Poisson's equation takes the form

$$u_{i-1} - 2u_i + u_{i+1} = v_i(\delta x)^2, \qquad ...10.68$$

for $i = 1, N$, where $v_i \equiv v(x_i)$. Furthermore, $u_0 = u_l$ and $u_{N+1} = u_h$.

It is helpful to regard the above set of discretised equations as a matrix equation. Let $u = (u_1, u_2, \cdots, u_N)$ be the vector of the u-values, and let

$$w = \left[v_1(\delta x)^2 - u_l, v_2(\delta x)^2, v_3(\delta x)^2, \cdots, v_{N-1}(\delta x)^2, v_N(\delta x)^2 - u_h\right] \qquad ...10.69$$

be the vector of the source terms. The discretised equations can be written as:

$$Mu = w. \qquad \text{...10.70}$$

The matrix M takes the form

$$M = \begin{pmatrix} -2 & 1 & 0 & 0 & 0 & 0 \\ 1 & -2 & 1 & 0 & 0 & 0 \\ 0 & 1 & -2 & 1 & 0 & 0 \\ 0 & 0 & 1 & -2 & 1 & 0 \\ 0 & 0 & 0 & 1 & -2 & 1 \\ 0 & 0 & 0 & 0 & 1 & -2 \end{pmatrix} \qquad \text{...10.71}$$

for $N = 6$. The generalisation to other N values is fairly obvious. Matrix M is termed a *tridiagonal* matrix, since only those elements which lie on the three leading diagonals are non-zero.

The formal solution to equation (10.70) is

$$u = M^{-1}w, \qquad \text{...10.72}$$

where M^{-1} is the inverse matrix to M. Unfortunately, the most efficient general purpose algorithm for inverting an $N \times N$ matrix—namely, Gauss-Jordan elimination with partial pivoting—requires $O(N^3)$ arithmetic operations. It is fairly clear that this is a disastrous scaling for finite-difference solutions of Poisson's equation. Every time we doubled the resolution (*i.e.*, doubled the number of grid-points) the required cpu time would increase by a factor of about eight. Consequently, adding a second dimension (which effectively requires the number of grid-points to be squared) would be prohibitively expensive in terms of cpu time. Fortunately, there is a well-known trick for inverting an $N \times N$ *tridiagonal* matrix which only requires $O(N)$ arithmetic operations.

Consider a general $N \times N$ tridiagonal matrix equation $Mu = w.$. Let *a*, *b*, and *c* be the vectors of the left, centre and right diagonal elements of the matrix, respectively Note that a_1 and c_N are undefined, and can be conveniently set to zero. Our matrix equation can now be written

$$a_i u_{i-1} + b_i u_i + c_i u_{i+1} = w_i, \qquad \text{...10.73}$$

for $i = 1, N$. Let us search for a solution of the form

$$u_{i+1} = x_i u_i + y_i. \qquad \text{...10.74}$$

Substitution into equation (10.73) yields

$$a_i u_{i-1} + b_i u_i + c_i (x_i u_i + y_i) = w_i, \qquad \text{...10.75}$$

which can be rearranged to give

$$u_i = -\frac{a_i u_{i-1}}{b_i + c_i x_i} + \frac{w_i - c_i y_i}{b_i + c_i x_i}. \qquad \text{...10.76}$$

However, if equation (10.74) is general then we can write $u_i = x_{i-1}u_{i-1} + y_{i-1}$. Comparison with the previous equation yields

$$x_{i-1} = -\frac{a_i}{b_i + c_i x_i}, \qquad \text{...10.77}$$

and

$$y_{i-1} = \frac{w_i - c_i y_i}{b_i + c_i x_i}. \qquad \text{...10.78}$$

We can now solve our tridiagonal matrix equation in two stages. In the first stage, we scan *up* the leading diagonal from $i = N$ to 1 using equation (10.77) and (10.78). Thus,

$$x_{N-1} = -\frac{a_N}{b_N}, \; y_{N-1} = \frac{w_N}{b_N}, \qquad \text{...10.79}$$

since $c_N = 0$. Furthermore,

$$x_i = -\frac{a_{i+1}}{b_{i+1} + c_{i+1}x_{i+1}}, \qquad y_i = \frac{w_{i+1} - c_{i+1}y_{i+1}}{b_{i+1} + c_{i+1}x_{i+1}} \qquad \text{...10.80}$$

for $i = N-2,1$. Finally,

$$x_0 = 0, \; y_0 = \frac{w_1 - c_1 y_1}{b_1 + c_1 x_1}, \qquad \text{...10.81}$$

since $a_1 = 0$. We have now defined all of the x_i and y_i. In the second stage, we scan *down* the leading diagonal from $i = 0$ to $N-1$ using equation (10.74). Thus,

$$u_1 = y_0, \qquad \text{...10.82}$$

since $x_0 = 0$, and

$$u_{i+1} = x_i u_i + y_i \qquad \text{...10.83}$$

for $i = 1, N-1$. We have now inverted our tridiagonal matrix equation using $O(N)$ arithmetic operations.

Clearly, we can use the above algorithm to invert equation (10.70), with the source terms specified in equation (10.69), and the diagonals of matrix M given by $a_i = 1$ for $i = 2, N$, plus $b_i = -2$ for $i = 1, N$, and $c_i = 1$ for $i = 1, N-1$.

An Example Tridiagonal Matrix Solving Routine

Listed below is an example tridiagonal matrix solving routine which utilises the Blitz++ library.

```
// Tridiagonal.cpp
// Function to invert tridiagonal matrix equation.
// Left, centre, and right diagonal elements of matrix
```

```
// stored in arrays a, b, c, respectively.
// Right-hand side stored in array w.
// Solution written to array u.
// Matrix is NxN. Arrays a, b, c, w, u assumed to be of extent N+2,
// with redundant 0 and N+1 elements.
#include <blitz/array.h>
using namespace blitz;
void Tridiagonal (Array<double,1> a, Array<double,1> b, Array<double,1> c,
                  Array<double,1> w, Array<double,1>& u)
{
  // Find N. Declare local arrays.
  int N = a.extent(0) - 2;
  Array<double,1> x(N), y(N);
  // Scan up diagonal from i = N to 1
  x(N-1) = - a(N) / b(N);
  y(N-1) = w(N) / b(N);
  for (int i = N-2; i > 0; i—)
    {
      x(i) = - a(i+1) / (b(i+1) + c(i+1) * x(i+1));
     y(i) = (w(i+1) - c(i+1) * y(i+1)) / (b(i+1) + c(i+1) * x(i+1));
    }
  x(0) = 0.;
  y(0) = (w(1) - c(1) * y(1)) / (b(1) + c(1) * x(1));
  // Scan down diagonal from i = 0 to N-1
  u(1) = y(0);
  for (int i = 1; i < N; i++)
    u(i+1) = x(i) * u(i) + y(i);
}
```

1-d Problem with Mixed Boundary Conditions

Previously, we solved Poisson's equation in one dimension subject to Dirichlet boundary conditions, which are the simplest conceivable boundary conditions. Let us now consider the following much more general set of boundary conditions:

$$\alpha_l u(x) + \beta_l \frac{du(x)}{dx} = \gamma_l, \qquad ...10.84$$

at $x = x_l$, and

$$\alpha_h u(x) + \beta_h \frac{du(x)}{dx} = \gamma_h, \qquad \text{...10.85}$$

at $x = x_h$. Here, α_l, β_l, etc., are known constants. The above boundary conditions are termed *mixed,* since they are a mixture of Dirichlet and Neumann boundary conditions.

Using the previous notation, the discretised versions of equation (10.84) and (10.85) are:

$$\alpha_l, u_0 + \beta_l \frac{u_1 - u_0}{\delta x} = \gamma_l, \qquad \text{...10.86}$$

$$\alpha_h u_{N+1} + \beta_h \frac{u_{N+1} - u_N}{\delta x} = \gamma_h, \qquad \text{...10.87}$$

respectively. The above expressions can be rearranged to give

$$u_0 = \frac{\gamma_l \delta x - \beta_l u_1}{\alpha_l \delta x - \beta_l}, \qquad \text{...10.88}$$

$$u_{N+1} = \frac{\gamma_h \delta x + \beta_h u_N}{\alpha_h \delta x + \beta_h}. \qquad \text{...10.89}$$

Using equation (10.67), (10.88), and (10.89), the problem can be reduced to a tridiagonal matrix equation $Mu = w$, where the left, centre, and right diagonals of M possess the elements $a_i = 1$ for $i = 2, N$, with

$$b_1 = -2 - \frac{\beta_l}{\alpha_l \delta x - \beta_l}, \qquad \text{...10.90}$$

and $b_i = -2$ for $i = 2, N-1$, plus

$$b_N = -2 + \frac{\beta_h}{\alpha_h \delta x + \beta_h}, \qquad \text{...10.91}$$

and $c_i = 1$ for $i = 1, N-1$, respectively. The elements of the right-hand side are

$$w_1 = v_1 (\delta x)^2 - \frac{\gamma_l \delta x}{\alpha_l \delta x - \beta_l}, \qquad \text{...10.92}$$

with $w_i = v_i (\delta x)^2$ for $i = 2, N-1$, and

$$w_N = v_N (\delta x)^2 - \frac{\gamma_h \delta x}{\alpha_h \delta x + \beta_h}. \qquad \text{...10.93}$$

Our tridiagonal matrix equation can be inverted using the algorithm discussed previously.

An Example 1-d Poisson Solving Routine

Listed below is an example 1-d Poisson solving routine which utilises the previously listed tridiagonal matrix solver and the Blitz++ library.

```
// Poisson1D.cpp
// Function to solve Poisson's equation in 1-d:
//   d^2 u / dx^2 = v   for  xl <= x <= xh
//   alpha_l u + beta_l du/dx = gamma_l   at x=xl
//   alpha_h u + beta_h du/dx = gamma_h   at x=xh
// Arrays u and v assumed to be of extent N+2.
// Now, ith elements of arrays correspond to
//   x_i = xl + i * dx      i=0,N+1
// Here, dx = (xh - xl) / (N+1) is grid spacing.
#include <blitz/array.h>
using namespace blitz;
void Tridiagonal (Array<double,1> a, Array<double,1> b, Array<double,1> c,
                  Array<double,1> w, Array<double,1>& u);
void Poisson1D (Array<double,1>& u, Array<double,1> v,
                double alpha_l, double beta_l, double gamma_l,
                double alpha_h, double beta_h, double gamma_h,
                double dx)
{
  // Find N. Declare local arrays.
  int N = u.extent(0) - 2;
  Array<double,1> a(N+2), b(N+2), c(N+2), w(N+2);
  // Initialise tridiagonal matrix
  for (int i = 2; i <= N; i++) a(i) = 1.;
  for (int i = 1; i <= N; i++) b(i) = -2.;
  b(1) -= beta_l / (alpha_l * dx - beta_l);
  b(N) += beta_h / (alpha_h * dx + beta_h);
  for (int i = 1; i <= N-1; i++) c(i) = 1.;
  // Initialise right-hand side vector
  for (int i = 1; i <= N; i++)
      w(i) = v(i) * dx * dx;
  w(1) -= gamma_l * dx / (alpha_l * dx - beta_l);
  w(N) -= gamma_h * dx / (alpha_h * dx + beta_h);
  // Invert tridiagonal matrix equation
  Tridiagonal (a, b, c, w, u);
  // Calculate i=0 and i=N+1 values
  u(0) = (gamma_l * dx - beta_l * u(1)) /
    (alpha_l * dx - beta_l);
  u(N+1) = (gamma_h * dx + beta_h * u(N)) /
    (alpha_h * dx + beta_h);
}
```

An Example Solution of Poisson's Equation in 1-d

Let us now solve Poisson's equation in one dimension, with mixed boundary conditions, using the finite difference technique discussed above. We seek the solution of

$$\frac{d^2u(x)}{dx^2}=v(x), \qquad ...10.94$$

in the region $0 \le x \le 1$, with $v(x)=1-2x^2$. The boundary conditions at $x_l=0$ and $x_h=1$ take the mixed form specified in equation (10.84) and (10.85). Of course, we can solve this problem analytically. In fact,

$$u(x)=g+hx+\frac{x^2}{2}-\frac{x^4}{6}, \qquad ...10.95$$

where

$$g=\frac{\gamma_l(\alpha_h+\beta_h)-\beta_l\left[\gamma_h-(\alpha_h+\beta_h)/3\right]}{\alpha_l\alpha_h+\alpha_l\beta_h-\beta_l\alpha_h} \qquad ...10.96$$

$$h=\frac{\alpha_l\left[\gamma_h-(\alpha_h+\beta_h)/3\right]-\gamma_l\alpha_h}{\alpha_l\alpha_h+\alpha_l\beta_h-\beta_l\alpha_h}. \qquad ...10.97$$

Figure 10.15 shows a comparison between the analytic and finite difference solutions for $N=100$. It can be seen that the finite difference solution mirrors the analytic solution almost exactly.

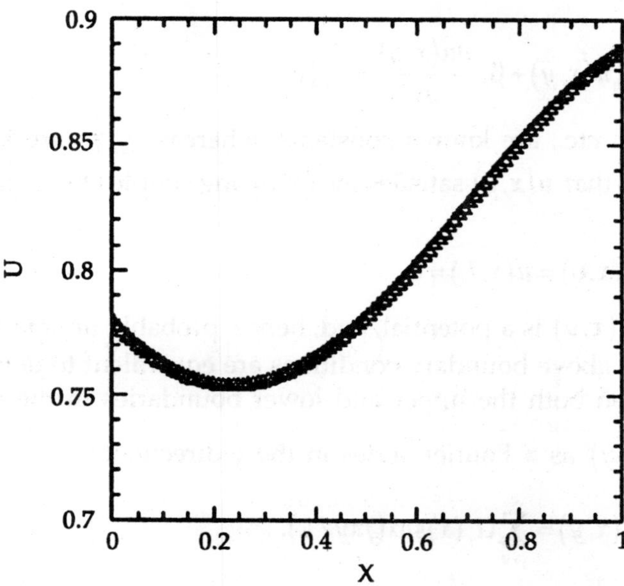

Figure 10.15: *Solution of Poisson's equation in one dimension with* $v=1-2x^2$, $\alpha_l=1$, $\beta_l=-1$, $\gamma_l=1$, $\alpha_h=1$, $\beta_h=1$, *and* $\gamma_h=1$. *The dotted curve (obscured) shows the analytic solution, whereas the open triangles show the finite difference solution for N=100*

2-d Problem with Dirichlet Boundary Conditions

Let us consider the solution of Poisson's equation in two dimensions. Suppose that

$$\frac{\partial^2 u(x,y)}{\partial x^2}+\frac{\partial^2 u(x,y)}{\partial y^2}=v(x,y), \qquad \text{...10.98}$$

for $x_l \leq x \leq x_h$, and $0 \leq y \leq L$. By direct analogy with our previous method of solution in the 1-d case, we could discretise the above 2-d problem using a second-order, central difference scheme in both the *x*- and *y*-directions. Unfortunately, such a discretisation scheme yields a set of equations which *cannot* be reduced to a simple tridiagonal matrix equation. In fact, all of the efficient numerical algorithms for solving this type of problem are *iterative* in nature. For instance, the *Jacobi* method, the *Gauss-Seidel* method, the *successive over-relaxation* method, and the *multi-grid* method.Regrettably, unless such iteration methods are extremely sophisticated (*e.g.*, the multi-grid method), and, hence, beyond the scope of this course, they tend to converge very poorly. In the following, rather than discuss iterative methods which do not work very well, we shall instead discuss a non-iterative method which works effectively for a restricted set of problems. The method in question is termed a *spectral method,* since it involves expanding *u* and *v* as truncated Fourier series in the *y*-direction.

Suppose that $u(x,y)$ satisfies mixed boundary conditions in the *x*-direction: i.e.,

$$\alpha_l u(x,y)+\beta_l \frac{\partial u(x,y)}{\partial x}=\gamma_l(y), \qquad \text{...10.99}$$

at $x=x_l$, and

$$\alpha_h u(x,y)+\beta_h \frac{\partial u(x,y)}{\partial x}=\gamma_h(y), \qquad \text{...10.100}$$

at $x=x_h$. Here, α_l, β_l, etc., are known constants, whereas γ_l, γ_h are known functions of *y*. Furthermore, suppose that $u(x,y)$ satisfies the following simple Dirichlet boundary conditions in the *y*-direction:

$$u(x,0)=u(x,L)=0. \qquad \text{...10.101}$$

Note that, since $u(x,y)$ is a potential, and, hence, probably undetermined to an arbitrary additive constant, the above boundary conditions are equivalent to demanding that take the *same* constant value on both the upper and lower boundaries in the *y*-direction.

Let us write $u(x,y)$ as a Fourier series in the *y*-direction:

$$u(x,y)=\sum_{j=0}^{\infty} U_j(x)\sin(j\pi y/L). \qquad \text{...10.102}$$

Note that the above expression for *u* automatically satisfies the boundary conditions in the *y*-direction. The $\sin(j\pi y/L)$ functions are *orthogonal,* and form a complete set, in the interval $y=0,L$. In fact,

$$\frac{2}{L}\int_0^L \sin(j\pi y/L)\sin(k\pi y/L)\,dy = \delta_{jk}. \qquad ...10.103$$

Thus, we can write the source term as

$$v(x,y) = \sum_{j=0}^{\infty} V_j(x)\sin(j\pi y/L), \qquad ...10.104$$

where

$$V_j(x) = \frac{2}{L}\int_0^L v(x,y)\sin(j\pi y/L)\,dy. \qquad ...10.105$$

Furthermore, the boundary conditions in the x-direction become

$$\alpha_l U_j(x) + \beta_l \frac{dU_j(x)}{dx} = \Gamma_{lj}, \qquad ...10.106$$

at $x = x_l$, and

$$\alpha_h U_j(x) + \beta_h \frac{dU_j(x)}{dx} = \Gamma_{hj}, \qquad ...10.107$$

at $x = x_h$, where

$$\Gamma_{lj} = \frac{2}{L}\int_0^L \gamma_l(y)\sin(j\pi y/L)\,dy, \qquad ...10.108$$

etc.

Substituting equation (10.102) and (10.104) into equation (10.98), and equating the coefficients of the $\sin(j\pi y/L)$ (since these functions are orthogonal), we obtain

$$\frac{d^2U_j(x)}{dx^2} - \frac{j^2\pi^2}{L^2}U_j(x) = V_j(x), \qquad ...10.109$$

for $j = 0,\infty$. Now, we can discretise the problem in the y-direction by truncating our Fourier expansion: i.e., by only solving the above equations for $j = 0, J$, rather than $j = 0,\infty$. This is essentially equivalent to discretisation in the y-direction on the equally-spaced grid-points $y_j = jL/J$. The problem is discretised in the x-direction by dividing the domain into equal segments, according to equation (10.66), and approximating d^2/dx^2 via the second-order, central difference scheme specified in equation (10.67). Thus, we obtain

$$U_{i-1,j} - (2 + j^2k^2)U_{ij} + U_{i+1,j} = V_{i,j}(\delta x)^2, \qquad ...10.110$$

for $i = 1, N$ and $j = 0, j$. Here, $U_{i,j} \equiv U_j(x_i)$, $V_{i,j} \equiv V_j(x_i)$, and $k = \pi\delta x/L$. The boundary conditions (10.106) and (10.107) discretise to give:

$$U_{0j} = \frac{\Gamma_{lj}\delta x - \beta_l U_{1j}}{\alpha_l \delta x - \beta_l}, \qquad ...10.111$$

$$U_{N+1,j} = \frac{\Gamma_{hj}\delta x + \beta_h U_{N,j}}{\alpha_h \delta x + \beta_h}, \quad ...10.112$$

for $j = 0, J$. equation (10.110), (10.111), and (10.112) constitute a set of $J+1$ *uncoupled* tridiagonal matrix equations (with one equation for each separate j value). These equations can be inverted to give the $U_{i,j}$. Finally, the $u(x_i, y_j)$ values can be reconstructed from equation (10.102). Hence, we have solved the problem.

2-d Problem with Neumann Boundary Conditions

Let us redo the above calculation, replacing the Dirichlet boundary conditions (10.101) with the following simple Neumann boundary conditions:

$$\frac{\partial u(x, y=0)}{\partial y} = \frac{\partial u(x, y=L)}{\partial y} = 0. \quad ...10.113$$

In this case, we can express $u(x,y)$ in the form

$$u(x,y) = \sum_{j=0}^{\infty} U_j(x)\cos(j\pi y/L), \quad ...10.114$$

which automatically satisfies the boundary conditions in the y-direction. Likewise, we can write the source term $v(x,y)$ as

$$v(x,y) = \sum_{j=0}^{\infty} V_j(x)\cos(j\pi y/L), \quad ...10.115$$

where

$$V_j(x) = \frac{2}{L}\int_0^L v(x,y)\cos(j\pi y/L)\,dy, \quad ...10.116$$

since

$$\frac{2}{L}\int_0^L \cos(j\pi y/L)\cos(k\pi y/L)\,dy = \delta_{jk}. \quad ...10.117$$

Finally, the boundary conditions in the x-direction become

$$\alpha_l U_j(x) + \beta_l \frac{dU_j(x)}{dx} = \Gamma_{lj}, \quad ...10.118$$

at $x = x_l$, and

$$\alpha_h U_j(x) + \beta_h \frac{dU_j(x)}{dx} = \Gamma_{hj}, \quad ...10.119$$

at $x = x_h$, where

$$\Gamma_{lj} = \frac{2}{L}\int_0^L \gamma_l(y)\cos(j\pi y/L)\,dy, \quad ...10.120$$

etc. Note, however, that the factor in front of the integrals in equation (10.116) and (10.120) takes the special value $1/L$ for the $j=0$ harmonic.

As before, we truncate the Fourier expansion in the y-direction, and discretise in the x-direction, to obtain the set of tridiagonal matrix equations specified in equation (10.110), (10.111), and (10.112). We can solve these equations to obtain the $U_{i,j}$, and then reconstruct $u(x_i, y_j)$ the from equation (10.114). Hence, we have solved the problem.

Fast Fourier Transform

The method outlined for solving Poisson's equation in 2-d with simple Dirichlet boundary conditions in the y-direction requires us to perform very many *Fourier-sine transforms*:

$$F_j^S = \frac{2}{J}\sum_{k=1}^{J-1} f_k \sin(jk\pi/J) \qquad \text{...10.121}$$

for $j=0, J$, and inverse Fourier-sine transforms:

$$f_j = \sum_{k=1}^{J-1} F_k^S \sin(jk\pi/J). \qquad \text{...10.122}$$

Here, f_j is the value of $f(y)$ at $y_j = jL/J$. Thus, equation (10.121) is analogous to equation (10.105) and (10.108), whereas equation (10.122) can be used to reconstruct the $u(x_i, y_j)$ from the $U_{i,j}$. Likewise, the method outlined in Sect. 5.8 for solving Poisson's equation in 2-d with simple Neumann boundary conditions in the y-direction requires us to perform very many *Fourier-cosine transforms*:

$$F_j^C = \frac{f_0}{J} + \frac{2}{J}\sum_{k=1}^{J-1} f_k \cos(jk\pi/J) + \frac{(-1)^j f_j}{J} \qquad \text{...10.123}$$

for $j=0, J$, and inverse Fourier-cosine transforms:

$$f_j = \sum_{k=0}^{J} F_k^C \cos(jk\pi/J). \qquad \text{...10.124}$$

Unfortunately, performing such transforms directly requires $O(J^2)$ arithmetic operations, which means that they are *extremely expensive* in terms of cpu resources. There is, however, an ingenious algorithm for performing Fourier transforms which only takes $O(J \ln J)$ arithmetic operations [which is much less than $O(J^2)$ operations when J is large]. This algorithm is known as the *fast Fourier transform* or FFT.

The details of the FFT algorithm lie beyond the scope of this course. Roughly speaking, the algorithm works by building up the transform in stages using 2, 4, 8, 16, etc. grid-points. In this course, we shall employ the best-known publicly available FFT library, called the fftw library, to perform all of our Fourier-sine and -cosine transforms. Unfortunately, the fftw library does not directly calculate Fourier-sine and -cosine transforms. Instead, it calculates *complex* Fourier transforms:

$$F_j = \frac{1}{2J}\sum_{k=0}^{2J-1} f_k \exp(-ijk\pi/J) \qquad ...10.125$$

for $j = 0, 2J-1$, and complex inverse Fourier transforms:

$$f_j = \sum_{k=0}^{2J-1} F_k \exp(ijk\pi/J). \qquad ...10.126$$

Note that f_j and F_j are *periodic* in j with period $2J$. Note, further, that the data-sets associated with complex Fourier transforms contain *twice* as many elements as the data-sets associated with sine and cosine transforms. However, we can easily convert a sine or cosine transform into a complex transform by *extending* its data-set. Thus, for a sine transform we write:

$$f_{2J-j} = -f_j, \qquad ...10.127$$

$$F_{2J-j} = -F_j, \qquad ...10.128$$

for $j = 1, J-1$, in which case

$$F_j^S = 2iF_j. \qquad ...10.129$$

Likewise, for a cosine transform we write:

$$f_{2J-j} = f_j, \qquad ...10.130$$

$$F_{2J-j} = F_j, \qquad ...10.131$$

for $j = 1, J-1$, in which case

$$F_j^C = 2F_j. \qquad ...10.132$$

Listed below are a set of wrapper routines which employ the fftw library to perform Fourier-sine and -cosine transforms.

```
// FFT.cpp
// Set of functions to calculate Fourier-cosine and -sine transforms
// of real data using fftw Fast-Fourier-Transform library.
// Input/ouput arrays are assumed to be of extent J+1.
// Uses version 2 of fftw library (incompatible with vs 3).
#include <fftw.h>
#include <blitz/array.h>
using namespace blitz;
// Calculates Fourier-cosine transform of array f in array F
void fft_forward_cos (Array<double,1> f, Array<double,1>& F)
{
```

```
   // Find J. Declare local arrays.
   int J = f.extent(0) - 1;
   int N = 2 * J;
   fftw_complex ff[N], FF[N];
   // Load and extend data
   c_re (ff[0]) = f(0); c_im (ff[0]) = 0.;
   c_re (ff[J]) = f(J); c_im (ff[J]) = 0.;
   for (int j = 1; j < J; j++)
     {
        c_re (ff[j]) = f(j); c_im (ff[j]) = 0.;
        c_re (ff[2*J-j]) = f(j); c_im (ff[2*J-j]) = 0.;
     }
   // Call fftw routine
   fftw_plan p = fftw_create_plan (N, FFTW_FORWARD, FFTW_ESTIMATE);
   fftw_one (p, ff, FF);
   fftw_destroy_plan (p);
   // Unload data
   F(0) = c_re (FF[0]); F(J) = c_re (FF[J]);
   for (int j = 1; j < J; j++)
     {
        F(j) = 2. * c_re (FF[j]);
     }
   // Normalise data
   F /= 2. * double (J);
}
// Calculates inverse Fourier-cosine transform of array F in array f
void fft_backward_cos (Array<double,1> F, Array<double,1>& f)
{
   // Find J. Declare local arrays.
   int J = f.extent(0) - 1;
   int N = 2 * J;
   fftw_complex ff[N], FF[N];
   // Load and extend data
```

```
    c_re (FF[0]) = F(0); c_im (FF[0]) = 0.;
    c_re (FF[J]) = F(J); c_im (FF[J]) = 0.;
    for (int j = 1; j < J; j++)
      {
        c_re (FF[j]) = F(j) / 2.; c_im (FF[j]) = 0.;
        FF[2*J-j] = FF[j];
      }
    // Call fftw routine
    fftw_plan p = fftw_create_plan (N, FFTW_BACKWARD, FFTW_ESTIMATE);
    fftw_one (p, FF, ff);
    fftw_destroy_plan (p);
    // Unload data
    f(0) = c_re (ff[0]); f(J) = c_re (ff[J]);
    for (int j = 1; j < J; j++)
      {
        f(j) = c_re (ff[j]);
      }
}
// Calculates Fourier-sine transform of array f in array F
void fft_forward_sin (Array<double,1> f, Array<double,1>& F)
{
    // Find J. Declare local arrays.
    int J = f.extent(0) - 1;
    int N = 2 * J;
    fftw_complex ff[N], FF[N];
    // Load and extend data
    c_re (ff[0]) = 0.; c_im (ff[0]) = 0.;
    c_re (ff[J]) = 0.; c_im (ff[J]) = 0.;
    for (int j = 1; j < J; j++)
      {
        c_re (ff[j]) = f(j); c_im (ff[j]) = 0.;
        c_re (ff[2*J-j]) = - f(j); c_im (ff[2*J-j]) = 0.;
      }
```

```
    // Call fftw routine
    fftw_plan p = fftw_create_plan (N, FFTW_FORWARD, FFTW_ESTIMATE);
    fftw_one (p, ff, FF);
    fftw_destroy_plan (p);
    // Unload data
    F(0) = 0.; F(J) = 0.;
    for (int j = 1; j < J; j++)
      {
        F(j) = - 2. * c_im (FF[j]);
      }
    // Normalise data
    F /= 2. * double (J);
}
// Calculates inverse Fourier-sine transform of array F in array f
void fft_backward_sin (Array<double,1> F, Array<double,1>& f)
{
    // Find J. Declare local arrays.
    int J = f.extent(0) - 1;
    int N = 2 * J;
    fftw_complex ff[N], FF[N];
    // Load and extend data
    c_re (FF[0]) = 0.; c_im (FF[0]) = 0.;
    c_re (FF[J]) = 0.; c_im (FF[J]) = 0.;
    for (int j = 1; j < J; j++)
      {
        c_re (FF[j]) = 0.; c_im (FF[j]) = - F(j) / 2.;
        c_re (FF[2*J-j]) = 0.; c_im (FF[2*J-j]) = F(j) / 2.;
      }
    // Call fftw routine
    fftw_plan p = fftw_create_plan (N, FFTW_BACKWARD, FFTW_ESTIMATE);
    fftw_one (p, FF, ff);
    fftw_destroy_plan (p);
    // Unload data
    f(0) = 0.; f(J) = 0.;
```

```
   for (int j = 1; j < J; j++)
      {
         f(j) = c_re (ff[j]);
      }
}
```

An Example 2-d Poisson Solving Routine

Listed below is an example 2-d Poisson solving routine which employs the previously listed tridiagonal matrix inversion and FFT wrapper routines, as well as the Blitz++ library.

```
// Poisson2d.cpp
// Function to solve Poisson's equation in 2-d:
//  d^2 u / dx^2 + d^2 u / dy^2 = v  for  xl <= x <= xh  and  0
<= y <= L
//  alphaL u + betaL du/dx = gammaL(y)   at x=xl
//  alphaH u + betaH du/dx = gammaH(y)   at x=xh
// In y-direction, either simple Dirichlet boundary conditions:
//  u(x,0) = u(x,L) = 0
// or simple Neumann boundary conditions:
//  du/dy(x,0) = du/dy(x,L) = 0
// Matrices u and v assumed to be of extent N+2, J+1.
// Arrays gammaL, gammaH assumed to be of extent J+1.
// Now, (i,j)th elements of matrices correspond to
//  x_i = xl + i * dx      i=0,N+1
//  y_j = j * L / J         j=0,J
// Here, dx = (xh - xl) / (N+1) is grid spacing in x-direction.
// Now, kappa = pi * dx / L
// Finally, Neumann=0/1 selects Dirichlet/Neumann bcs in y-direction.
#include <blitz/array.h>
using namespace blitz;
void fft_forward_cos (Array<double,1> f, Array<double,1>& F);
void fft_backward_cos (Array<double,1> F, Array<double,1>& f);
void fft_forward_sin (Array<double,1> f, Array<double,1>& F);
void fft_backward_sin (Array<double,1> F, Array<double,1>& f);
```

```
void Tridiagonal (Array<double,1> a, Array<double,1> b, Array<double,1> c,
                  Array<double,1> w, Array<double,1>& u);
void Poisson2D (Array<double,2>& u, Array<double,2> v,
                double alphaL, double betaL, Array<double,1> gammaL,
                double alphaH, double betaH, Array<double,1> gammaH,
                double dx, double kappa, int Neumann)
{
  // Find N and J. Declare local arrays.
  int N = u.extent(0) - 2;
  int J = u.extent(1) - 1;
  Array<double,2> V(N+2, J+1), U(N+2, J+1);
  Array<double,1> GammaL(J+1), GammaH(J+1);
  // Fourier transform boundary conditions
  if (Neumann)
    {
      fft_forward_cos (gammaL, GammaL);
      fft_forward_cos (gammaH, GammaH);
    }
  else
    {
      fft_forward_sin (gammaL, GammaL);
      fft_forward_sin (gammaH, GammaH);
    }
  // Fourier transform source term
  for (int i = 1; i <= N; i++)
    {
      Array<double,1> In(J+1), Out(J+1);
      for (int j = 0; j <= J; j++) In(j) = v(i, j);
      if (Neumann)
        fft_forward_cos (In, Out);
      else
```

```
          fft_forward_sin (In, Out);
        for (int j = 0; j <= J; j++) V(i, j) = Out(j);
    }
// Solve tridiagonal matrix equations
if (Neumann)
    {
        for (int j = 0; j <= J; j++)
            {
              Array<double,1> a(N+2), b(N+2), c(N+2), w(N+2), uu(N+2);
              // Initialise tridiagonal matrix
              for (int i = 2; i <= N; i++) a(i) = 1.;
              for (int i = 1; i <= N; i++)
              b(i) = -2. - double (j * j) * kappa * kappa;
              b(1) -= betaL / (alphaL * dx - betaL);
              b(N) += betaH / (alphaH * dx + betaH);
              for (int i = 1; i <= N-1; i++) c(i) = 1.;
              // Initialise right-hand side vector
              for (int i = 1; i <= N; i++)
                 w(i) = V(i, j) * dx * dx;
              w(1) -= GammaL(j) * dx / (alphaL * dx - betaL);
              w(N) -= GammaH(j) * dx / (alphaH * dx + betaH);
              // Invert tridiagonal matrix equation
              Tridiagonal (a, b, c, w, uu);
              for (int i = 1; i <= N; i++) U(i, j) = uu(i);
            }
    }
else
    {
        for (int j = 1; j < J; j++)
            {
              Array<double,1> a(N+2), b(N+2), c(N+2), w(N+2), uu(N+2);
              // Initialise tridiagonal matrix
```

```
            for (int i = 2; i <= N; i++) a(i) = 1.;
            for (int i = 1; i <= N; i++)
              b(i) = -2. - double (j * j) * kappa * kappa;
            b(1) -= betaL / (alphaL * dx - betaL);
            b(N) += betaH / (alphaH * dx + betaH);
            for (int i = 1; i <= N-1; i++) c(i) = 1.;
            // Initialise right-hand side vector
            for (int i = 1; i <= N; i++)
              w(i) = V(i, j) * dx * dx;
            w(1) -= GammaL(j) * dx / (alphaL * dx - betaL);
            w(N) -= GammaH(j) * dx / (alphaH * dx + betaH);
            // Invert tridiagonal matrix equation
            Tridiagonal (a, b, c, w, uu);
            for (int i = 1; i <= N; i++) U(i, j) = uu(i);
          }
        for (int i = 1; i <= N   ; i++)
          {
            U(i, 0) = 0.; U(i, J) = 0.;
          }
      }
  // Reconstruct solution via inverse Fourier transform
  for (int i = 1; i <= N; i++)
      {
        Array<double,1> In(J+1), Out(J+1);
        for (int j = 0; j <= J; j++) In(j) = U(i, j);
        if (Neumann)
          fft_backward_cos (In, Out);
        else
          fft_backward_sin (In, Out);
        for (int j = 0; j <= J; j++) u(i, j) = Out(j);
      }

  // Calculate i=0 and i=N+1 values
  for (int j = 0; j <= J; j++)
      {
```

```
        u(0, j) = (gammaL(j) * dx - betaL * u(1, j)) /
          (alphaL * dx - betaL);
        u(N+1, j) = (gammaH(j) * dx + betaH * u(N, j)) /
          (alphaH * dx + betaH);
    }
}
```

An Example Solution of Poisson's Equation in 2-d

Let us now use the techniques discussed above to solve Poisson's equation in two dimensions. Suppose that the source term is

$$v(x,y) = 6xy(1-y) - 2x^3 \qquad \text{...10.133}$$

for $0 \le x \le 1$ and $0 \le y \le 1$. The boundary conditions at $x = 0$ are $\alpha_l = 1$, $\beta_l = 0$, and $\gamma_l = 0$, whereas the boundary conditions at $x = 1$ are $\alpha_h = 1$, $\beta_h = 0$, and $\gamma_h = y(1-y)$. The simple Dirichlet boundary conditions $u(x,0) = u(x,1) = 0$ are applied at $y = 0$ and $y = 1$. Of course, this problem can be solved analytically to give

$$u(x,y) = y(1-y)x^3. \qquad \text{...10.134}$$

Figures 10.16 and 10.17 show comparisons between the analytic and finite difference solutions for $N = J = 64$. It can be seen that the finite difference solution mirrors the analytic solution almost exactly.

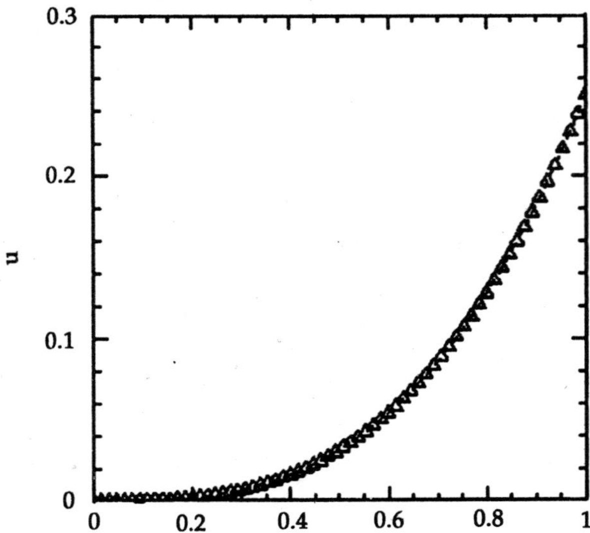

Figure 10.16: Solution of Poisson's equation in two dimensions with simple Dirichlet boundary conditions in the y-direction. The solution is plotted versus x at $y = 0.5$. The dotted curve (obscured) shows the analytic solution, whereas the open triangles show the finite difference solution for $N = j = 64$

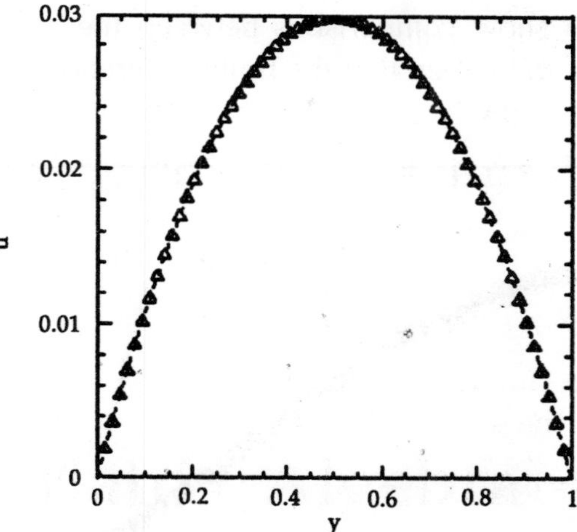

***Figure 10.17:** Solution of Poisson's equation in two dimensions with simple Dirichlet boundary conditions in the y-direction. The solution is plotted versus y at* $x = 0.5$ *. The dotted curve (obscured) shows the analytic solution, whereas the open triangles show the finite difference solution for* $N = J = 64$

As a second example, suppose that the source term is

$$v(x,y) = -2(2y^3 - 3y^2 + 1) + 6(1 - x^2)(2y - 1) \quad ...10.135$$

for $0 \le x \le 1$ and $0 \le y \le 1$. The boundary conditions at $x = 0$ are $\alpha_l = 1$, $\beta_l = 0$, and $\gamma_l = 2y^3 - 3y^2 + 1$, whereas the boundary conditions at $x = 1$ are $\alpha_h = 1$, $\beta_h = 0$, and $\gamma_h = 0$. The simple Neumann boundary conditions $\partial u(x,0)/\partial y = \partial u(x,1)/\partial y = 0$ are applied at $y = 0$ and $y = 1$. Of course, this problem can be solved analytically to give

$$u(x,y) = (1 - x^2)(2y^3 - 3y^2 + 1). \quad ...10.136$$

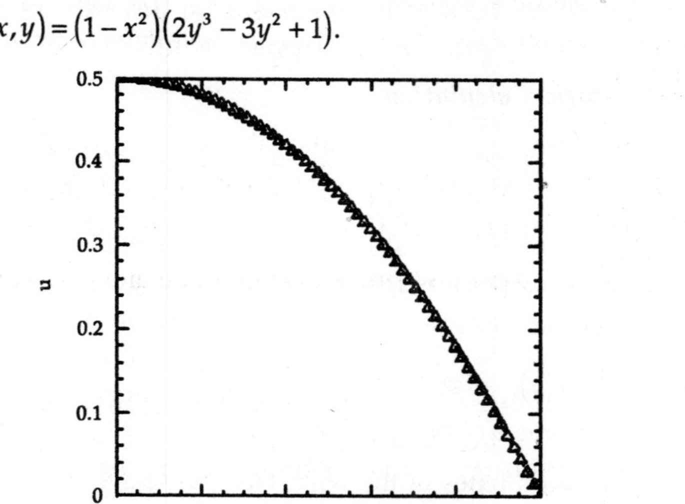

***Figure 10.18:** Solution of Poisson's equation in two dimensions with simple Neumann boundary conditions in the y-direction. The solution is plotted versus x at* $y = 0.5$ *. The dotted curve (obscured) shows the analytic solution, whereas the open triangles show the finite difference solution for* $N = J = 64$

Figures 10.18 and 10.19 show comparisons between the analytic and finite difference solutions for $N = J = 64$. It can be seen that the finite difference solution mirrors the analytic solution almost exactly.

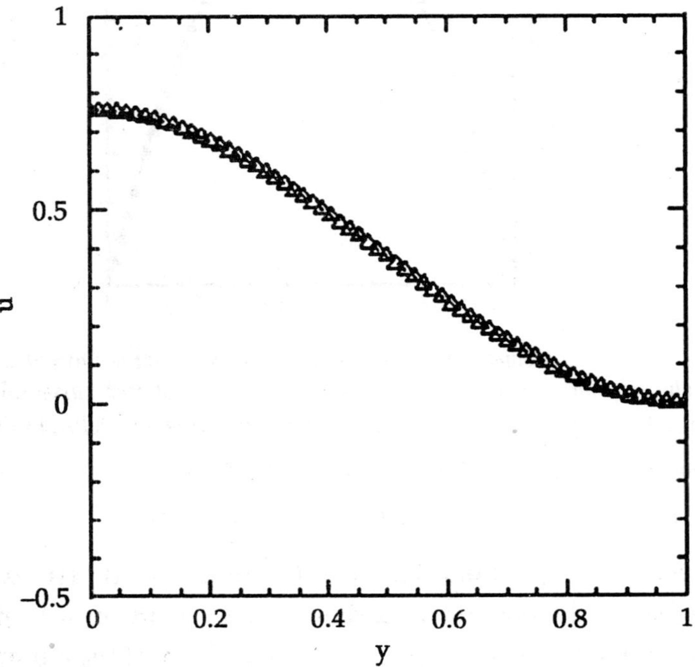

Figure 10.19: Solution of Poisson's equation in two dimensions with simple Neumann boundary conditions in the y-direction. The solution is plotted versus y at $x = 0.5$ *. The dotted curve (obscured) shows the analytic solution, whereas the open triangles show the finite difference solution for* $N = j = 64$

Example 2-d Electrostatic Calculation

Let us perform an example 2-d electrostatic calculation. Consider a charged wire running parallel to the axis of a uniform, hollow, rectangular, conducting channel. Suppose that the vertices of the channel lie at $(x,y)=(0,0)$, $(0,1)$, $(1,0)$, and $(1,1)$. Suppose, further, that the wire carries a uniform charge per unit length of magnitude unity. The electric potential $\phi(x,y)$ inside the channel satisfies

$$\frac{\partial^2\phi(x,y)}{\partial x^2}+\frac{\partial^2\phi(x,y)}{\partial y^2}=v(x,y)=-\delta(x-x_0)\delta(y-y_0), \qquad ...10.137$$

where (x_0,y_0) are the coordinates of the wire. Here, we have conveniently normalised our units such that the factor ϵ_0 is absorbed into the normalisation. Assuming that the box is grounded, the potential is subject to the Dirichlet boundary conditions $\phi = 0$ at $x = 0$, $x = 1$, $y = 0$, and $y = 1$. We require the solution in the region $0 \le x \le 1$ and $0 \le y \le 1$.

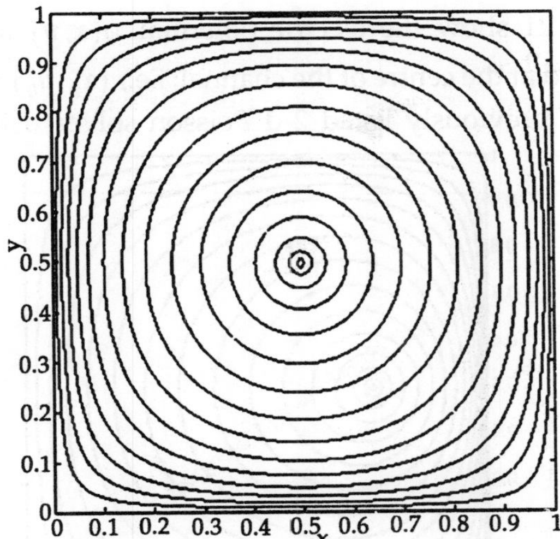

Figure 10.20: Contour plot of the electric potential generated by a charged wire placed at the centre of a grounded rectangular channel. The wire is located at $(x,y)=(0.5,0.5)$*, whereas the channel walls are at* $x=0$*,* $x=1$*,* $y=0$*, and* $y=1$*. Calculation performed with* $N=J=128$

Note that when discretising equation (10.137) the right-hand side becomes

$$v(x_i,y_i)=\frac{1}{\delta x\delta y} \qquad ...10.138$$

on the grid-point closest to the wire, with $v(x_i,y_i)=0$ on the remaining grid-points. Here, δx and δy are the grid spacings in the x- and y- directions, respectively.

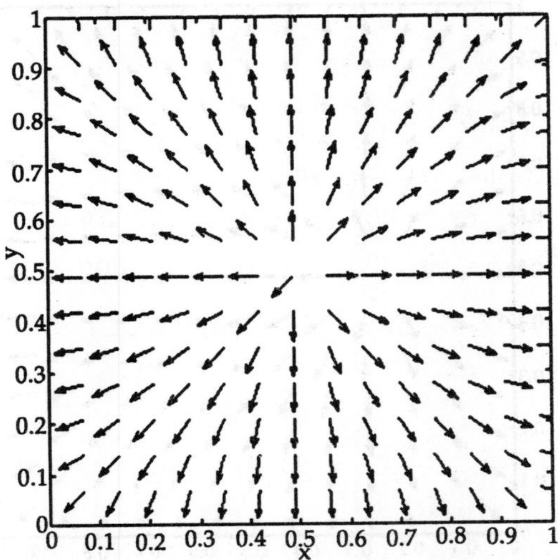

Figure 10.21: Vector plot showing the direction of the electric field generated by a charged wire placed at the centre of a grounded rectangular channel. The wire is located at $(x,y)=(0.5,0.5)$*, whereas the channel walls are at* $x=0$*,* $x=1$*,* $y=0$*, and* $y=1$*. Calculation performed with* $N=J=128$

Figures 10.20 and 10.21 show the electric potential $\phi(x,y)$ and electric field $E=-\nabla\phi$ generated by a wire placed at the centre of the channel: i.e., $(x_0,y_0)=(0.5,0.5)$. The calculation was performed with the previously listed 2-d Poisson solver using $N=J=128$.

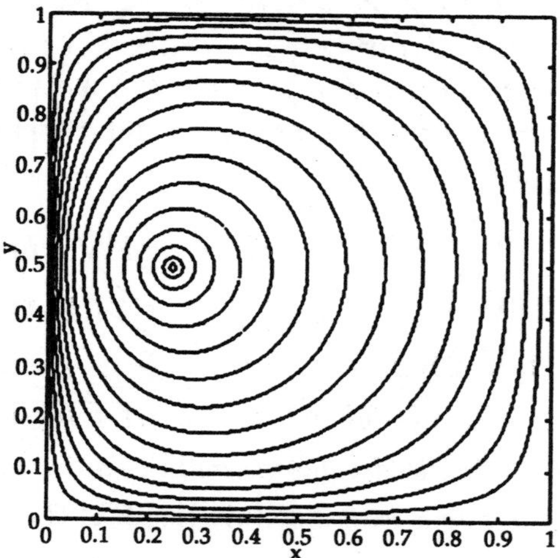

Figure 10.22: Contour plot of the electric potential generated by a charged wire offset from the centre of a grounded rectangular channel. The wire is located at $(x,y)=(0.25,0.5)$, *whereas the channel walls are at* $x=0$, $x=1$, $y=0$, *and* $y=1$. *Calculation performed with* $N=J=128$

Figures 10.22 and 10.23 show the electric potential $\phi(x,y)$ and electric field $E=\nabla\phi$ generated by a wire offset from the centre of the channel: i.e., $(x_0,y_0)=(0.25,0.5)$. The calculation was performed with the previously listed 2-d Poisson solver using $N=J=128$.

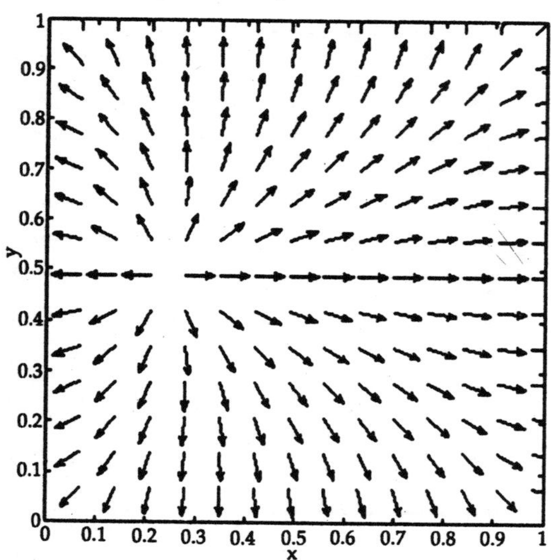

Figure 10.23: Vector plot showing the direction of the electric field generated by a charged wire offset from the centre of a grounded rectangular channel. The wire is located at $(x,y)=(0.25,0.5)$, *whereas the channel walls are at* $x=0$, $x=1$, $y=0$, *and* $y=1$. *Calculation performed with* $N=J=128$

3-d Problems

The techniques for solving Poisson's equation in two dimensions with a restricted class of boundary conditions can easily be generalised to three dimensions. In the 3-d case, it is necessary to Fourier transform in *two* directions (the y and z directions, say) in order to reduce the problem to a system of uncoupled tridiagonal matrix equations. These equations can be inverted in the usual manner, and the solution can then be reconstructed via a double inverse Fourier transform.

Diffusion Equation

The diffusion equation

$$\frac{\partial T(r,t)}{\partial t} = D\nabla^2 T(r,t), \qquad \text{...10.139}$$

where $D > 0$ is the (uniform) coefficient of diffusion, describes many interesting physical phenomena. For instance, in heat conduction we can write

$$q = -k\nabla T, \qquad \text{...10.140}$$

where q is the heat flux, T the temperature, and k the coefficient of thermal conductivity. The above equation merely states that heat flows down a temperature gradient. In the absence of sinks or sources of heat, energy conservation requires that

$$-\frac{\partial Q}{\partial t} = \int q \cdot dS, \qquad \text{...10.141}$$

where Q is the thermal energy contained in some volume V bounded by a closed surface S. The above equation states that the rate of decrease of the thermal energy content of volume equals the instantaneous heat flux flowing across its boundary. However, $Q = \int cTdV$, where is the heat capacity per unit volume. Making use of the previous equations, as well as the divergence theorem, we obtain the following diffusion equation for the temperature:

$$\frac{\partial T}{\partial t} = D\nabla^2 T, \qquad \text{...10.142}$$

where $D = k/c$. In a typical heat conduction problem, we are given the temperature $T(r,t_0)$ at some initial time t_0, and then asked to evaluate $T(r,t)$ at all subsequent times. Such a problem is known as an *initial value problem*. The spatial boundary conditions can be either of type Dirichlet (*i.e.*, T specified on the boundary), type Neumann (*i.e.*, ∇T specified on the boundary), or some combination.

1-d Problem with Mixed Boundary Conditions

Consider the solution of the diffusion equation in one dimension. Suppose that

$$\frac{\partial T(x,t)}{\partial t} = D\frac{\partial^2 T(x,t)}{\partial x^2}, \qquad \text{...10.143}$$

for $x_l \le x \le x_h$, subject to the mixed spatial boundary conditions

$$\alpha_l(t)T(x,t)+\beta_l(t)\frac{\partial T(x,t)}{\partial x}=\gamma_l(t), \qquad ...10.144$$

at $x = x_l$, and

$$\alpha_h(t)T(x,t)+\beta_h(t)\frac{\partial T(x,t)}{\partial x}=\gamma_h(t), \qquad ...10.145$$

at $x = x_h$. Here, α_l, β_l, etc., are known functions of time. Of course, $T(x, t_0)$ must be specified at some initial time t_0.

Equation (10.143) needs to be discretised in both time and space. In time, we discretise on the equally spaced grid

$$t_n = t_0 + n\delta t, \qquad ...10.146$$

where δt is the *time-step*. Adopting a simple first-order differencing scheme, equation (10.143) becomes

$$\frac{T(x,t_{n+1})-T(x,t_n)}{\delta t}=D\frac{\partial^2 T(x,t_n)}{\partial x^2}+O(\delta t). \qquad ...10.147$$

In space, we discretise on the usual equally spaced grid-points specified in equation (10.66), and approximate d^2/dx^2 via the second-order, central difference scheme introduced in equation (10.67). The spatial boundary conditions are discretised in a similar manner to equation (10.86) and (10.87). Thus, equation (10.147) yields

$$\frac{T_i^{n+1}-T_i^n}{\delta t}=D\frac{T_{i-1}^n-2T_i^n+T_{i+1}^n}{(\delta x)^2}, \qquad ...10.148$$

or

$$T_i^{n+1}=T_i^n+C\left(T_{i-1}^n-2T_i^n+T_{i+1}^n\right) \qquad ...10.149$$

for $i = 1, N$, where $T_i^n \equiv T(x_i, t_n)$, and $C = D\delta t/(\delta x)^2$. The discretised boundary conditions take the form

$$T_0^n=\frac{\gamma_l^n\delta x-\beta_l^n T_1^n}{\alpha_l^n\delta x-\beta_l^n}, \qquad ...10.150$$

$$T_{N+1}^n=\frac{\gamma_h^n\delta x+\beta_h^n T_N^n}{\alpha_h^n\delta x+\beta_h^n}, \qquad ...10.151$$

where $\gamma_l^n \equiv \gamma_l(t_n)$, etc. The discretisation scheme outlined above is termed first-order in time and second-order in space.

Equations (10.149)-(10.151) constitute a fairly straightforward iterative scheme which can be used to evolve the $T(x,t)$ in time.

An Example 1-d Diffusion Equation Solver

Listed below is an example 1-d diffusion equation solving routine which makes use of the Blitz++ library.

```
// Diffusion1D.cpp
// Function to evolve diffusion equation in 1-d:
//   dT / dt = D d^2 T / dx^2   for   xl <= x <= xh
//   alpha_l T + beta_l dT/dx = gamma_l   at x=xl
//   alpha_h T + beta_h dT/dx = gamma_h   at x=xh
// Array T assumed to be of extent N+2.
// Now, ith element of array corresponds to
//   x_i = xl + i * dx      i=0,N+1
// Here, dx = (xh - xl) / (N+1) is grid spacing.
// Function evolves T by single time-step.
// C = D dt / dx^2, where dt is time-step.
// Uses explicit scheme.
#include <blitz/array.h>
using namespace blitz;
void Diffusion1D (Array<double,1>& T,
                  double alpha_l, double beta_l, double gamma_l,
                  double alpha_h, double beta_h, double gamma_h,
                  double dx, double C)
{
  // Set N. Declare local array.
  int N = T.extent(0) - 2;
  Array<double,1> T0(N+2);
  // Evolve T
  T0 = T;
  for (int i = 1; i <= N; i++)
    T(i) += C * (T0(i-1) - 2. * T0(i) + T0(i+1));
  // Set boundary conditions
  T(0) = (gamma_l * dx - beta_l * T(1)) / (alpha_l * dx - beta_l);
  T(N+1) = (gamma_h * dx + beta_h * T(N)) / (alpha_h * dx + beta_h);
}
```

An Example 1-d Solution of the Diffusion Equation

Let us now solve the diffusion equation in 1-d using the finite difference technique discussed above. We seek the solution of equation (10.143) in the region $-x_0 \le x \le x_0$, subject to the initial condition

$$T(x,t_0) = \exp\left(\frac{-x^2}{4Dt_0}\right), \quad ...10.152$$

where $t_0 > 0$. The spatial boundary conditions are

$$T(\pm x_0, t) = \sqrt{\frac{t_0}{t}} \exp\left(\frac{-x_0^2}{4Dt}\right). \quad ...10.153$$

Of course, we can solve this problem analytically to give

$$T(x,t) = \sqrt{\frac{t_0}{t}} \exp\left(\frac{-x^2}{4Dt}\right). \quad ...10.154$$

Note that the above equation describes a Gaussian pulse which gradually decreases in height and broadens in width in such a manner that its area is conserved. The width of the pulse varies approximately as

$$\Delta x \sim \sqrt{Dt}. \quad ...10.155$$

Moreover, the pulse approaches a δ-function as $t \to 0$.

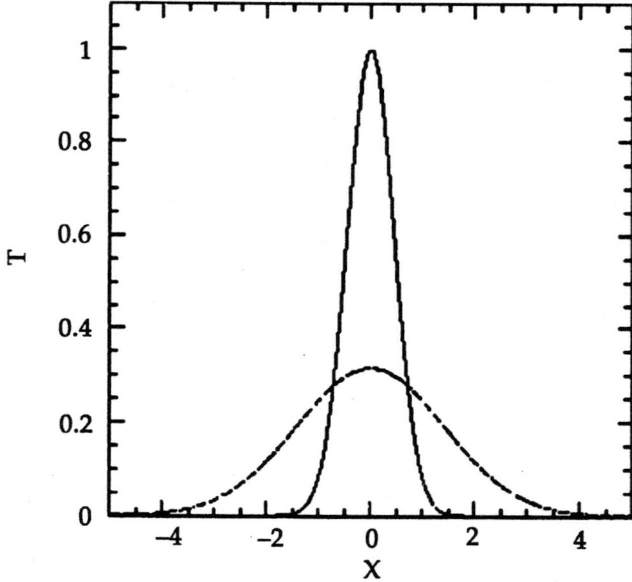

Figure 10.24: *Diffusive evolution of a 1-d Gaussian pulse. Numerical calculation performed using* $D = 1$, $x_0 = 5$, $\delta t = 4 \times 10^{-3}$, *and* $N = 100$. *The pulse is evolved from* $t = 0.1$ *to* $t = 1$. *The solid curve shows the initial condition at* $t = 0.1$, *the dashed curve the numerical solution at* $t = 1$, *and the dotted curve (obscured by the dashed curve) the analytic solution at* $t = 1$

Figure 10.24 shows a comparison between the analytic and numerical solutions for a calculation performed using $D = 1$, $x_0 = 5$, $t_0 = 0.1$, $\delta t = 4 \times 10^{-3}$, and $N = 100$. It can be seen that the analytic and numerical solutions are in excellent agreement.

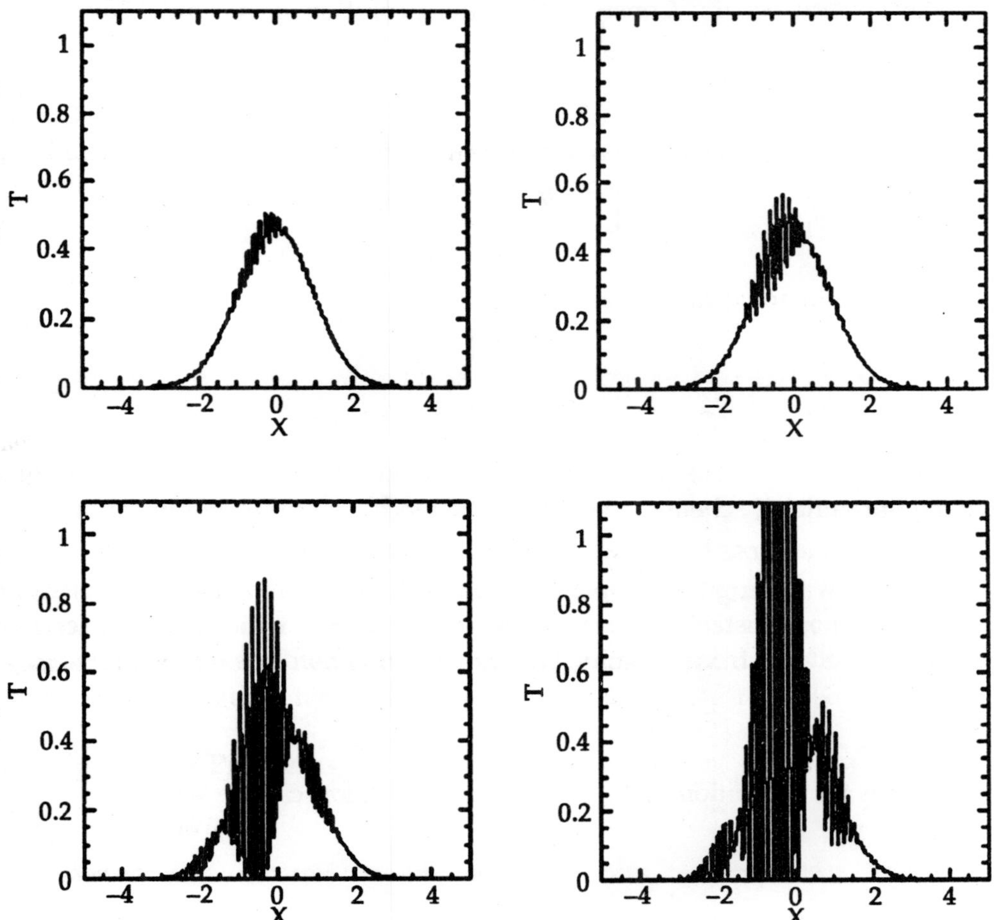

Figure 10.25: *Diffusive evolution of a 1-d Gaussian pulse. Numerical calculation performed using* $D = 1$, $x_0 = 5$, $\delta t = 4 \times 10^{-3}$, *and* $N = 125$. *The simulation is started at* $t = 0.1$. *The top-left, top-right, bottom-left, and bottom-right panels show the solution at* $t = 0.45$, $t = 0.46$, $t = 0.47$, *and* $t = 0.48$, *respectively*

It is reasonable to expect that as N increases at fixed δt (*i.e.*, the spatial resolution increases at fixed temporal resolution) the numerical solution should become more and more accurate. This is indeed the case—at least, until N exceeds a critical value. Beyond this value, there is a catastrophic breakdown in the numerical solution.

This breakdown is illustrated in figure 10.25. It can be seen that the solution develops rapidly growing short-wavelength oscillations. Indeed, the solution eventually becomes effectively infinite. Let us investigate this unusual and rather disturbing phenomenon.

Von Neumann Stability Analysis

Clearly, our simple finite difference algorithm for solving the 1-d diffusion equation is subject to a numerical instability under certain circumstances. Let us try to establish when this instability occurs. Consider the time evolution of a single Fourier mode of wave-number k:

$$T(x,t)=\hat{T}(t)e^{ikx}. \qquad ...10.156$$

Substitution of the above expression into our finite difference scheme (10.149) yields

$$\hat{T}^{n+1}e^{ikx_n}=\hat{T}^{n}e^{ikx_n}\left[1+C\left(e^{-ik\delta x}-2+e^{+ik\delta x}\right)\right], \qquad ...10.157$$

or

$$\hat{T}^{n+1}=A\hat{T}^{n}, \qquad ...10.158$$

where

$$A=1-2C(1-\cos k\delta x)=1-4C\sin^2(k\delta x/2). \qquad ...10.159$$

Thus, the amplitude of the Fourier mode is amplified by a factor A at each time-step. In order for the differencing scheme to be stable, the modulus of this amplification factor must be *less than unity* for *all* possible values of k. Now, the largest possible value of $\sin^2(k\delta x/2)$. is unity: hence, the wavelength corresponding to this value is that of the most unstable Fourier mode. In fact, the most unstable mode possesses a wavelength which is *half* the grid-spacing: i.e., $\lambda=\delta x/2$. It follows from equation (10.159) that this mode is stable provided

$$C<\frac{1}{2}. \qquad ...10.160$$

Finally, from the definition of C, our stability condition can be written

$$\delta t<\frac{(\delta x)^2}{2D}. \qquad ...10.161$$

Note that $C=0.408$ for the stable calculation shown in figure 10.24, whereas C=0.635 for the unstable calculation shown in figure 10.25. Incidentally, the type of stability analysis outlined above is called *von Neumann stability analysis*. Note that the neglect of the spatial boundary conditions in the above calculation is justified because the unstable modes vary on very small length-scales which are typically of order the grid spacing.

According to equation (10.161), our finite difference scheme for solving the 1-d diffusion equation is only stable provided that the time-step remains below some critical value. Note that this critical value scales like the *square* of the grid-spacing. This is a very unfavourable scaling, since it implies that a doubling of the spatial resolution requires a simultaneous reduction in the time-step by a factor of four in order to maintain numerical stability. Certainly, the above constraint limits us to absurdly small time-steps in high resolution calculations. Is there any way of overcoming this constraint?

Crank-Nicholson Scheme

Let us revisit our temporal differencing scheme:

$$\frac{T(x,t_{n+1})-T(x,t_n)}{\delta t}=D\frac{\partial^2 T(x,t_n)}{\partial x^2}+O(\delta t). \qquad ...10.162$$

Note that the right-hand side is evaluated entirely at the *start* of the time-step: i.e., at t_n. This type of temporal differencing scheme is termed an *explicit* scheme. Now, explicit schemes are very straightforward to implement, but are also notoriously prone to numerical instabilities. Fortunately, we can often overcome these instabilities by making our differencing scheme *implicit* in nature. An implicit scheme is one in which the right-hand side is evaluated partly (or wholly) at the *end* of the time-step: i.e., at t_{n+1}. Unfortunately, implicit schemes are generally a great deal more complicated to implement than explicit schemes.

The well-known Crank-Nicholson implicit method for solving the diffusion equation involves taking the average of the right-hand side between the beginning and end of the time-step. In other words,

$$\frac{T(x,t_{n+1})-T(x,t_n)}{\delta t}=\frac{D}{2}\frac{\partial^2 T(x,t_n)}{\partial x^2}+\frac{D}{2}\frac{\partial^2 T(x,t_{n+1})}{\partial x^2}+O(\delta t)^2. \qquad ...10.163$$

As indicated by the error term, this method is actually *second-order* in time.

Adopting our usual spatial differencing scheme, the above expression yields

$$T_i^{n+1}-\frac{C}{2}\left(T_{i-1}^{n+1}-2T_i^{n+1}+T_{i+1}^{n+1}\right)=T_i^n+\frac{C}{2}\left(T_{i-1}^n-2T_i^n+T_{i+1}^n\right). \qquad ...10.164$$

When we perform a von Neumann stability analysis of the above scheme, we obtain the following expression for the amplification factor:

$$A=\frac{1-2C\sin^2(k\delta x/2)}{1+2C\sin^2(k\delta x/2)}. \qquad ...10.165$$

Note that $|A|<1$ for all values of . It follows that the Crank-Nicholson scheme is *unconditionally stable*. Unfortunately, equation (10.164) constitutes a tridiagonal matrix equation linking the T_i^{n+1} and the T_i^n. Thus, the price we pay for the high accuracy and unconditional stability of the Crank-Nicholson scheme is having to invert a tridiagonal matrix equation at each time-step. Usually, this price is well worth paying.

An Improved 1-d Diffusion Equation Solver

Listed below is an improved 1-d diffusion equation solver which uses the Crank-Nicholson scheme, as well as the previous listed tridiagonal matrix solver and the Blitz++ library. Note the great structural similarity between this solver and the previously listed 1-d Poisson solver.

```
// CrankNicholson1D.cpp
// Function to evolve diffusion equation in 1-d:
//   dT / dt = D d^2 T / dx^2   for   xl <= x <= xh
```

```
//   alpha_l T + beta_l dT/dx = gamma_l   at x=xl
//   alpha_h T + beta_h dT/dx = gamma_h   at x=xh
// Array T assumed to be of extent N+2.
// Now, ith element of array corresponds to
//   x_i = xl + i * dx      i=0,N+1
// Here, dx = (xh - xl) / (N+1) is grid spacing.
// Function evolves T by single time-step.
// C = D dt / dx^2, where dt is time-step.
// Uses Crank-Nicholson implicit scheme.
#include <blitz/array.h>
using namespace blitz;
void Tridiagonal (Array<double,1> a, Array<double,1> b, Array<double,1> c,
                  Array<double,1> w, Array<double,1>& u);
void CrankNicholson1D (Array<double,1>& T,
                       double alpha_l, double beta_l, double gamma_l,
                       double alpha_h, double beta_h, double gamma_h,
                       double dx, double C)
{
  // Find N. Declare local arrays.
  int N = T.extent(0) - 2;
  Array<double,1> a(N+2), b(N+2), c(N+2), w(N+2);
  // Initialise tridiagonal matrix
  for (int i = 2; i <= N; i++) a(i) = - 0.5 * C;
  for (int i = 1; i <= N; i++) b(i) = 1. + C;
  b(1) += 0.5 * C * beta_l / (alpha_l * dx - beta_l);
  b(N) -= 0.5 * C * beta_h / (alpha_h * dx + beta_h);
  for (int i = 1; i <= N-1; i++) c(i) = - 0.5 * C;
  // Initialise right-hand side vector
  for (int i = 1; i <= N; i++)
    w(i) = T(i) + 0.5 * C * (T(i-1) - 2. * T(i) + T(i+1));
  w(1) += 0.5 * C * gamma_l * dx / (alpha_l * dx - beta_l);
  w(N) += 0.5 * C * gamma_h * dx / (alpha_h * dx + beta_h);
  // Invert tridiagonal matrix equation
  Tridiagonal (a, b, c, w, T);
  // Calculate i=0 and i=N+1 values
```

```
  T(0) = (gamma_l * dx - beta_l * T(1)) /
     (alpha_l * dx - beta_l);
  T(N+1) = (gamma_h * dx + beta_h * T(N)) /
     (alpha_h * dx + beta_h);
}
```

An Improved 1-d Solution of the Diffusion Equation

Let us now solve the simple diffusion problem introduced with the above listed Crank-Nicholson routine. Figure 10.26 shows a comparison between the analytic and numerical solutions for a calculation performed using $D=1$, $x_0=5$, $t_0=0.1$, $\delta t=0.1$, and $N=100$. It can be seen that the analytic and numerical solutions are in excellent agreement. Note, however, that the time-step used in this calculation (*i.e.,* $\delta t=0.1$) is much larger than that used in our previous calculation (*i.e.,* $\delta t=4\times10^{-3}$), which employed an explicit differencing scheme. According to equation (10.161), an explicit scheme is limited to time-steps less than about 5×10^{-3} for the problem under investigation.

Thus, we have been able to exceed this limit by a factor of 20 with our implicit scheme, yet still maintain numerical stability. Note that our Crank-Nicholson scheme is able to obtain accurate results with a time-step as large as 0.1 because it is *second-order* in time.

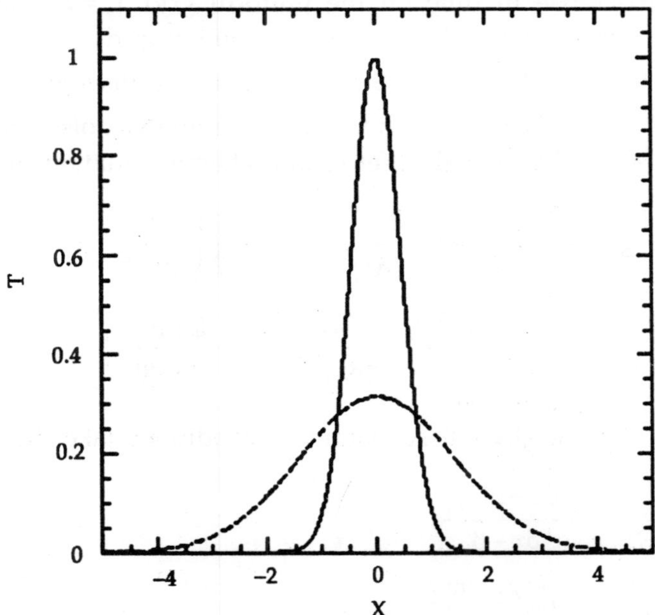

Figure 10.26: *Diffusive evolution of a 1-d Gaussian pulse. Numerical calculation performed using* $D=1$, $x_0=5$, $\delta t=0.1$, *and* $N=100$. *The pulse is evolved from* $t=0.1$ *to* $t=1$. *The solid curve shows the initial condition at* $t=0.1$, *the dashed curve the numerical solution at* $t=1$, *and the dotted curve (obscured by the dashed curve) the analytic solution at* $t=1$

2-d Problem with Dirichlet Boundary Conditions

Let us consider the solution of the diffusion equation in two dimensions. Suppose that

$$\frac{\partial T(x,y,t)}{\partial t}=D\frac{\partial^2 T(x,y,t)}{\partial x^2}+D\frac{\partial^2 T(x,y,t)}{\partial y^2}, \qquad \text{...10.166}$$

for $x_l \le x \le x_h$, and $0 \le y \le L$. Suppose that $T(x,y,t)$ satisfies mixed boundary conditions in the x-direction:

$$\alpha_l(t)T(x,y,t)+\beta_l(t)\frac{\partial T(x,y,t)}{\partial x}=\gamma_l(y,t), \qquad \text{...10.167}$$

at $x=x_l$, and

$$\alpha_h(t)T(x,y,t)+\beta_h(t)\frac{\partial T(x,y,t)}{\partial x}=\gamma_h(y,t), \qquad \text{...10.168}$$

at $x=x_h$. Here, α_l, β_l, etc., are known functions of t, whereas γ_l, γ_h are known functions of y and t. Furthermore, suppose that $T(x,y,t)$ satisfies the following simple Dirichlet boundary conditions in the y-direction:

$$T(x,0,t)=T(x,L,t)=0. \qquad \text{...10.169}$$

As before, we discretise in time on the uniform grid $t_n = t_0 + n\delta t$, for $n=0,1,2,\cdots$. Furthermore, in the x-direction, we discretise on the uniform grid $x_i = x_l + i\delta x$, for $i=0, N+1$, where $\delta x=(x_h - x_l)/(N+1)$. Finally, in the y-direction, we discretise on the uniform grid $y_j = j\delta y$, for $j=0, J$, where $\delta y = L/J$. Adopting the Crank-Nicholson temporal differencing scheme and the second-order spatial differencing scheme, equation (10.166) yields

$$\frac{T_{i,j}^{n+1}-T_{i,j}^{n}}{\delta t}-\frac{D}{2}\frac{T_{i-1j}^{n+1}-2T_{i,j}^{n+1}+T_{i+1,j}^{n+1}}{(\delta x)^2}-\frac{D}{2}\left(\frac{\partial^2 T}{\partial y^2}\right)_{i,j}^{n+1}=$$

$$\frac{D}{2}\frac{T_{i-1,j}^{n}-2T_{i,j}^{n}+T_{i+1,j}^{n}}{(\delta x)^2}+\frac{D}{2}\left(\frac{\partial^2 T}{\partial y^2}\right)_{i,j}^{n}, \qquad \text{...10.170}$$

where $T_{i,j}^n \equiv T(x_i, y_j, t_n)$. The discretised boundary conditions take the form

$$T_{0,j}^n=\frac{\gamma_{lj}^n\delta x-\beta_l^n T_{1,j}^n}{\alpha_l^n\delta x-\beta_l^n}, \qquad \text{...10.171}$$

$$T_{N+1,j}^n=\frac{\gamma_{h,j}^n\delta x+\beta_h^n T_{N,j}^n}{\alpha_h^n\delta x+\beta_h^n}, \qquad \text{...10.172}$$

plus

$$T_{i,0}^n=T_{i,j}^n=0. \qquad \text{...10.173}$$

Here, $\alpha_l^n \equiv \alpha_l(t_n)$, etc., and $\gamma_{lj}^n \equiv \gamma_l(y_j, t_n)$, etc.

Adopting the Fourier method, we write the $T_{i,j}^n$ in terms of their Fourier-sine harmonics:

$$T_{i,j}^n = \sum_{k=0}^{J} \hat{T}_{i,k}^n \sin(jk\pi / J), \qquad ...10.174$$

which automatically satisfies the boundary conditions (10.173). The above expression can be inverted to give

$$\hat{T}_{i,j}^n = \frac{2}{J}\sum_{k=0}^{J} T_{i,k}^n \sin(jk\pi / J). \qquad ...10.175$$

When equation (10.170) is written in terms of the $\hat{T}_{i,j}^n$, it reduces to

$$-\frac{C}{2}\hat{T}_{i-1j}^{n+1} + \{1 + C(1 + j^2k^2/2)\}\hat{T}_{i,j}^{n+1} - \frac{C}{2}\hat{T}_{i+1,j}^{n+1} = \frac{C}{2}\hat{T}_{i-1,j}^n + \{1 - C(1 + j^2k^2/2)\}\hat{T}_{i,j}^n + \frac{C}{2}\hat{T}_{i+1,j}^n, \qquad ...10.176$$

for $i = 1, N$, and $j = 0, J$. Here, $C = D\delta t / (\delta x)^2$, and $k = \pi\delta x / L$. Moreover, the boundary conditions (10.171) and (10.172) yield

$$\hat{T}_{0,j}^n = \frac{\Gamma_{lj}^n \delta x - \beta_l^n \hat{T}_{1,j}^n}{\alpha_l^n \delta x - \beta_l^n}, \qquad ...10.177$$

$$\hat{T}_{N+1,j}^n = \frac{\Gamma_{hj}^n \delta x + \beta_h^n \hat{T}_{Nj}^n}{\alpha_l^n \delta x + \beta_h^n}, \qquad ...10.178$$

where

$$\hat{\Gamma}_{lj}^n = \frac{2}{J}\sum_{k=0}^{J} \gamma_{l,k}^n \sin(jk\pi / J), \qquad ...10.179$$

etc. Equations (10.176)—(10.178) constitute a set of $J+1$ uncoupled tridiagonal matrix equations for the $T_{i,j}^{n+1}$, with one equation for each separate value of *j*.

In order to advance our solution by one time-step, we first Fourier transform the $T_{i,j}^n$ and the boundary conditions, according to equation (10.175) and (10.179). Next, we invert the $J+1$ tridiagonal equations (10.176)—(10.178) to obtain the $T_{i,j}^{n+1}$. Finally, we reconstruct the $T_{i,j}^n$ via equation (10.174).

2-d Problem with Neumann Boundary Conditions

Let us replace the Dirichlet boundary conditions (10.173) by the following simple Neumann boundary conditions:

$$\frac{\partial T(x,0,t)}{\partial y} = \frac{\partial T(x,L,t)}{\partial y} = 0. \qquad ...10.180$$

The method of solution outlined in the previous section is unaffected, except that the Fourier-sine transforms are replaced by Fourier-cosine transforms. An example 2-d diffusion equation solver.

Listed below is an example 2-d diffusion equation solver which uses the Crank-Nicholson scheme, as well as the previous listed tridiagonal matrix solver and the Blitz++ library. Note the great structural similarity between this solver and the previously listed 2-d Poisson solver.

```
// CrankNicholson2D.cpp
// Function to evolve diffusion equation in 2-d:
// dT / dt = D d^2 T / dx^2 +  D d^2 T / dy^2   for  xl <= x <= xh
// and  0 <= y <= L
//   alphaL T + betaL dT/dx = gammaL(y)   at x=xl
//   alphaH T + betaH dT/dx = gammaH(y)   at x=xh
// In y-direction, either simple Dirichlet boundary conditions:
//   T(x,0) = T(x,L) = 0
// or simple Neumann boundary conditions:
//   dT/dy(x,0) = dT/dy(x,L) = 0
// Matrix T assumed to be of extent N+2, J+1.
// Arrays gammaL, gammaH assumed to be of extent J+1.
// Now, (i,j)th elements of matrices correspond to
//   x_i = xl + i * dx      i=0,N+1
//   y_j = j * L / J         j=0,J
// Here, dx = (xh - xl) / (N+1) is grid spacing in x-direction.
// Now, C =  D dt / dx^2, and kappa = pi * dx / L
// Finally, Neumann=0/1 selects Dirichlet/Neumann bcs in y-direction.
// Uses Crank-Nicholson scheme.
#include <blitz/array.h>
using namespace blitz;
void fft_forward_cos (Array<double,1> f, Array<double,1>& F);
void fft_backward_cos (Array<double,1> F, Array<double,1>& f);
void fft_forward_sin (Array<double,1> f, Array<double,1>& F);
void fft_backward_sin (Array<double,1> F, Array<double,1>& f);
void Tridiagonal (Array<double,1> a, Array<double,1> b, Array<double,1> c,
                  Array<double,1> w, Array<double,1>& u);
void CrankNicholson2D (Array<double,2>& T,
                  double alphaL, double betaL, Array<double,1> gammaL,
                  double alphaH, double betaH, Array<double,1> gammaH,
                  double dx, double C, double kappa, int Neumann)
{
```

```
   // Find N and J. Declare local arrays.
   int N = T.extent(0) - 2;
   int J = T.extent(1) - 1;
   Array<double,2> TT(N+2, J+1), V(N+2, J+1);
   Array<double,1> GammaL(J+1), GammaH(J+1);
   // Fourier transform T
   for (int i = 0; i <= N+1; i++)
     {
       Array<double,1> In(J+1), Out(J+1);
       for (int j = 0; j <= J; j++) In(j) = T(i, j);
       if (Neumann)
         fft_forward_cos (In, Out);
       else
         fft_forward_sin (In, Out);
       for (int j = 0; j <= J; j++) TT(i, j) = Out(j);
     }
   // Fourier transform boundary conditions
   if (Neumann)
     {
       fft_forward_cos (gammaL, GammaL);
       fft_forward_cos (gammaH, GammaH);
     }
   else
     {
       fft_forward_sin (gammaL, GammaL);
       fft_forward_sin (gammaH, GammaH);
     }
   // Construct source term
   for (int i = 1; i <= N; i++)
     for (int j = 0; j <= J; j++)
       V(i, j) =
         0.5 * C * TT(i-1, j) +
         (1. - C * (1. + 0.5 * double (j * j) * kappa * kappa))
* TT(i, j) +
         0.5 * C * TT(i+1, j);
   // Solve tridiagonal matrix equations
   if (Neumann)
     {
```

```
        for (int j = 0; j <= J; j++)
          {
             Array<double,1> a(N+2), b(N+2), c(N+2), w(N+2), u(N+2);
             // Initialise tridiagonal matrix
             for (int i = 2; i <= N; i++) a(i) = - 0.5 * C;
             for (int i = 1; i <= N; i++)
            b(i) = 1. + C * (1. + 0.5 * double (j * j) * kappa * kappa);
             b(1) += 0.5 * C * betaL / (alphaL * dx - betaL);
             b(N) -= 0.5 * C * betaH / (alphaH * dx + betaH);
             for (int i = 1; i <= N-1; i++) c(i) = - 0.5 * C;
             // Initialise right-hand side vector
             for (int i = 1; i <= N; i++)
               w(i) = V(i, j);
           w(1) += 0.5 * C * GammaL(j) * dx / (alphaL * dx - betaL);
           w(N) += 0.5 * C * GammaH(j) * dx / (alphaH * dx + betaH);
             // Invert tridiagonal matrix equation
             Tridiagonal (a, b, c, w, u);
             for (int i = 1; i <= N; i++) TT(i, j) = u(i);
          }
      }
    else
      {
        for (int j = 1; j < J; j++)
          {
             Array<double,1> a(N+2), b(N+2), c(N+2), w(N+2), u(N+2);
             // Initialise tridiagonal matrix
             for (int i = 2; i <= N; i++) a(i) = - 0.5 * C;
             for (int i = 1; i <= N; i++)
           b(i) = 1. + C * (1. + 0.5 * double (j * j) * kappa * kappa);
             b(1) -= betaL / (alphaL * dx - betaL);
             b(N) += betaH / (alphaH * dx + betaH);
             for (int i = 1; i <= N-1; i++) c(i) = - 0.5 * C;
             // Initialise right-hand side vector
             for (int i = 1; i <= N; i++)
               w(i) = V(i, j);
           w(1) += 0.5 * C * GammaL(j) * dx / (alphaL * dx - betaL);
           w(N) += 0.5 * C * GammaH(j) * dx / (alphaH * dx + betaH);
             // Invert tridiagonal matrix equation
```

```
                Tridiagonal  (a,  b,  c,  w,  u);
                for  (int  i  =  1;  i  <=  N;  i++)  TT(i,  j)  =  u(i);
            }
          for  (int  i  =  1;  i  <=  N   ;  i++)
            {
                TT(i,  0)  =  0.;  TT(i,  J)  =  0.;
            }
      }
    //  Reconstruct  solution
    for  (int  i  =  1;  i  <=  N;  i++)
      {
          Array<double,1>  In(J+1),  Out(J+1);
          for  (int  j  =  0;  j  <=  J;  j++)  In(j)  =  TT(i,  j);
          if  (Neumann)
             fft_backward_cos  (In,  Out);
          else
             fft_backward_sin  (In,  Out);
          for  (int  j  =  0;  j  <=  J;  j++)  T(i,  j)  =  Out(j);
      }
    //  Calculate  i=0  and  i=N+1  values
    for  (int  j  =  0;  j  <=  J;  j++)
      {
          T(0,  j)  =  (gammaL(j)  *  dx  -  betaL  *  T(1,  j))  /
             (alphaL  *  dx  -  betaL);
          T(N+1,  j)  =  (gammaH(j)  *  dx  +  betaH  *  T(N,  j))  /
             (alphaH  *  dx  +  betaH);
      }
}
```

An Example 2-d Solution of the Diffusion Equation

Let us now solve the diffusion equation in 2-d using the finite difference technique discussed above. We seek the solution of equation (10.166) in the region $0 \le x \le 1$ and $0 \le y \le 1$, subject to the following initial condition at $t = 0$:

$$T(x,y,0) = 1 \quad \text{for } |x - 0.5| < 0.1 \text{ and } |y - 0.5| < 0.1,$$
$$T(x,y,0) = 0 \quad \text{otherwise.} \qquad \text{...10.181}$$

The boundary conditions are simply $T(0,y) = T(1,y) = T(x,0) = T(x,1) = 0$.

3-d Problems

The techniques discussed above for solving the diffusion equation in two dimensions with a restricted class of boundary conditions can easily be generalised to three dimensions. In the 3-d case, it is necessary to Fourier transform in *two* directions (the y and z directions, say) in order to reduce the problem to a system of uncoupled tridiagonal matrix equations. These equations can be inverted in the usual manner, and the solution can then be reconstructed via a double inverse Fourier transform.

11

Particle-in-cell Codes

The Particle-in-Cell (PIC) method refers to a technique used to solve a certain class of partial differential equations. In this method, individual particles (or fluid elements) in a Lagrangian frame are tracked in continuous phase space, whereas moments of the distribution such as densities and currents are computed simultaneously on Eulerian (stationary) mesh points.

PIC methods were already in use as early as 1955, even before the first Fortran compilers were available. The method gained popularity for plasma simulation in the late 1950s and early 1960s by Buneman, Dawson, Hockney, Birdsall, Morse and others. In plasma physics applications, the method amounts to following the trajectories of charged particles in self-consistent electromagnetic (or electrostatic) fields computed on a fixed mesh.

Consider an unmagnetised, uniform, 1-dimensional plasma consisting of N electrons and N unit-charged ions. Now, ions are much more massive than electrons. Hence, on short time-scales, we can treat the ions as a static neutralising background, and only consider the motion of the electrons. Let r_i be the x-coordinate of the ith electron. The equations of motion of the ith electron are written:

$$\frac{dr_i}{dt} = v_i, \qquad \text{...11.1}$$

$$\frac{dv_i}{dt} = -\frac{eE(r_i)}{m_e}, \qquad \text{...11.2}$$

where $e > 0$ is the magnitude of the electron charge, m_e the electron mass, and $E(x)$ the x-component of the electric field-strength at position x. Now, the electric field-strength can be expressed in terms of an electric potential:

$$E(x) = -\frac{d\phi}{dx}. \qquad ...11.3$$

Furthermore, from the Poisson-Maxwell equation, we have

$$\frac{d^2\phi(x)}{dx^2} = -\frac{e}{\epsilon_0}\{n_0 - n(x)\}, \qquad ...11.4$$

where ϵ_0 is the permittivity of free-space, $n(x)$ the electron number density (*i.e.*, $n(x)dx$ is the number of electrons in the interval x to $x+dx$), and n_0 the uniform ion number density. Of course, the average value of $n(x)$ is equal to n_0, since there are equal numbers of ions and electrons.

Let us consider an initial electron distribution function consisting of two counter-propagating Maxwellian beams of mean speed v_b and thermal spread v_{th}: i.e.,

$$f(x,v) = \frac{n_0}{2}\left\{\frac{1}{\sqrt{2\pi}\,v_{th}} e^{-(v-v_b)^2/2v_{th}^2} + \frac{1}{\sqrt{2\pi}\,v_{th}} e^{-(v+v_b)^2/2v_{th}^2}\right\}. \qquad ...11.5$$

Here, $f(x,v)\,dx\,dv$ is the number of electrons between x and $x+dx$ with velocities in the range v to $v+dv$. Of course, $n(x) = \int_{-\infty}^{\infty} f(x,v)\,dv$.. The beam temperature T is related to the thermal velocity via $v_{th} = \sqrt{k_B T/m_e}$, where k_B is the Boltzmann constant. It is well-known that if v_b is significantly larger than v_{th} then the above distribution is unstable to a plasma instability called the *two-stream instability*. Let us investigate this instability numerically.

Normalisation Scheme

It is convenient to normalise time with respect to ω_p^{-1}, where

$$\omega_p^2 = \frac{n_0 e^2}{\epsilon_0 m_e} \qquad ...11.6$$

is the so-called *plasma frequency*: i.e., the typical frequency of electrostatic electron oscillations. Likewise it is convenient to normalise length with respect to the so-called *Debye length*:

$$\lambda_D = \frac{v_{th}}{\omega_p},$$

which is the length-scale above which the electrons exhibit collective (*i.e.*, plasma-like) effects, instead of acting like individual particles.

Our normalised equations take the form:

$$\frac{\partial x_i}{dt} = v_i, \qquad ...11.7$$

$$\frac{\partial v_i}{dt} = -E(x_i), \qquad ...11.8$$

$$E(x) = -\frac{d\phi(x)}{dx}, \quad \text{...11.9}$$

$$\frac{d^2\phi(x)}{dx^2} = \frac{n(x)}{n_0} - 1. \quad \text{...11.10}$$

whereas our initial distribution function becomes

$$f(x,v) = \frac{n_0}{2}\left\{\frac{1}{\sqrt{2\pi}}e^{-(v-v_b)^2/2} + \frac{1}{\sqrt{2\pi}}e^{-(v+v_b)^2/2}\right\}. \quad \text{...11.11}$$

Note that $v_{th} = 1$ in normalised units.

Let us solve the above system of equations in the domain $0 \le x \le L$. Furthermore, for the sake of simplicity, let us adopt *periodic* boundary conditions: i.e., let us identify the left and right boundaries of our solution domain. It follows that $n(0) = n(L)$, $\phi(0) = \phi(L)$, and $E(0) = E(L)$. Moreover, any electron which crosses the right boundary of the solution domain must reappear at the left boundary with the same velocity, and *vice versa*.

Solution of Electron Equations of Motion

We can solve the electron equations of motion, (11.7) and (11.8), as a set of $2N$ coupled first-order ODEs using the RK4 methods. However, in order to evaluate the right-hand sides of these equations we need to know the electric field $E(x)$ at each time-step. We can achieve this by solving Poisson's equation, (11.10), every time-step. However, in order to determine the source term for this equation we need to calculate the electron number density $n(x)$, which is, of course, a function of the instantaneous electron locations.

Evaluation of Electron Number Density

In order to obtain the electron number density $n(x)$ from the electron coordinates r_i we adopt a so-called *particle-in-cell* (PIC) approach. Let us define a set of J equally spaced spatial grid-points located at coordinates

$$x_j = j\delta x, \quad \text{...11.12}$$

for $j = 0, J-1$, where $\delta x = L/J$. Let $n_j \equiv n(x_j)$. Suppose that the ith electron lies between the j th and $(j+1)$ th grid-points: i.e., $x_j < r_i < x_{j+1}$. We let

$$n_j \to n_j + \left(\frac{x_{j+1} - r_i}{x_{j+1} - x_j}\right) / \delta x, \quad \text{...11.13}$$

and

$$n_{j+1} \to n_{j+1}\left(\frac{r_i - x_j}{x_{j+1} - x_j}\right) / \delta_x. \quad \text{...11.14}$$

Thus, $n_j\delta x$ increases by 1 if the electron is at the j th grid-point, $n_{j+1}\delta x$ increases by 1 if the electron is at the $(j+1)$ th grid-point, and $n_j\delta x$ and $n_{j+1}\delta x$ both increase by $1/2$ if the electron is halfway between the two grid-points, etc. Performing a similar assignment for each electron in turn allows us to build up the n_j from the electron coordinates (assuming that all the n_j are initialised to zero at the start of this process).

Solution of Poisson's equation

Consider the solution of Poisson's equation:

$$\frac{d^2\phi(x)}{dx^2} = \rho(x), \qquad ...11.15$$

where $\rho(x) = n(x)/n_0 - 1$. Note that $n_0 = N/L$ in normalised units. Let $\phi_j \equiv \phi(x_j)$ and $\rho_j \equiv \rho(x_j)$. We can write

$$\phi_j = \sum_{j'=0,J-1} \hat{\phi}_{j'} e^{ijj'2\pi/J}, \qquad ...11.16$$

$$\rho_j = \sum_{j'=0,J-1} \hat{\rho}_{j'} e^{ijj'2\pi/J}, \qquad ...11.17$$

which automatically satisfies the periodic boundary conditions $\phi_J = \phi_0$ and $\rho_J = \rho_0$. Note that $\hat{\rho}_0 = 0$, since $\int_0^L n(x)\,dx = n_0$. The other $\hat{\rho}_j$ are obtained from

$$\hat{\rho}_j = \frac{1}{J} \sum_{j'=0,J-1} \rho_{j'} e^{-ijj'2\pi/J}, \qquad ...11.18$$

for $j = 1, J-1$. The Fourier transformed version of Poisson's equation yields

$$\hat{\phi}_0 = 0 \qquad ...11.19$$

and

$$\hat{\phi}_j = -\frac{\hat{\rho}_j}{j^2k^2} \qquad ...11.20$$

for $j = 1, J/2$, where $k = 2\pi/L$. Finally,

$$\hat{\phi}_j = \hat{\phi}^*_{J-j} \qquad ...11.21$$

for $j = J/2+1$ to $J-1$, which ensures that the ϕ_j remain real. The discretised version of equation (11.9) is

$$E_j = \frac{\phi_{j-1} - \phi_{j+1}}{2\delta x}. \qquad ...11.22$$

Of course, $j=0$ and $j=j-1$ are special cases which can be resolved using the periodic boundary conditions. Finally, suppose that the coordinate of the ith electron lies between the j th and $(j+1)$ th grid-points: i.e., $x_j < r_i < x_{j+1}$. We can then use linear interpolation to evaluate the electric field seen by the ith electron:

$$E(r_i) = \left(\frac{x_{j+1} - r_i}{x_{j+1} - x_j}\right) E_j + \left(\frac{r_i - x_j}{x_{j+1} - x_j}\right) E_{j+1}. \quad ...11.23$$

An Example: 1D PIC Code

The following code is an implementation of the ideas developed above.

The main function reads in the calculation parameters, checks that they are sensible, initialises the electron coordinates, and then evolves the electron equations of motion from $t=0$ to some specified t_{max}, using a fixed step RK4 routine with some specified time-step δt. Information on the electron phase-space coordinates and the electric field is periodically written to various data-files.

```
// 1-d PIC code to solve plasma two-stream instability problem.
#include <stdlib.h>
#include <stdio.h>
#include <math.h>
#include <time.h>
#include <blitz/array.h>
#include <fftw.h>
using namespace blitz;
void Output (char* fn1, char* fn2, double t,
           Array<double,1> r, Array<double,1> v);
void Density (Array<double,1> r, Array<double,1>& n);
void Electric (Array<double,1> phi, Array<double,1>& E);
void Poisson1D (Array<double,1>& u, Array<double,1> v, double kappa);
void rk4_fixed (double& x, Array<double,1>& y,
          void (*rhs_eval)(double, Array<double,1>, Array<double,1>&),
           double h);
void rhs_eval (double t, Array<double,1> y, Array<double,1>& dydt);
void Load (Array<double,1> r, Array<double,1> v, Array<double,1>& y);
void UnLoad (Array<double,1> y, Array<double,1>& r, Array<double,1>& v);
double distribution (double vb);
```

```
double L; int N, J;
int main()
{
                    // Parameters
  L;                // Domain of solution 0 <= x <= L (in Debye lengths)
  N;                // Number of electrons
  J;                // Number of grid points
  double vb;        // Beam velocity
  double dt;        // Time-step (in inverse plasma frequencies)
  double tmax;      // Simulation run from t = 0. to t = tmax
                    // Get parameters
  printf ("Please input N:  "); scanf ("%d", &N);
  printf ("Please input vb:  "); scanf ("%lf", &vb);
  printf ("Please input L:  "); scanf ("%lf", &L);
  printf ("Please input J:  "); scanf ("%d", &J);
  printf ("Please input dt:  "); scanf ("%lf", &dt);
  printf ("Please input tmax:  "); scanf ("%lf", &tmax);
  int skip = int (tmax / dt) / 10;
  if ((N < 1) || (J < 2) || (L <= 0.) || (vb <= 0.)
      || (dt <= 0.) || (tmax <= 0.) || (skip < 1))
    {
      printf ("Error - invalid input parameters\n");
      exit (1);
    }
  // Set names of output files
  char* phase[11]; char* data[11];
 phase[0] = "phase0.out";phase[1] = "phase1.out";phase[2] = "phase2.out";
 phase[3] = "phase3.out";phase[4] = "phase4.out";phase[5] = "phase5.out";
 phase[6] = "phase6.out";phase[7] = "phase7.out";phase[8] = "phase8.out";
 phase[9] = "phase9.out";phase[10] = "phase10.out";data[0] = "data0.out";
  data[1] = "data1.out"; data[2] = "data2.out"; data[3] = "data3.out";
  data[4] = "data4.out"; data[5] = "data5.out"; data[6] = "data6.out";
  data[7] = "data7.out"; data[8] = "data8.out"; data[9] = "data9.out";
```

```
data[10] = "data10.out";
// Initialise solution
double t = 0.;
int seed = time (NULL); srand (seed);
Array<double,1> r(N), v(N);
for (int i = 0; i < N; i++)
  {
    r(i) = L * double (rand ()) / double (RAND_MAX);
    v(i) = distribution (vb);
  }
Output (phase[0], data[0], t, r, v);
// Evolve solution
Array<double,1> y(2*N);
Load (r, v, y);
for (int k = 1; k <= 10; k++)
  {
    for (int kk = 0; kk < skip; kk++)
      {
        // Take time-step
        rk4_fixed (t, y, rhs_eval, dt);

        // Make sure all coordinates in range 0 to L.
        for (int i = 0; i < N; i++)
          {
            if (y(i) < 0.) y(i) += L;
            if (y(i) > L) y(i) -= L;
          }
        printf ("t = %11.4e\n", t);
      }
    printf ("Plot %3d\n", k);

    // Output data
    UnLoad (y, r, v);
```

```
            Output(phase[k], data[k], t, r, v);
        }

    return 0;
}
```

The following routine outputs the simulation data to various data-files.

```
// Write data to output files
void Output (char* fn1, char* fn2, double t,
            Array<double,1> r, Array<double,1> v)
{
    // Write phase-space data
    FILE* file = fopen (fn1, "w");
    for (int i = 0; i < N; i++)
        fprintf (file, "%e %e\n", r(i), v(i));
    fclose (file);
    // Write electric field data
    Array<double,1> ne(J), n(J), phi(J), E(J);
    Density (r, ne);
    for (int j = 0; j < J; j++)
        n(j) = double (J) * ne(j) / double (N) - 1.;
    double kappa = 2. * M_PI / L;
    Poisson1D (phi, n, kappa);
    Electric (phi, E);

    file = fopen (fn2, "w");
    for (int j = 0; j < J; j++)
        {
            double x = double (j) * L / double (J);
            fprintf (file, "%e %e %e %e\n", x, ne(j), n(j), E(j));
        }
    double x = L;
    fprintf (file, "%e %e %e %e\n", x, ne(0), n(0), E(0));
    fclose (file);
}
```

The following routine returns a random velocity distributed on a double Maxwellian distribution function corresponding to two counter-streaming beams. The algorithm used to achieve this is called the *rejection method,* and will be discussed later in this course.

```
// Function to distribute electron velocities randomly so as
// to generate two counter propagating warm beams of thermal
// velocities unity and mean velocities +/- vb.
// Uses rejection method.
double distribution (double vb)
{
    // Initialise random number generator
    static int flag = 0;
    if (flag == 0)
      {
        int seed = time (NULL);
        srand (seed);
        flag = 1;
      }
    // Generate random v value
    double fmax = 0.5 * (1. + exp (-2. * vb * vb));
    double vmin = - 5. * vb;
    double vmax = + 5. * vb;
   double v = vmin + (vmax - vmin) * double (rand ()) / double (RAND_MAX);
    // Accept/reject value
    double f = 0.5 * (exp (-(v - vb) * (v - vb) / 2.) +
                exp (-(v + vb) * (v + vb) / 2.));
    double x = fmax * double (rand ()) / double (RAND_MAX);
    if (x > f) return distribution (vb);
    else return v;
}
```

The routine below evaluates the electron number density on an evenly spaced mesh given the instantaneous electron coordinates.

```
// Evaluates electron number density n(0:J-1) from
// array r(0:N-1) of electron coordinates.
void Density (Array<double,1> r, Array<double,1>& n)
```

```
{
   // Initialise
   double dx = L / double (J);
   n = 0.;
   // Evaluate number density.
   for (int i = 0; i < N; i++)
      {
         int j = int (r(i) / dx);
         double y = r(i) / dx - double (j);
         n(j) += (1. - y) / dx;
         if (j+1 == J) n(0) += y / dx;
         else n(j+1) += y / dx;
      }
}
```

The following functions are wrapper routines for using the fftw library with periodic functions.

```
// Functions to calculate Fourier transforms of real data
// using fftw Fast-Fourier-Transform routine.
// Input/ouput arrays are assumed to be of extent J.
// Calculates Fourier transform of array f in arrays Fr and Fi
void fft_forward (Array<double,1>f, Array<double,1>&Fr,
         Array<double,1>& Fi)
{
   fftw_complex ff[J], FF[J];
   // Load data
   for (int j = 0; j < J; j++)
      {
         c_re (ff[j]) = f(j); c_im (ff[j]) = 0.;
      }
   // Call fftw routine
   fftw_plan p = fftw_create_plan (J, FFTW_FORWARD, FFTW_ESTIMATE);
   fftw_one (p, ff, FF);
   fftw_destroy_plan (p);
   // Unload data
```

```
  for (int j = 0; j < J; j++)
    {
      Fr(j) = c_re (FF[j]); Fi(j) = c_im (FF[j]);
    }
  // Normalise data
  Fr /= double (J);
  Fi /= double (J);
}
// Calculates inverse Fourier transform of arrays Fr and Fi in array f
void fft_backward (Array<double,1> Fr, Array<double,1> Fi,
        Array<double,1>& f)
{
  fftw_complex ff[J], FF[J];
  // Load data
  for (int j = 0; j < J; j++)
    {
      c_re (FF[j]) = Fr(j); c_im (FF[j]) = Fi(j);
    }
  // Call fftw routine
  fftw_plan p = fftw_create_plan (J, FFTW_BACKWARD, FFTW_ESTIMATE);
  fftw_one (p, FF, ff);
  fftw_destroy_plan (p);
  // Unload data
  for (int j = 0; j < J; j++)
      f(j) = c_re (ff[j]);
```

The following routine solves Poisson's equation in 1-D to find the instantaneous electric potential on a uniform grid.

```
// Solves 1-d Poisson equation:
//      d^u / dx^2 = v    for  0 <= x <= L
// Periodic boundary conditions:
//      u(x + L) = u(x),   v(x + L) = v(x)
// Arrays u and v assumed to be of length J.
// Now, jth grid point corresponds to
```

```
//       x_j = j dx   for j = 0,J-1
// where dx = L / J.
// Also,
//       kappa = 2 pi / L
void Poisson1D (Array<double,1>& u, Array<double,1> v, double kappa)
{
   // Declare local arrays.
   Array<double,1> Vr(J), Vi(J), Ur(J), Ui(J);
   // Fourier transform source term
   fft_forward (v, Vr, Vi);
   // Calculate Fourier transform of u
   Ur(0) = Ui(0) = 0.;
   for (int j = 1; j <= J/2; j++)
      {
         Ur(j) = - Vr(j) / double (j * j) / kappa / kappa;
         Ui(j) = - Vi(j) / double (j * j) / kappa / kappa;
      }
   for (int j = J/2; j < J; j++)
      {
         Ur(j) = Ur(J-j);
         Ui(j) = - Ui(J-j);
      }
   // Inverse Fourier transform to obtain u
   fft_backward (Ur, Ui, u);
}
```

The following function evaluates the electric field on a uniform grid from the electric potential.

```
// Calculate electric field from potential
void Electric (Array<double,1> phi, Array<double,1>& E)
{
   double dx = L / double (J);
   for (int j = 1; j < J-1; j++)
      E(j) = (phi(j-1) - phi(j+1)) / 2. / dx;
   E(0) = (phi(J-1) - phi(1)) / 2. / dx;
   E(J-1) = (phi(J-2) - phi(0)) / 2. / dx;
}
```

The following routine is the right-hand side routine for the electron equations of motion. Is is designed to be used with the fixed-step RK4 solver described earlier in this course.

```
// Electron equations of motion:
//      y(0:N-1)    = r_i
//      y(N:2N-1)   = dr_i/dt
void rhs_eval (double t, Array<double,1> y, Array<double,1>& dydt)
{
   // Declare local arrays
   Array<double,1> r(N), v(N), rdot(N), vdot(N), r0(N);
   Array<double,1> ne(J), rho(J), phi(J), E(J);
   // Unload data from y
   UnLoad (y, r, v);
   // Make sure all coordinates in range 0 to L
   r0 = r;
   for (int i = 0; i < N; i++)
      {
         if (r0(i) < 0.) r0(i) += L;
         if (r0(i) > L) r0(i) -= L;
      }
   // Calculate electron number density
   Density (r0, ne);
   // Solve Poisson's equation
   double n0 = double (N) / L;
   for (int j = 0; j < J; j++)
      rho(j) = ne(j) / n0 - 1.;
   double kappa = 2. * M_PI / L;
   Poisson1D (phi, rho, kappa);
   // Calculate electric field
   Electric (phi, E);
   // Equations of motion
   for (int i = 0; i < N; i++)
      {
         double dx = L / double (J);
```

```
            int j = int (r0(i) / dx);
            double y = r0(i) / dx - double (j);
            double Efield;
            if (j+1 == J)
                Efield = E(j) * (1. - y) + E(0) * y;
            else
                Efield = E(j) * (1. - y) + E(j+1) * y;
            rdot(i) = v(i);
            vdot(i) = - Efield;
        }

    // Load data into dydt
    Load (rdot, vdot, dydt);
}
```

The following functions load and unload the electron phase-space coordinates into the solution vector y used by the RK4 routine.

```
// Load particle coordinates into solution vector
void Load (Array<double,1> r, Array<double,1> v, Array<double,1>& y)
{
    for (int i = 0; i < N; i++)
        {
            y(i) = r(i);
            y(N+i) = v(i);
        }
}
// Unload particle coordinates from solution vector
void UnLoad (Array<double,1> y, Array<double,1>& r, Array<double,1>& v)
{
    for (int i = 0; i < N; i++)
        {
            r(i) = y(i);
            v(i) = y(N+i);
        }
}
```

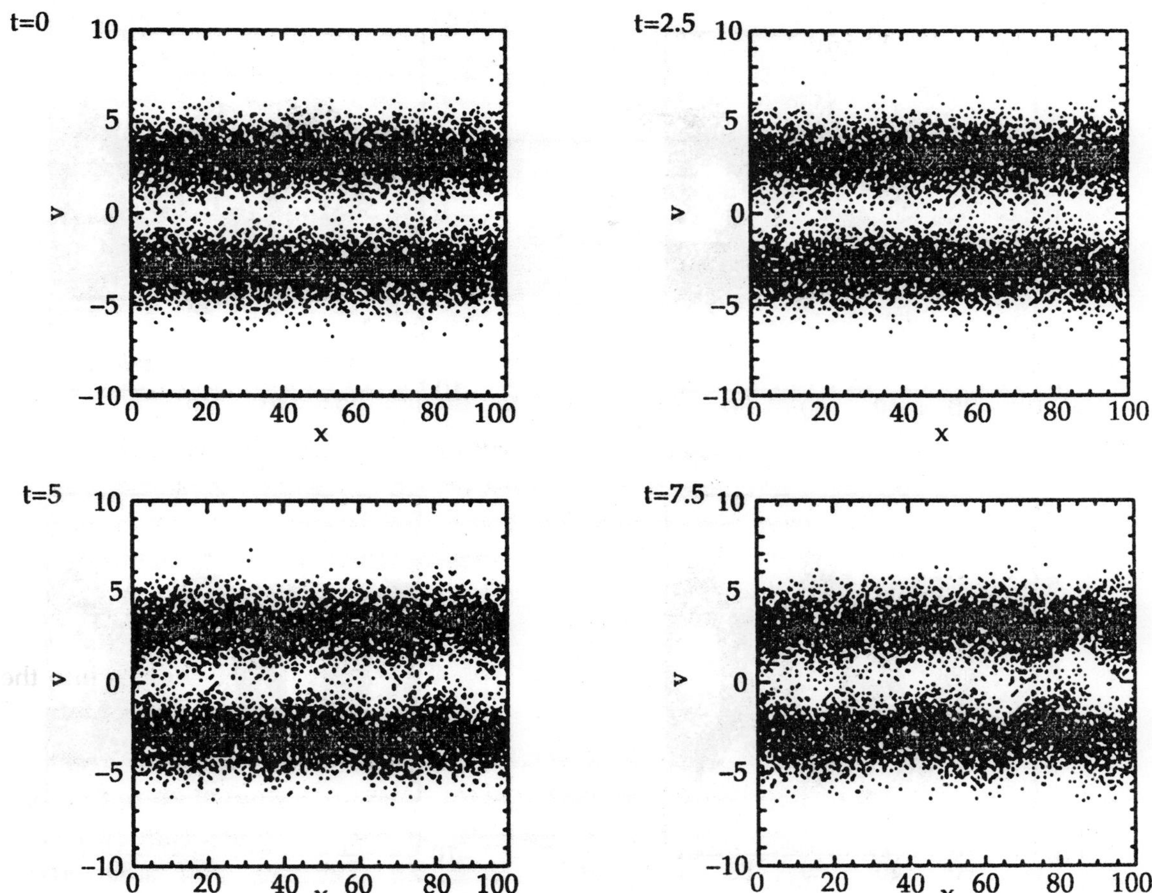

Figure 11.1: *The electron phase-space distribution evaluated at various times for a 1-dimensional two-stream instability calculation performed with* $N = 200$, $J = 1000$, $L = 100$, $v_b = 3$, *and* $\delta t = 0.1$

Figures 11.1 and 11.2 show the electron phase-space distributions evaluated at equally spaced times for a two-stream instability calculation performed with 2×10^4 electrons. It can be seen that the distribution initially takes the form of two uniform bands, corresponding to two counter-streaming electron beams.

However, as time progresses, the bands spontaneously develop structure which grows in magnitude and eventually converts the phase-space distribution into a set of connected vortices. In this final state, the electrons are basically bouncing backwards and forwards in a quasi-periodic electric potential generated by non-uniformities in the electron density.

In other words, the instability effectively destroys the two beams. For this reason, the two-stream instability is of major concern in particle accelerators, which often consist of counter-propagating charged particle beams.

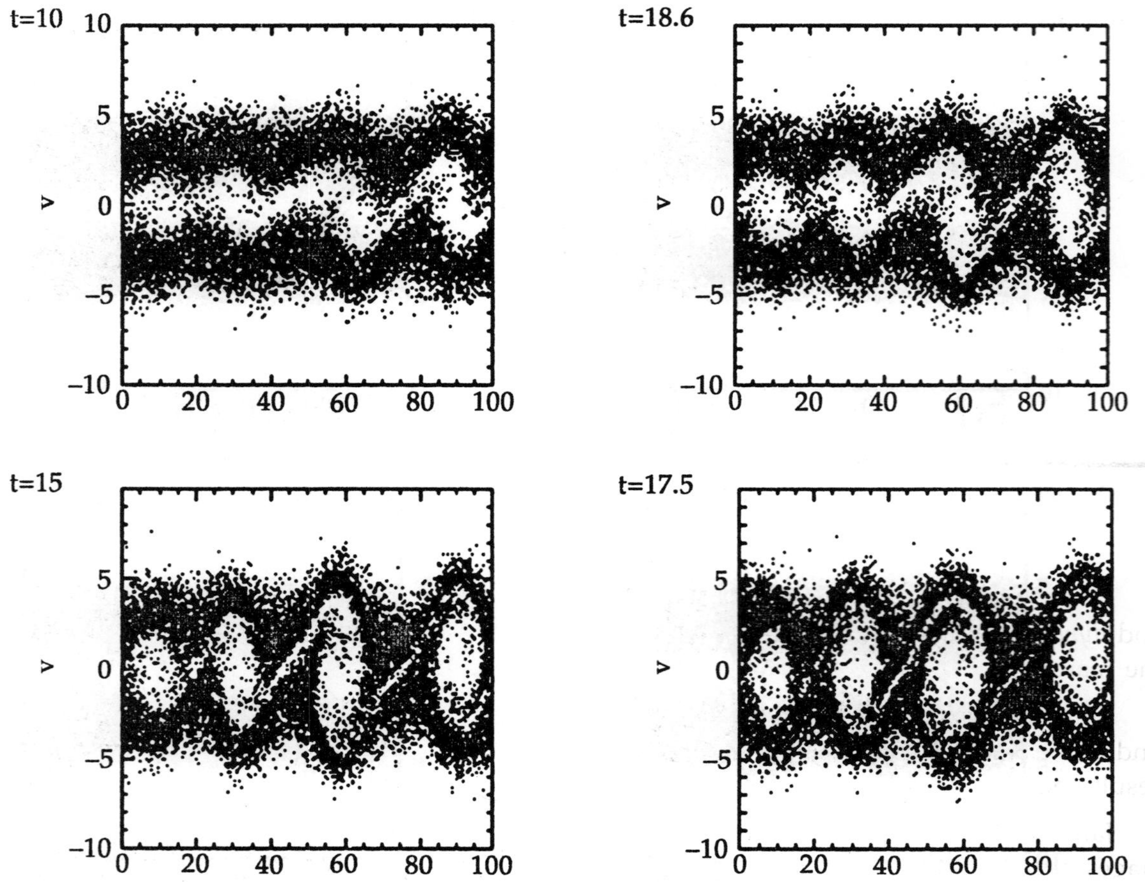

Figure 11.2: The electron phase-space distribution evaluated at various times for a 1-dimensional two-stream instability calculation performed with $N = 20000$, $J = 1000$, $L = 100$, $v_b = 3$, *and* $\delta t = 0.1$

Obviously, the ideas discussed above could be generalised in a fairly straight-forward manner to deal with the evolution of two and three-dimensional charged particle distributions. PIC codes have the advantage that they are reasonably straight-forward to write. Unfortunately, PIC codes also have a number of disadvantages. The first is that PIC codes suffer from high levels of statistical noise, since they generally only deal with a relatively small number of particles (typically, $\leq 10^6$). Real physical systems do not exhibit anything like the same level of statistical noise, since they generally contain of order Avogadro's number ($\sim 10^{24}$) of interacting particles. Another problem with PIC codes is that they do not handle charged particle collisions very well. The reason for this is that there are generally a large number of particles in each cell (for practical reasons), and the short range Coulomb fields of these particles tend to cancel one another out (recall that the electric field is only calculated at the cell vertices).

12

Short-time Critical Dynamics

A spin model has both *static* properties (e.g., the critical exponent of the magnetisation) and *dynamic* properties. The latter characterise the behaviour of the system over time, e.g., the manner in which it decays from an initial state of magnetisation 1.

This chapter obtains simulation results concerning short-time dynamics in the Ising model and in the 3-state Potts model, compares them with results in the literature, and gives a new result concerning the 4-state Potts model.

Janssen *et al.* (1989), using a renormalisation group approach, discovered that in the early stages of the relaxation process in an Ising or a q-state Potts model at critical temperature with small initial magnetisation one may expect to find (via Monte Carlo studies) an increase in the magnetisation before it crosses over to the long-term relaxation behaviour. This initial increase is known as "the critical initial slip" and is characterised by a new universal dynamic critical exponent θ.

Based on the work of Janssen *et al.* various authors have noted the following scaling relation for Ising and Potts spin models:

$$M^{(k)}\ (t,\ \tau,\ L,\ m_0) = b^{-k\beta/\nu}\ .\ M^{(k)}\ (b^{-z}t,\ b^{1/\nu}\tau,\ b^{-1}L,\ b^{x}{}_0{\cdot}m_0)$$

where $M^{(k)}$ is the k-th moment of the magnetisation, t is time, τ is the reduced temperature, $(T-T_c)/T_c$, L is the lattice size, m_0 is the initial magnetisation, z is the dynamic critical exponent, ν is the critical exponent of the correlation length, x_0 is a critical exponent called "the scaling dimension of the initial magnetisation" and b is a scaling factor.

As shown by Okano *et al.* 1997 this has the consequence for the time evolution of the magnetisation M(t) in the initial stage of the relaxation process that at the critical temperature $M(t) \sim m_0 t^{\theta}$ where $q = (x_0 - \beta/\nu)/z$ (with β denoting the critical exponent of the magnetisation) is a new universal dynamic critical exponent.

A second consequence is that for zero initial magnetisation the 2nd moment of the magnetisation $M^{(2)}(t) \sim t^{(d-2\beta/\nu)/z}$ where d is the dimension, β and ν are static critical exponents and z is the dynamic critical exponent.

Thirdly, for zero initial magnetisation, the autocorrelation $A(t) \sim t^{-d/z+\theta}$ where d, z and θ are as above.

Thus, for a given d, it is possible, by means of a spin model study, to obtain the following (by measuring the slope of log-log plots of the quantity under consideration against time):

(a) An estimate of θ by monitoring the magnetisation.

(b) An estimate of z by monitoring the autocorrelation.

(c) An estimate of β/ν by monitoring the 2nd moment of the magnetisation.

We reproduce some measurements for the Ising and the 3-state Potts model given by Okano *et al.* 1997. Having established that the simulation software gives correct results in these cases we then go on to obtain a new result for θ for the 4-state Potts model and an estimate for z.

Ising Model

Magnetisation

The magnetisation (per spin) in the Ising model is defined as $\Sigma_i(S_i/N)$ where N is the number of spins and S_i = +1 or -1 for each spin S_i.

Okano *et al.* performed simulations using a square lattice of size 128x128 and initial magnetisations, M_0, in the range 0.08 to 0.02 (using both Metropolis and Glauber algorithms). In each case magnetisation from timepoints 0 - 120 was measured, and the magnetisation against the time was plotted using a log-log plot. Okano *et al.* were able to obtain a least-squares fit to the data, and the slope of the line gives the value of the short-time dynamic critical exponent θ (for the particular initial magnetisation used). By means of reciprocal-size projection Okano *et al.* then obtained an estimate for M_0 of zero, namely 0.191(1), which is the final estimate for θ for the Ising model (using Glauber dynamics) by Okano *et al.*

For M_0 = 0.02, Glauber dyamics and 300,000 samples Okano *et al.* obtained a value of 0.187(1) for θ. This experiment was repeated using the simulation programme, with 15,000 samples, and the result (with error bars) is shown in Figure 12.1.

The error bars in the logarithm values of the magnetisation in Figure 12.1 were obtained from the values of the standard deviation of the magnetisation as follows, where M = mean magnetisation (at a given timepoint) and sd = standard deviation of the measurements of the magnetisation (at that timepoint):

$$\text{Positive error bar: } \ln(1+sd/M)$$

$$\text{Negative error bar: } -\ln(1-sd/M)$$

This follows from the algebraic relations:

$$\ln(M+sd) = \ln(M) + \ln(1+sd/M)$$

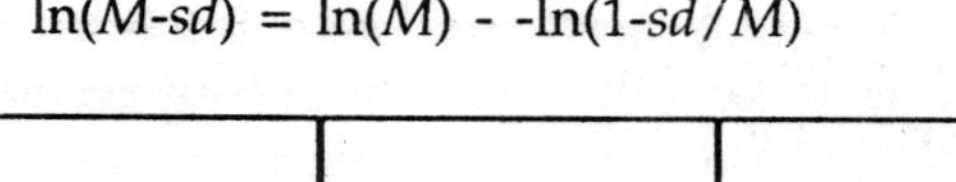

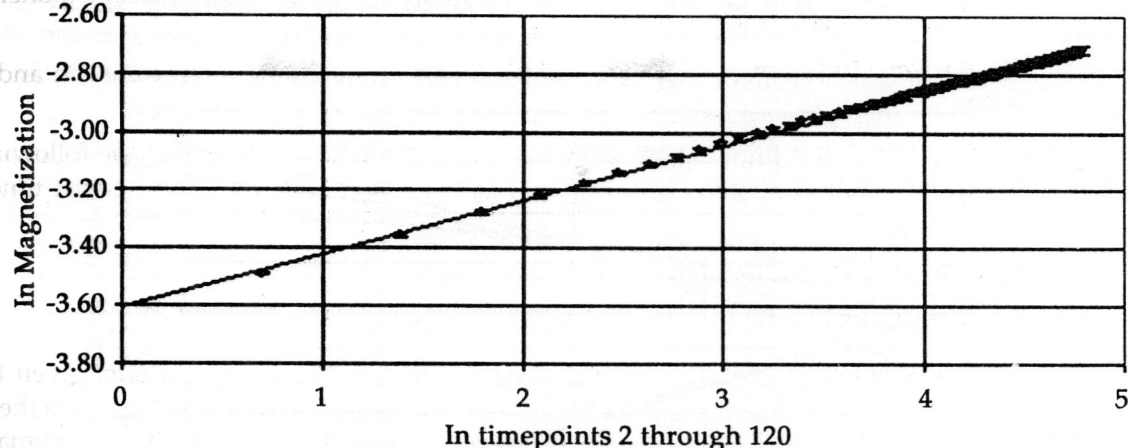

Figure 12.1: *Ising model, 2d square lattice, 128×128, critical temperature, Glauber dynamics, initial magnetisation=0.02, 15,000 samples*

Running the FIT programme on the 60 magnetisation values, with error bars, yields:

```
60 data lines read.
Standard deviations present.
Data fitted to y(x) = a + bx.

        a = -3.6129,  siga = 0.0042
        b = 0.1888,  sigb = 0.0012
chi^2 = 10.6210,        Q = 1.0000
```

We may conclude that θ for this case is 0.189(1). This is between the value for initial magnetisation 0.02 of 0.187(1) obtained by Okano *et al.* and their final estimate of 0.191(1).

Autocorrelation

Autocorrelation is a measure of the correlation of the spins of a lattice with their initial state, or more exactly (for the Ising model):

$$A_{\text{Ising}}(t) = 2.\{\Sigma_i[\delta(S_{i,0},S_{i,t})]/N\}-1$$

where N is the number of spins and for the ith spin $S_{i,0}$ is its initial value, $S_{i,t}$ is its value at timepoint t and δ() is the Kronecker delta function: $\delta(x,y) = 1$ if and only if $x = y$ (otherwise 0). At the initial state the correlation is 1. As spins are flipped the correlation decreases, becoming zero when the spins are totally uncorrelated with their initial state.

Okano *et al.* performed a simulation using a square lattice of size 256x256 and zero initial magnetisation (using both the Metropolis and the Glauber algorithms and 35,000 samples). The autocorrelation from timepoints 30 - 100 was measured, and plotted against time using a log-log plot. Okano *et al.* were able to obtain a least-squares fit to the data, and the slope of the line gives the value of $-d/z + \theta$, where d is the dimension (in this case 2) and z is the dynamic critical exponent. With Glauber dynamics Okano *et al.* obtained a value of -0.737(1) for $-d/z + \theta$.

This experiment was repeated using the simulation programme, with 2,800 samples, and the result is shown in Figure 12.2 (with error bars).

The error bars in the logarithm values of the autocorrelation in Figure 12.2 were obtained from the standard deviation of the autocorrelation as described above.

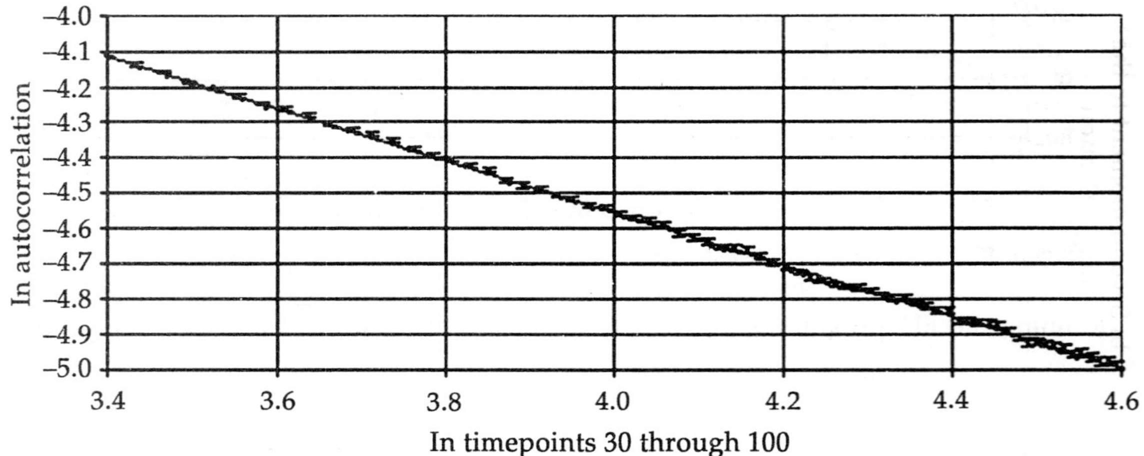

Figure 12.2: *Ising model, square lattice, 256×256, critical temperature, Glauber dynamics, zero initial magnetisation, 2800 samples*

Running the FIT programme on the autocorrelation log values, with error bars, yields:

```
71 data lines read.
Standard deviations present.
Data fitted to y(x) = a + bx.
     a = -1.5956,  siga = 0.0091
     b = -0.7391,  sigb = 0.0023
chi^2 = 39.4005,        Q = 0.9984
```

From this we may conclude that $-d/z + \theta$ is -0.739(2), which is almost the same as the -0.737(1) obtained by Okano *et al.*

Second Moment of the Magnetisation

In the Ising model the 2nd moment of the magnetisation, $M^{(2)}$, is simply the mean square magnetisation (or rather, for a given timepoint t, the mean of the squares of the magnetisations measured at t averaged over many samples), i.e.,

$$M^{(2)} = [(\Sigma_i S_i)/N]^2$$

In the simulation performed by Okano *et al.* to measure the autocorrelation the 2nd moment of the magnetisation was also measured, over timepoints 60 - 150. After plotting $M^{(2)}$ against time using a log-log plot Okano *et al.* obtained a least-squares fit to the data, and the slope of the line gave the value of $(d\text{-}2\beta/\nu)/z$, where β is the critical exponent of the magnetisation and ν is the critical exponent of the correlation length. With Glauber dynamics Okano *et al.* obtained a value of 0.817(7) for $(d\text{-}2\beta/\nu)/z$.

In the same repetition, using the simulation software, of the experiment in which the autocorrelation was measured the 2nd moment of the magnetisation was also measured, and the result is shown in Figure 12.3.

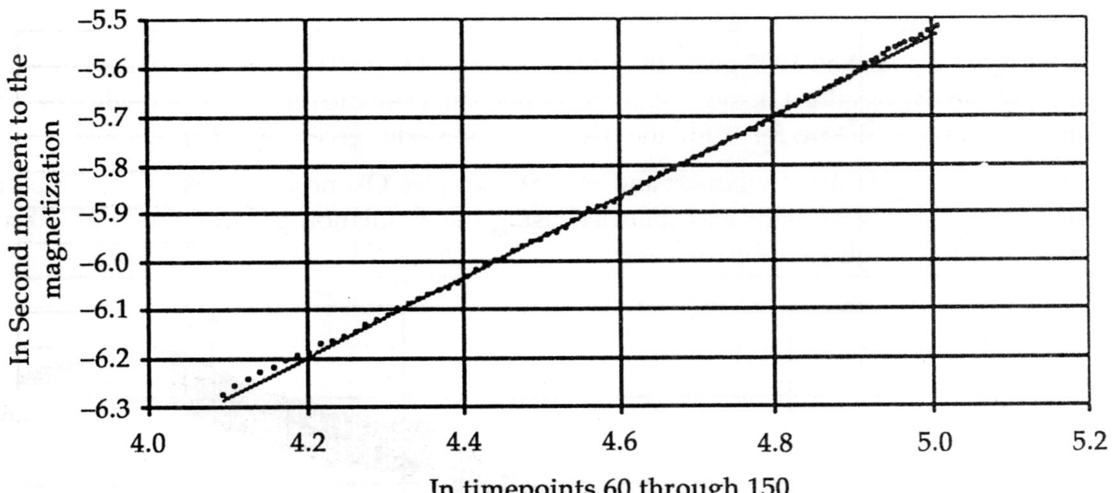

Figure 12.3: *Ising model, square lattice, 256×256, critical temperature, Glauber dynamics, zero initial magnetisation, 2,800 samples*

Running the FIT programme on the ln $M^{(2)}$ values yields:

```
91 data lines read.
Standard deviations absent.
Data fitted to y(x)  = a + bx.
      a = -9.6546,  siga = 0.0112
      b = +0.8234,  sigb = 0.0024
chi^2 = 0.0032
```

From this we may conclude that $(d\text{-}2\beta/\nu)/z$ is 0.823(3), which is consistent with the value of 0.817(7) obtained by Okano *et al.*

3-state Potts Model

Magnetisation

For the q-state Potts model magnetisation (per spin) is defined as follows:

$$M = \{q/[(q-1).N]\}.\Sigma_i[\delta(S_i,1)-1/q]$$

where $\delta(S_i,1)$, the Kronecker delta, is 1 if $S_i = 1$ and 0 otherwise. Thus if all spins have spin value +1 then M = 1, if all have a spin value other than +1 then $M = -1/(q-1)$, and if the spin values 1, ..., q are distributed randomly among the N spins (so that N/q have spin value +1) then $M = 0$.

Okano *et al.* performed simulations with the 3-state Potts model similar to those described. To measure the value of the short-time dynamic critical exponent θ a square lattice of linear size 72 and initial magnetisations, M_0, in the range 0.08 to 0.02 (using both the Metropolis and the Glauber algorithms) were used.

Table 3 of Okano *et al.* 1997 gives the results for M_0 of 0.08, 0.06, 0.04 and 0.02. By means of reciprocal-size projection Okano *et al.* then obtained an estimate for $M_0 = 0$, namely 0.075(3), which is the final estimate for θ for the 3-state Potts model given by Okano *et al.*

For $M_0 = 0.02$, Glauber dynamics and 600,000 samples Okano *et al.* obtained a value of 0.084(3) for θ. This experiment was repeated using the simulation programme, with 50,000 samples, and the result (with error bars).

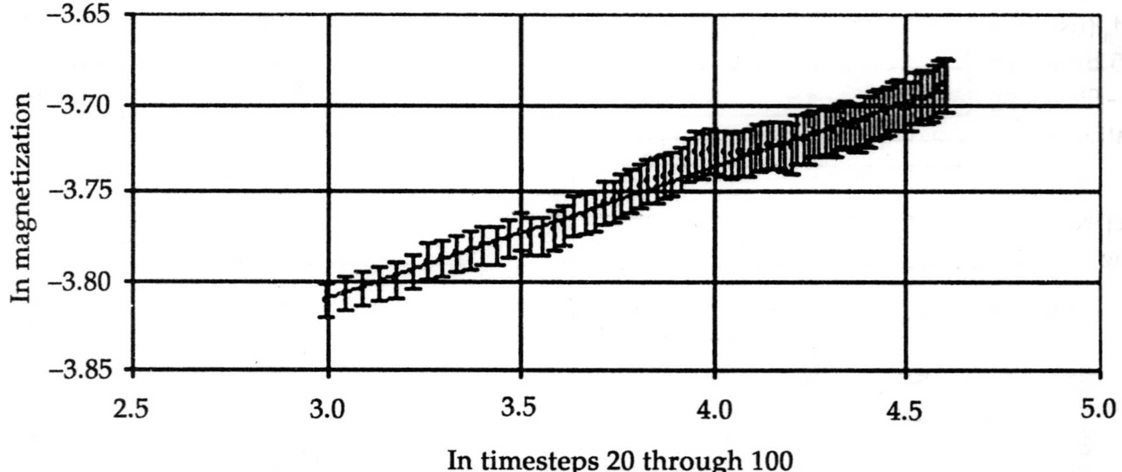

***Figure 12.4:** 3-state Potts mode, square lattice 72×72, critical temperature, Glauber dynamics, initial magnetisation=0.02, 50,000 samples*

Running the FIT programme on the magnetisation log values (for timepoints 20 through 100, the same as used by Okano *et al.*) with error bars yields:

```
81 data lines read.

Standard deviations present.

Data fitted to y(x) = a + bx.

      a = -4.0331, siga = 0.0114

      b = +0.0742, sigb = 0.0029

chi^2 = 5.8651,       Q = 1.0000
```

From this we may conclude that θ for this case is 0.074(3). This is lower than the value 0.084(3) for initial magnetisation of 0.02 obtained by Okano *et al.* but is close to their final estimate for the square 3-state Potts q of 0.075(3).

Autocorrelation

The definition of autocorrelation for the q-state Potts model is:

$$A_{\text{Potts}}(t)=\left\{\Sigma_i\left[\delta\left(S_{i,0},S_{i,t}\right)-1/q\right]-1/q\right\}/N$$

that is,

$$A_{\text{Potts}}(t)=\left(\left\{\Sigma_i\left[\delta\left(S_{i,0},S_{i,t}\right)\right]\right\}/N\right)-1/q$$

At the initial state the correlation is $(q\text{-}1)/q$. As spins are flipped over time the correlation decreases, becoming zero when the spins are totally uncorrelated with their initial state.

Okano *et al.* performed simulations with square lattices of size 144, 288 and 576 and zero M_0 (using both the Metropolis and the Glauber algorithms). The autocorrelation over the first 85 timepoints was measured. Okano *et al.* plotted autocorrelation against time for timepoints 5 - 50 using a log-log plot and averaged the values of the slopes of the plots for the three lattice sizes to obtain a value for $-d/z+\theta$ of -0.836(2).

For lattice size 144 Okano *et al.* obtained -0.839(1) for $-d/z+\theta$. This experiment was repeated using the simulation programme, with 6000 samples, and the result is shown in Figure 12.5 (with error bars).

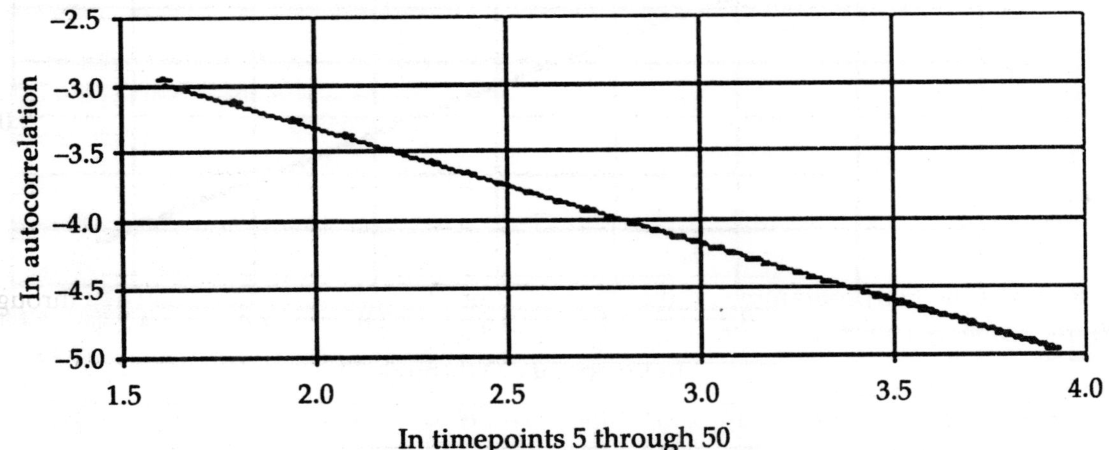

Figure 12.5: *3-state Potts model, square lattice, 144×144, critical temperature, Glauber dynamics, zero initial magnetisation, 6000 samples*

Running the FIT programme on the autocorrelation log values, with error bars, yields:

```
46 data lines read.

Standard deviations present.
```

```
Data fitted to y(x) = a + bx.

     a = -1.5643,  siga = 0.0012
     b = -0.8714,  sigb = 0.0005
chi^2 = 914.9231,       Q = 0.0000
```

From this we might conclude that $-d/z + \theta$ is -0.871(1). However, the Q value is zero, so we must question this result. The problem appears to come from choosing a range of timepoints (namely, 5 - 50) which begins too early. Okano *et al.* write:

> Even though universal behaviour is expected in the *macroscopic early time*, in general, the dynamic system needs a certain microscopic time period to get rid of the microscopic short-wave effects and to enter the macroscopic quasi-stable state. Such a time period is called the microscopic time scale t_{mic}. It is only for $t > t_{mic}$ that the exponents extracted from the short-time dynamics will be universal, i.e. will be independent of the algorithms and other microscopic details.

If we do a log-log plot of autocorrelation against time for timepoints 20 - 70, rather than 5 - 50, we obtain Figure 12.6 (with error bars):

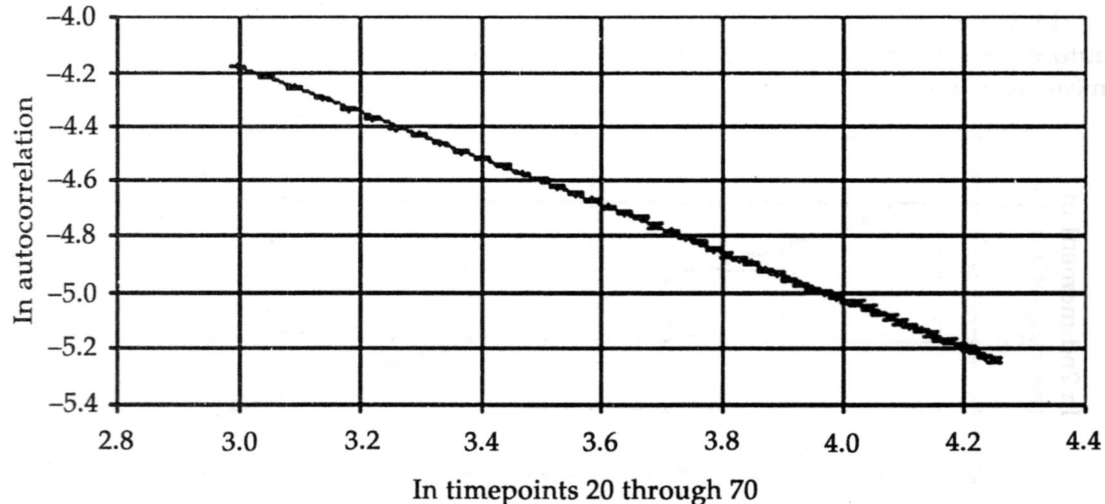

Figure 12.6: *3-state Potts model, square lattice, 144×144, critical temperature, Glauber dynamics, zero initial magnetisation, 6000 samples*

Running the FIT programme on the autocorrelation log values, with error bars, yields:

```
51 data lines read.
Standard deviations present.
Data fitted to y(x) = a + bx.
```

```
      a  =  -1.6582,  siga  =  0.0061
      b  =  -0.8417,  sigb  =  0.0017
chi^2  =  30.9428,       Q  =  0.9795
```

Now $Q > 0.1$ and we may conclude that $-d/z + \theta$ is -0.842(2). This is consistent with the result for the lattice of size 144 of -0.839(1) obtained by Okano *et al.*

Second Moment of the Magnetisation

Following the definition given by Okano *et al.* 1997 of the 2nd moment of the magnetisation, $M^{(2)}$, for the 3-state Potts model we define $M^{(2)}$ for the q-state Potts model as:

$$M^{(2)} = \{q/[(q-1).N]\}^2.\left[\left(\Sigma_r\left\{\Sigma_i\left[\delta(S_i,r)-1/q\right]\right\}^2\right)/q\right]$$

where for $1 \le r \le q$, $\delta(S_i,r)$ is 1 if $S_i = r$ and 0 otherwise.

In the simulation performed by Okano *et al.* to measure the autocorrelation, $M^{(2)}$ was also measured, over timepoints 35 - 100. After plotting $M^{(2)}$ against time using a log-log plot Okano *et al.* obtained a least-squares fit to the data, and the slope of the line gave the value of $(d\text{-}2\beta/\nu)/z$, where β and ν are as before. Okano *et al.* averaged over lattice sizes 144 and 288, using Glauber dynamics, to obtain a value of 0.788(1) for $(d\text{-}2\beta/\nu)/z$. For lattice size 144 Okano *et al.* obtained a value of 0.789(2).

In the repetition, with the simulation software, of the experiment in which the autocorrelation was measured (using 6000 samples and a lattice of size 144) $M^{(2)}$ was also measured; the result is shown in Figure 12.7.

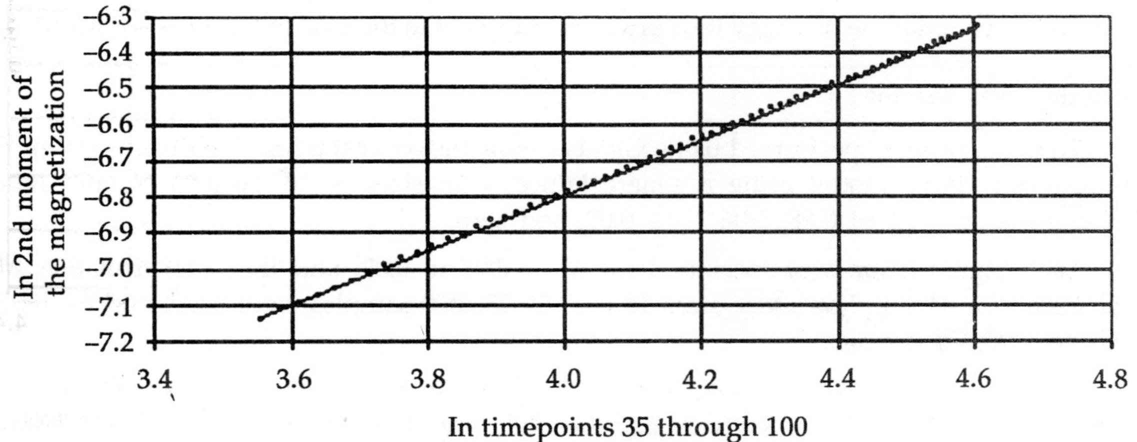

Figure 12.7: *3-state Potts model, square lattice, 144×144, critical temperature, Glauber dynamics, zero initial magnetisation, 6000 samples*

Running the FIT programme on the ln $M^{(2)}$ values yields:

```
66 data lines read.
Standard deviations absent.
```

```
Data  fitted  to  y(x)  =  a  +  bx.
       a  =  -9.8497,  siga  =  0.0046
       b  =  +0.7644,  sigb  =  0.0011
chi^2  =  0.0005
```

from which we may conclude that $(d\text{-}2\beta/\nu)/z$ is 0.764(1), which is smaller than the value of 0.789(2) obtained by Okano *et al.*

The value one obtains for $(d\text{-}2\beta/\nu)/z$ depends on the range of timepoints. By fitting a line to the ln $M^{(2)}$ values over timepoints 20 - 60 one obtains:

```
41  data  lines  read.
Standard  deviations  absent.
Data  fitted  to  y(x)  =  a  +  bx.
       a  =  -9.9446,  siga  =  0.0061
       b  =  +0.7890,  sigb  =  0.0017
chi^2  =  0.0004
```

from which one could also conclude that $(d\text{-}2\beta/\nu)/z$ is 0.789(2), which is the same as the value obtained by Okano *et al.*

The results from the simulation programme which have been described for the Ising model and the 3-state Potts model on a square lattice, when compared to the results obtained by Okano *et al.*, show that we are justified in having some degree of confidence in the simulation programme. We may thus proceed to apply it to the case of the 4-state Potts model.

4-state Potts Model

Simulations were performed using Glauber dynamics over the first 100 timepoints for the pure 4-state Potts model using a square lattice, with sizes of 16, 32 and 64, and initial magnetisations, M_0, of 0.08, 0.06, 0.04, 0.02, and zero.

The magnetisation was measured for M_0 = 0.08 to 0.02, and the autocorrelation was measured for M_0 = 0. For lattice sizes 16 and 32, 250,000 samples were used, and for lattice size 64, 100,000 samples.

The average time for simulations (on a 330 MHz Intel PC) was 6, 22 and 33 hours for lattice sizes 16, 32 and 64 respectively, and the total computing time was 305 hours (12.7 days).

Figures 12.8 and 12.9 are examples of data from simulations with a particular size and initial magnetisation, and show log-log plots against time (timepoints 30 through 100) for a square lattice of:

- the magnetisation (with lattice size 64, initial magnetisation 0.02 and 100,000 samples)
- the autocorrelation (with lattice size 32, zero initial magnetisation and 250,000 samples).

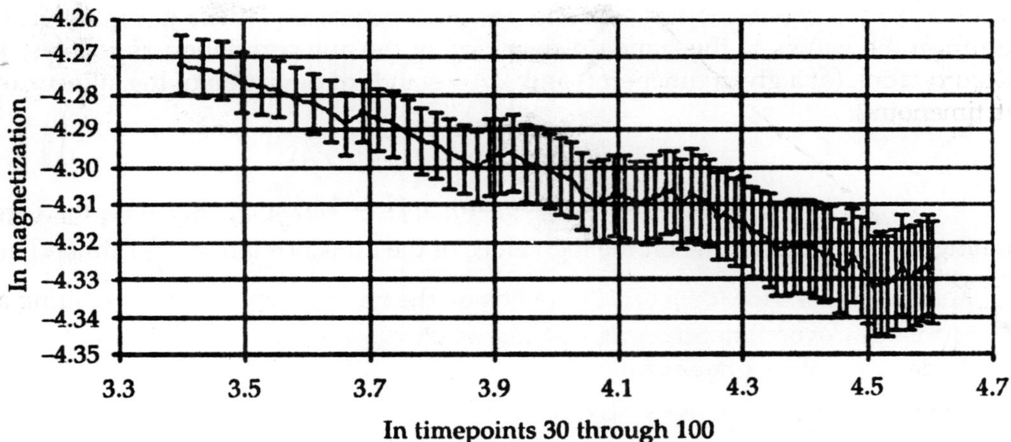

Figure 12.8: *4-state Potts, 64×64, initial magnetisation=0.02, 100,000 samples*

The error bars in the logarithm values of the magnetisation in Figure 12.8.

Running the FIT programme on the log values of the magnetisation over timepoints 30 - 100 gives:

```
71 data lines read.
Standard deviations present.
Data fitted to y(x) = a + bx.
     a = -4.1058, siga = 0.0157
     b = -0.0490, sigb = 0.0039
chi^2 = 2.5558,       Q = 1.0000
```

Thus the slope of the least-squares fit in this case (a first approximation to the 4-state Potts θ) is -0.0490(39), or rather -0.049(4).

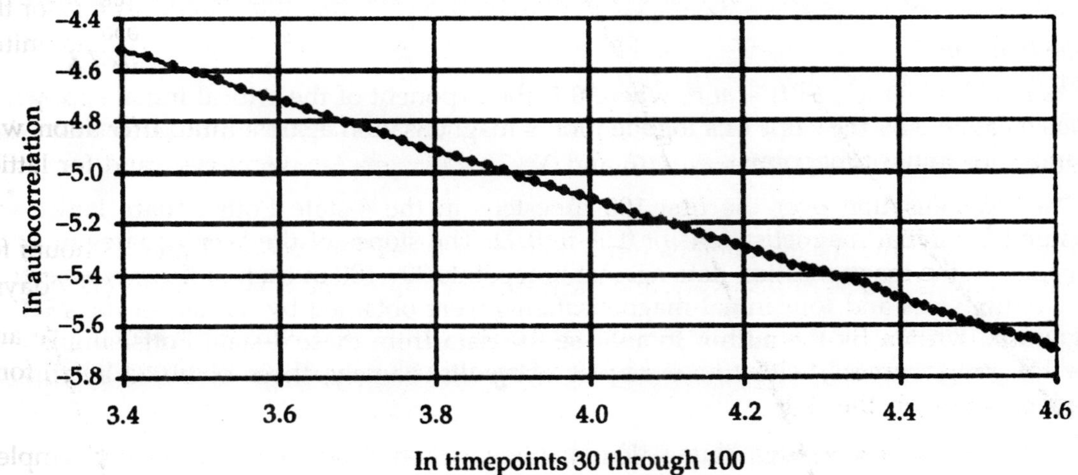

Figure 12.9: *4-state Potts, 32×32, zero initial magnetisation 250,000 samples*

The error bars in the logarithm values of the autocorrelation values in Figure 12.9 were obtained from the values of the standard deviation of the autocorrelation as follows, where A = autocorrelation (at a given timepoint) and sdA = standard deviation of the autocorrelation (at that timepoint):

$$\text{Positive error bar: } \ln(1+sdA/A)$$

$$\text{Negative error bar: } -\ln(1-sdA/A)$$

Running the FIT programme on the log values of the autocorrelation over timepoints 30 - 100 gives:

```
71 data lines read.
Standard deviations present.
Data fitted to y(x) = a + bx.
      a = -1.2223, siga = 0.0067
      b = -0.9705, sigb = 0.0017
chi^2 = 38.2004,        Q = 0.9990
```

Thus the slope of the least-squares fit in this case (a first approximation to $-d/z + \theta$) is -0.9705(17).

The strategy here adopted is as follows:

- An estimate for the 4-state Potts θ will be arrived at by measuring, for each of the three lattice sizes 16, 32 and 64, the slope of the log-log plot of magnetisation values over timepoints 30 - 100 for initial magnetisations, M_0, of 0.08, 0.06, 0.04 and 0.02, and projecting the slopes to $M_0 = 0$, then projecting these values to 1/size = 0.
- An estimate for $-d/z + \theta$ will be arrived at by measuring, for each lattice size, with $M_0 = 0$, the slope of the log-log plot of autocorrelation values over timepoints 30 - 100 and projecting to 1/size = 0.
- From (i) and (ii) we hope to obtain the 4-state Potts dynamic critical exponent, z.

Magnetisation

As noted in earlier, $M(t) \sim m_0 t^\theta$, where θ is the exponent of the critical initial slip, which is thus measured by the slope of a log-log plot of magnetisation against time (after discarding a number of initial timepoints).

The magnetisation over the first 100 timesteps in the 4-state Potts square lattice was measured for initial magnetisations of 0.08 to 0.02. The slopes of the least-squares fit to the log values of the magnetisation data against timepoints 30 - 100 in each of these twelve cases (three lattice sizes and four initial magnetisations) were obtained by the use of the 4STPFIT programme written by the author to analyse the data from these 4-state Potts simulations. "The FIT Programmes", with two examples of results, namely, those obtained by running the programme on the data :

- for lattice size 64 and $M_0 = 0.02$
- for lattice size 32 and $M_0 = 0$.

The slopes of the log-log plots of magnetisation against time (for timepoints 30 through 100) are given in Table 12.10 and are shown in Figure 12.10.

***Table 12.1:** 4-state Potts, short-time dynamics*
Slope of magnetisation (timepoints 30-100)

Initial magnetisation	0.08	0.06	0.04	0.02
Slope				
16	-0.0093	-0.0155	-0.0438	-0.0530
32	-0.0040	-0.0132	-0.0334	-0.0468
64	-0.0052	-0.0141	-0.0270	-0.0490
Errors				
16	0.0024	0.0033	0.0050	0.0107
32	0.0012	0.0016	0.0024	0.0050
64	0.0009	0.0012	0.0019	0.0039
Q				
16	1.0000	1.0000	1.0000	1.0000
32	1.0000	1.0000	1.0000	1.0000
64	1.0000	1.0000	1.0000	1.0000

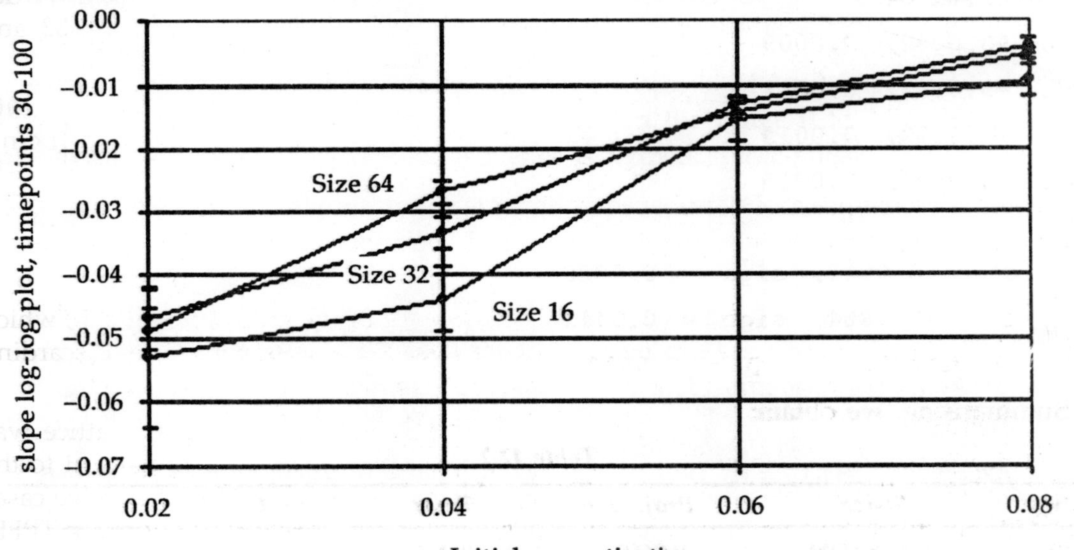

***Figure 12.10:** 4-state Potts magnetisation*

The FIT programme was then used on the data for each of the lattice sizes 16, 32 and 64 to obtain a least-squares fit and so to project the values to $M_0 = 0$, to obtain the following (the input data is displayed if the /data command-line parameter is used):

Lattice size 16:

```
0.08,  -0.0093,  0.0024
0.06,  -0.0155,  0.0033
0.04,  -0.0438,  0.0050
0.02,  -0.0530,  0.0107
        a  =  -0.0660,  siga  =  0.0076
        b  =  +0.7308,  sigb  =  0.1103
chi^2  =  6.6040,        Q  =  0.0368
```

Lattice size 32:

```
0.08,  -0.0040,  0.0012
0.06,  -0.0132,  0.0016
0.04,  -0.0334,  0.0024
0.02,  -0.0468,  0.0050
        a  =  -0.0579,  siga  =  0.0037
        b  =  +0.6866,  sigb  =  0.0533
chi^2  =  7.3314,        Q  =  0.0256
```

Lattice size 64:

```
0.08,  -0.0052,  0.0009
0.06,  -0.0141,  0.0012
0.04,  -0.0270,  0.0019
0.02,  -0.0490,  0.0039

        a  =  -0.0513,  siga  =  0.0028
        b  =  +0.5864,  sigb  =  0.0412
chi^2  =  9.6777,        Q  =  0.0079
```

Summarising, we obtain:

Table 12.2

Size	*1/size*	*Proj. zero*	*Error*	*Q*
16	0.062500	-0.0660	0.0076	0.0368
32	0.031250	-0.0579	0.0037	0.0256
64	0.015625	-0.0513	0.0028	0.0079

We can now plot the projected slopes for $M_0 = 0$ against the reciprocal of the lattice size to obtain the slope for M_0 in the ideal case of the infinite lattice:

```
0.0625,      -0.0660,  0.0076
0.03125,    -0.0579,  0.0037
0.015625,  -0.0513,  0.0028
        a  =  -0.0464,  siga  =  0.0045
        b  =  -0.3352,  sigb  =  0.1614
chi^2  =  0.1225,        Q  =  0.7263
```

Thus we arrive at an estimate for the 4-state Potts q of -0.046(5). Okano *et al.* 1997 remark that "not all the models will have a positive θ. For example, for the Potts model with $q = 4$ the exponent θ is likely negative or very close to zero." This result confirms that expectation.

Autocorrelation

As noted earlier, $A(t) \sim t^{-d/z+\theta}$, where θ is as above, d is the dimensionality (in this case 2) and z is the dynamic critical exponent. Thus by measuring the slope of a log-log plot of autocorrelation against time (after discarding a number of initial timepoints) we can obtain z if we know θ.

The autocorrelation was measured for $M_0 = 0$. The slopes of the least-squares fits to the log values of the autocorrelation data against timepoints 30 - 100 for each of the lattice sizes were obtained by the use of the 4STPFIT programme and are given in Table 12.3.

Table 12.3: Autocorrelation (zero initial magnetisation)

Size	*1/size*	*Slope A*	*Error*	*Q*
16	0.062500	-1.0337	0.0032	0.9931
32	0.031250	-0.9705	0.0017	0.9991
64	0.015625	-0.9706	0.0011	0.7991

We can now plot the autocorrelation slopes for $M_0 = 0$ against the reciprocal of the lattice size to obtain the slope in the ideal case of the infinite lattice:

```
0.0625,      -1.0337,  0.0032
0.03125,    -0.9705,  0.0017
0.015625,  -0.9706,  0.0011
        a  =  -0.9499,  siga  =  0.0018
        b  =  -1.0887,  sigb  =  0.0673
chi^2  =  97.8233,        Q  =  0.0000
```

Unfortunately the Q-value is zero, a sign that something is not right.

Figure 12.11 shows the plot for the slopes of A for each of the sizes 16, 32 and 64.

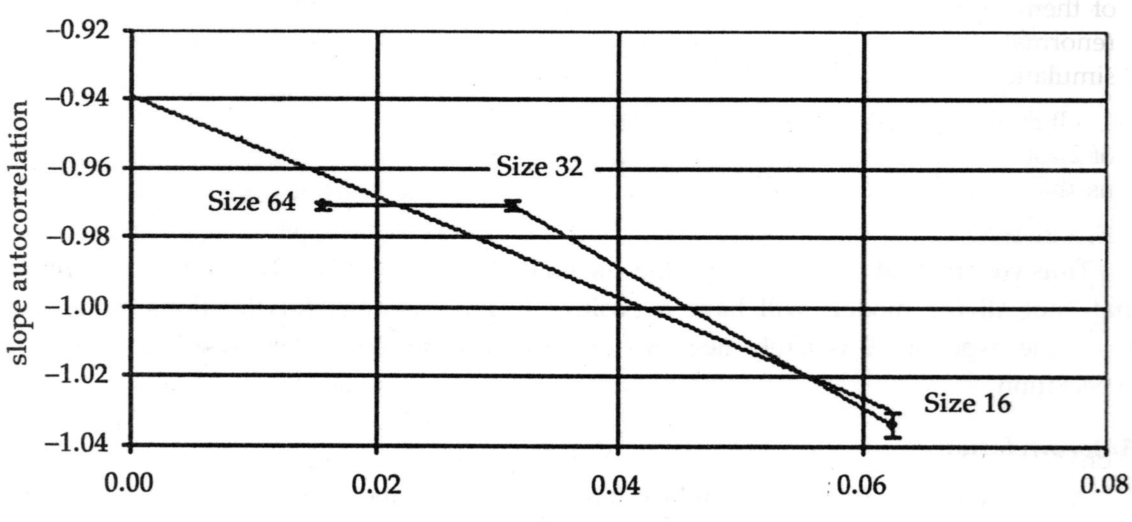

Figure 12.11: *4-state Potts, autocorrelation*

The reason for the zero Q-value seems to be that the errors have been underestimated. We can calculate a least-squares fit without reference to the error bars, obtaining (with the use of the FIT programme):

```
0.0625,     -1.0337
0.03125,    -0.9705
0.015625,  -0.9706
       a  =  -0.9390,  siga  =  0.0208
       b  =  -1.4427,  sigb  =  0.5020
chi^2  =  0.0003
```

thus taking -0.939(21) as our estimate for $-d/z + \theta$.

The Dynamic Critical Exponent z

Let slopeA = $-d/z + \theta$, then since d = 2 we have z = $2/(\theta$ - slope$A)$. In this chapter, we arrived at a value of -0.046(5) for θ, so by using the value obtained for slope$A$, we arrive at an estimate of z, namely 2.240(67), or 2.24(7).

The errors are obtained by noting the extreme values for $2/(\theta$ - slope$A)$ based on the errors for slope$A$ and for θ.

Okano *et al.* (1997) obtained z = 2.155(3) for the Ising model and 2.196(8) for the 3-state Potts model. For the 3-state Potts model Arcangelis and Jan (1986) obtained 2.43(15), and Schulke and Zheng (1995) obtained z = 2.198(8).

Only two published values for the 4-state Potts z were found in the literature, neither of them recent. Arcangelis and Jan (1986) obtained 2.36(20) using a dynamic Monte Carlo renormalisation group method, and Jain (1987) obtained 2.94(29) based on Monte Carlo simulations on the square lattice. Both of these are probably too large.

If there were a linear relation between q and z for the q-state Potts model then our estimate of 2.24(7) for the 4-state Potts z would be consistent with it because the FIT programme tells us that the following least squares fit gives a Q-value of 0.9884:

```
2,  2.155,  0.003
3,  2.197,  0.008
4,  2.240,  0.067
          a  =  +2.0709,  siga  =  0.0179
          b  =  +0.0420,  sigb  =  0.0083
chi^2  =  0.0002,         Q  =  0.9884
```

13

Monte Carlo Methods and Simulation

Monte Carlo

The term *Monte-Carlo* refers to a group of methods in which physical or mathematical problems are simulated by using random numbers. Sometimes this is done at a very simple level. For example, calculations of radiation damage in humans have been studied by simulating the firing of random particles into human tissue and randomly carrying out the various possible processes. After a lot of averaging one arrives at the likely damage due to different forms of incident radiation. Similar methods are used to simulate the tracks left in particle physics experiments. Here we will concentrate on 3 different types of calculations using random numbers.

Random Number Generators

Before discussing the uses of random numbers it is useful to have some idea of how random numbers are generated on a computer. Most methods depend on a *chaotic* sequence. The commonest is the *multiplicative congruential method* which relies on prime numbers. Consider the sequence

$$x_{n+1} = (a * x_n)\% b \qquad \text{...13.1}$$

where $x\%y$ refers to the remainder on dividing x by y and a and b are large integers which have no common factors (often both are prime numbers). This process generates all integers less than in an apparently random order. After all b integers have been generated the series will repeat itself. Thus one important question to ask about any random number generator is how frequently it repeats itself.

It is worth noting that the sequence is completely deterministic: if the same initial *seed*, x_0 is chosen the same sequence will be generated. This property is extremely useful for debugging purposes, but can be a problem when averaging has to be done over several runs. In such cases it is usual to initialise the *seed* from the system clock.

Routines exist which generate a sequence of integers, as described, or which generate floating point numbers in the range 0 to 1. Most other distributions can be derived from these. There are also very efficient methods for generating a sequence of random bits. The *Numerical Algorithms Group* library contains routines to generate a wide range of distributions.

A difficult but common case is the *Gaussian* distribution. One method simply averages several (say 10) uniform $(0 \to 1)$ random numbers and relies on the central limit theorem. Another method uses the fact that a distribution of complex numbers with both real and imaginary parts Gaussian distributed can also be represented as a distribution of amplitude and phase in which the amplitude has a *Poisson* distribution and the phase is uniformly distributed between 0 and 2π.

As an example of generating another distribution we consider the *Poisson* case,

$$p(y) = \exp(-y). \qquad ...13.2$$

Then, if $p'(x)$ and $p(y)$ are the probability distributions of x and y respectively,

$$\int_{-\infty}^{y=f(x)} p(y')dy' = \int_{-\infty}^{x} p'(x')dx' \qquad ...13.3$$

must be true for all x, as the probability of finding any $y' < y$ must be the same as that for finding any $x' < x$. It follows that

$$p(y) = p'(x)\left|\frac{dx}{dy}\right|. \qquad ...13.4$$

If x is uniformly distributed between 0 and 1, i.e. $p'(x) = 1$, then

$$p(y) = \frac{dx}{dy} = \exp(-y) \quad \rightarrow \quad y = -\ln x. \qquad ...13.5$$

Hence, a *Poisson* distribution is generated by taking the logarithm of numbers drawn from a uniform distribution.

Monte-Carlo Integration

Often we are faced with integrals which cannot be done analytically. Especially in the case of multidimensional integrals the simplest methods of discretisation can become prohibitively expensive. For example, the error in a trapezium rule calculation of a *d*-dimensional integral falls as $N^{-2/d}$, where N is the number of different values of the integrand used. In a Monte-Carlo calculation the error falls as $N^{-1/2}$ independently of the dimension. Hence for $d > 4$ Monte-Carlo integration will usually converge faster.

We consider the expression for the average of a *statistic*, $f(x)$ when x is a random number distributed according to a distribution $p(x)$, then

$$\langle f(x) \rangle = \int p(x) f(x) dx, \qquad ...13.6$$

which is just a generalisation of the well known results for (e.g.) $\langle x \rangle$ or $\langle x^2 \rangle$, where we are using the notation <> to denote averaging. Now consider an integral of the sort which might arise while using Laplace transforms.

$$\int_0^\infty \exp(-x) f(x) dx. \qquad ...13.7$$

This integral can be evaluated by generating a set of N random numbers, $\{x_1, ..., x_N\}$, from a Poisson distribution, $p(x) = \exp(-x)$, and calculating the mean of $f(x)$ as

$$\frac{1}{N} \sum_{i=1}^{N} f(x_i). \qquad ...13.8$$

The error in this mean is evaluated as usual by considering the corresponding *standard error of the mean*

$$\sigma^2 = \frac{1}{N-1} \left(\langle f^2(x) \rangle - \langle f(x) \rangle^2 \right). \qquad ...13.9$$

As mentioned earlier, Monte-Carlo integration can be particularly efficient in the case of multi-dimensional integrals. However this case is particularly susceptible to the flaws in random number generators. It is a common feature that when a random set of coordinates in a d-dimensional space is generated (i.e. a set of d random numbers), the resulting distribution contains (hyper)planes on which the probability is either significantly higher or lower than expected.

Metropolis Algorithm

In statistical mechanics we commonly want to evaluate thermodynamic averages of the form

$$\langle y \rangle = \frac{\sum_i y_i e^{\beta E_i}}{\sum_i e^{-\beta E_1}} \qquad ...13.10$$

where E_i is the energy of the system in state i and $\beta = 1/k_B T$. Such problems can be solved using the Metropolis et al. (1953) algorithm.

Let us suppose the system is initially in a particular state i and we change it to another state j. The detailed balance condition demands that in equilibrium the flow from i to j must be balanced by the flow from j to i. This can be expressed as

$$p_i T_{i \to j} = p_j T_{j \to i} \qquad ...13.11$$

where p_i is the probability of finding the system in state i and $T_{i \to j}$ is the probability (or rate) that a system in state i will make a transition to state j. (13.11) can be rearranged to read

$$\frac{T_{i \to j}}{T_{j \to i}} = \frac{p_j}{p_i} \qquad ...13.12$$

$$= e^{-\beta(E_j - E_i)}. \qquad ...13.13$$

Generally the right-hand-side of (13.3) is known and we want to generate a set of states which obey the distribution p_i. This can be achieved by choosing the transition rates such that

$$T_{i \to j} = \begin{cases} 1 & \text{if } p_j > p_i \text{ or } E_j < E_i \\ \dfrac{p_j}{p_i} \text{ or } e^{-\beta(E_j - E_i)} & \text{if } p_j < p_i \text{ or } E_j > E_i. \end{cases} \quad ...13.14$$

In practice if $p_j < p_i$ a random number, r, is chosen between 0 and 1 and the system is moved to state j only if r is less than p_j / p_i or $e^{-\beta(E_j - E_i)}$.

This method is not the only way in which the condition can be fulfilled, but it is by far the most commonly used.

An important feature of the procedure is that it is never necessary to evaluate the partition function, the denominator in (13.10) but only the relative probabilities of the different states. This is usually much easier to achieve as it only requires the calculation of the change of energy from one state to another.

Note that, although we have derived the algorithm in the context of thermodynamics, its use is by no means confined to that case. See for example the quantum Monte-Carlo methods.

Ising Model

As a simple example of the Metropolis Method we consider the Ising model of a ferromagnet

$$H = -J \sum_{ij} S_i S_j \quad ...13.15$$

where J is a positive energy, $S = \pm\frac{1}{2}$, and i and j are nearest neighbours on a lattice. In this case we change from one state to another by flipping a single spin and the change in energy is simply

$$\Delta E_i = -J \sum_j S_j \quad ...13.16$$

where the sum is only over the nearest neighbours of the flipped spin.

The simulation proceeds by choosing a spin (usually at random) and testing whether the energy would be increased or decreased by flipping the spin. If it is decreased the rules say that the spin should definitely be flipped. If, on the other hand, the energy is increased, a uniform random number, r, between 0 and 1 is generated and compared with $e^{-\beta \Delta E}$. If it is smaller the spin is flipped, otherwise the spin is unchanged.

Further information can be found in the Ising Model project.

Thermodynamic Averages

To average over a thermodynamic quantity it suffices to average over the values for the sequence of states generated by the Metropolis algorithm. However it is usually wise to carry out a number of Monte-Carlo steps before starting to do any averaging. This is to guarantee that the system is in thermodynamic equilibrium while the averaging is carried out.

The sequence of random changes is often considered as a sort of time axis. In practice we think (e.g.) about the time required to reach equilibrium. Sometimes, however, the transition rate becomes very low and the system effectively gets stuck in a non-equilibrium state. This is often the case at low temperatures when almost every change causes an increase in energy.

Quantum Monte-Carlo

The term *Quantum Monte-Carlo* does not refer to a particular method but rather to any method using Monte-Carlo type methods to solve quantum (usually many-body) problems. As an example consider the evaluation of the energy of a trial wave function $\Psi(r)$ where r is a $3N$ dimensional position coordinate of N particles,

$$E = \frac{\int \mathrm{d}r \Psi * (r) \hat{H} \Psi(r)}{\int \mathrm{d}r \Psi * (r) \Psi(r)} \qquad ...13.17$$

where $\hat{H}$ is the Hamiltonian operator. This can be turned into an appropriate form for Monte-Carlo integration by rewriting as

$$E = \int \mathrm{d}r \left\{ \frac{|\psi(r)|^2}{\int \mathrm{d}r |\psi(r)|^2} \right\} \left[\psi(r)^{-1} \hat{H} \psi(r) \right] \qquad ...13.18$$

such that the quantity in braces ({}) has the form of a probability distribution. This integral can now easily be evaluated by Monte-Carlo integration. Typically a sequence of r's is generated using the Metropolis algorithm, so that it is not even necessary to normalise the trial wave function, and the quantity in square brackets [] is averaged over the r's.

This method, *variational quantum Monte-Carlo*, presupposes we have a good guess for the wave function and want to evaluate an integral over it. It is only one of several different techniques which are referred to as *quantum Monte Carlo*. Others include, Diffusion Monte-Carlo, Green's function Monte-Carlo and World Line Monte-Carlo.

Molecular Dynamics

An alternative approach to studying the behaviour of large systems is simply to solve the equations of motion for a reasonably large number of molecules. Usually this is done by using one of the methods described for Ordinary Differential Equations to solve the coupled equations of motion for the atoms or molecules to be described. Molecular solids in particular can be studied by considering the molecules as rigid objects and using phenomenological classical equations of motion to describe the interaction between them. As a simple example the *Lennard-Jones* potential

$$\Phi(r) = \epsilon \left(\left(\frac{a}{r} \right)^{12} - 2 \left(\frac{a}{r} \right)^{6} \right) \qquad ...13.19$$

has been successfully used to describe the thermodynamics of noble gases.

General Principles

Consider a set of particles interacting through a 2-body potential, $\Phi(r)$. The equations of motion can be written in the form

$$\frac{dr_i}{dt} = v_i \qquad ...13.20$$

$$\frac{dv_i}{dt} = -\frac{1}{m_i}\frac{\partial}{\partial r_i}\sum_{j\neq i}\Phi(|r_i - r_j|). \qquad ...13.21$$

In practice it is better to use the scalar force $F(r) = \partial\Phi/\partial r$ to avoid unnecessary numerical differentiation.

A common feature of such problems is that the time derivative of one variable only involves the other variable as with r and v in the above equations of motion. In such circumstances a leap-frog like method suggests itself as the most appropriate. Hence we write

$$r_i^n = r_i^{n-2} + 2\delta t v_i^{n-1} \qquad ...13.22$$

$$v_i^{n+1} = v_i^{n-1} + \frac{2\delta t}{m}\sum_{j\neq i} F(|r_i^n - r_j^n|)\frac{r_i^n - r_j^n}{|r_i^n - r_i^n|}. \qquad ...13.23$$

This method has the advantage of simplicity as well as the merit of being properly conservative.

The temperature is defined from the kinetic energy of N particles via

$$\tfrac{3}{2}k_b T = \frac{1}{N}\left\langle \sum_i \tfrac{1}{2} m_i v_i^2 \right\rangle \qquad ...13.24$$

where the averages <> are taken with respect to time.

As an example of another thermodynamic quantity consider the specific heat at constant volume C_v. This can be calculated by changing the total energy by multiplying all the velocities by a constant amount ($\alpha \approx 1$) and running for some time to determine the temperature. Note that, as the temperature is defined in terms of a time average, multiplying all the velocities by α does not necessarily imply a simple change of temperature, $T \mapsto \alpha^2 T$. At a 1st order phase transition, for example, the system might equilibrate to the same temperature as before with the additional energy contributing to the latent heat.

When the specific heat is known for a range of temperatures it becomes possible, at least in principle, to calculate the entropy from the relationship

$$T\left.\frac{\partial \delta}{\partial T}\right)_v = \left.\frac{\partial U}{\partial T}\right)_v \qquad ...13.25$$

The pressure is rather more tricky as it is defined in terms of the free energy, $F = U - TS$, using

$$p = -\left.\frac{\partial F}{\partial V}\right)_T \qquad ...13.26$$

and hence has a contribution from the entropy as well as the internal energy. Nevertheless methods exist for calculating this.

It is also possible to define modified equations of motion which, rather than conserving energy and volume, conserve temperature or pressure. These are useful for describing isothermal or isobaric processes.

Note that for a set of mutually attractive particles it may not be necessary to constrain the volume but for mutually repulsive particles it certainly is necessary.

Ising Model

The Ising model for a ferromagnet is not only a very simple model which has a phase transition, but it can also be used to describe phase transitions in a whole range of other physical systems. The model is defined using the equation

$$E = -\sum_{ij} J_{ij} S_i \cdot S_j \qquad \text{...13.27}$$

where the i and j designate points on a lattice S and takes the values $\pm\frac{1}{2}$. The various different physical systems differ in the definition and sign of the various J_{ij}'s.

Model and Method

Here we will consider the simple case of a 2 dimensional square lattice with interactions only between nearest neighbours. In this case

$$E = -J \sum_{i,ji} S_i \cdot S_{ji} \qquad \text{...13.28}$$

where j_i is only summed over the 4 nearest neighbours of i.

This model can be studied using the Metropolis method as described in the notes, where the state can be changed by flipping a single spin. Note that the change in energy due to flipping the kth spin from ↓ to ↑ is given by

$$\Delta E_k = -J \sum_{jk} S_{jk}. \qquad \text{...13.29}$$

The only quantity which actually occurs in the calculation is

$$Z_k = \exp(-\Delta E_k / k_B T), \qquad \text{...13.30}$$

and this can only take one of five different values given by the number of neighbouring ↑ spins. Hence it is sensible to store these in a short array before starting the calculation. Note also that there is really only 1 parameter in the model, $J / k_B T$, so that it would make sense to write your programme in terms of this single parameter rather than J and T separately.

The calculation should use periodic boundary conditions, in order to avoid spurious effects due to boundaries. There are several different ways to achieve this. One of the most efficient is to think of the system as a single line of spins wrapped round a torus. This way it is possible to avoid a lot of checking for the boundary. For an $N \times N$ system of spins define an array of $2N^2$ elements using the shortest sensible variable type: char in C(++). It is easier to use 1 for spin $\uparrow$ and 0 for spin $\downarrow$, as this makes the calculation of the number of neighbouring $\uparrow$ spins easier. In order to map between spins in a 2d space S_r and in the 1d array S_k the following mapping can be used.

$$S_{r+\delta x} \mapsto S_{k+1}, \; S_{r-\delta x} \mapsto S_{k\_N^2-1}, \; S_{r+\delta y} \mapsto S_{k+N}, \; S_{r-\delta y} \mapsto S_{k+N^2-N} \quad \text{...13.31}$$

where the 2nd N^2 elements of the array are always maintained equal to the 1st N^2. This way it is never necessary to check whether one of the neighbours is over the edge. It is important to remember to change S_{k+N^2} whenever S_k is changed.

The calculation proceeds as follows:

1. Initialise the spins, either randomly or aligned.
2. Choose a spin to flip. It is better to choose a spin at random rather than systematically as systematic choices can lead to spurious temperature gradients across the system.
3. Decide whether to flip the spin by using the Metropolis condition.
4. If the spin is to be flipped, do so but remember to flip its mirror in the array.
5. Update the energy and magnetisation.
6. Add the contributions to the required averages.
7. Return to step 2 and repeat.

In general it is advisable to run the programme for some time to allow it to reach equilibrium before trying to calculate any averages. Close to a phase transition it is often necessary to run for much longer to reach equilibrium. The behaviour of the total energy during the run is usually a good guide to whether equilibrium has been reached. The total energy, E, and the magnetisation can be calculated from (13.27) and

$$M = \frac{1}{N}\sum_i S_i \quad \text{...13.32}$$

It should be possible to calculate these as you go along, by accumulating the changes rather than by recalculating the complete sum after each step. A 10×10 lattice should suffice for most purposes and certainly for testing, but you may require a much bigger lattice close to a transition.

A useful trick is to use the final state at one temperature as the initial state for the next slightly different temperature. That way the system won't need so long to reach equilibrium.

It should be possible to calculate the specific heat and the magnetic susceptibility. The specific heat could be calculated by differentiating the energy with respect to temperature. This is a numerically questionable procedure however. Much better is to use the relationship

$$C_v \propto \frac{1}{N}\left(\frac{J}{k_B T}\right)^2 \left(\langle E^2\rangle - \langle E\rangle^2\right) \quad \text{...13.33}$$

Similarly, in the paramagnetic state, the susceptibility can be calculated using

$$\chi \propto \frac{1}{N}\frac{J}{k_B T}\left(\langle S^2\rangle - \langle S\rangle^2\right) \qquad ...13.34$$

where $S = \sum_i S_i$ and the averages are over different states, i.e. can be calculated by averaging over the different Metropolis steps. Both these quantities are expected to diverge at the transition, but the divergence will tend to be rounded off due to the small size of the system. Note however that the fact that (13.33) & (13.34) have the form of variances, and that these diverge at the transition, indicates that the average energy and magnetisation will be subject to large fluctuations around the transition.

Finally a warning. A common error made in such calculations is to add a contribution to the averages only when a spin is flipped. In fact this is wrong as the fact that it isn't flipped means that the original state has a higher probability of occupation.

Quantum Monte Carlo Calculation

This project is to use the variational quantum Monte Carlo method to calculate the ground state energy of the He atom. The He atom is a two electron problem which cannot be solved analytically and so numerical methods are necessary. Quantum Monte Carlo is one of the more interesting of the possible approaches (although there are better methods for this particular problem).

The Schrödinger equation for the He atom in atomic units is,

$$\left(-\frac{1}{2}\nabla_{r_1}^2 - \frac{1}{2}\nabla_{r_2}^2 - \frac{2}{r_1} - \frac{2}{r_2} + \frac{1}{r_{12}}\right)\Psi(r_1, r_2) = E\Psi(r_1, r_2), \qquad ...13.35$$

where r_1 and r_2 are the position vectors of the two electrons, $r_1 = |r_1|, r_2 = |r_2|$, and $r_{12} = |r_1 - r_2|$. Energies are in units of the Hartree energy (1 Hartree = 2 Rydbergs) and distances are in units of the Bohr radius. The ground state spatial wavefunction is symmetric under exchange of the two electrons (the required anti-symmetry is taken care of by the spin part of the wavefunction, which we can forget about otherwise).

The expression for the energy expectation value of a particular trial wavefunction, $\Phi_T(r_1, r_2)$, is,

$$E_T = \frac{\int ... \int \Phi_T^* \hat{H} \Phi_T d^3r_1 d^3r_2}{\int ... \int \Phi_T^* \Phi_T d^3r_1 d^3r_2}. \qquad ...13.36$$

In the variational Monte Carlo method, this equation is rewritten in the form,

$$E_T = \int ... \int \left(\frac{1}{\Phi_T} \hat{H} \Phi_T\right) f(r_1, r_2) d^3r_1 d^3r_2, \qquad ...13.37$$

where

$$f(r_1,r_2)=\frac{\Phi_T^*(r_1,r_2)\Phi_T(r_1,r_2)}{\int...\int\Phi_T^*\Phi_T d^3r_1 d^3r_2} \qquad ...13.38$$

is interpreted as a probability density which is sampled using the Metropolis algorithm. Note that the Metropolis algorithm only needs to know *ratios* of the probability density at different points, and so the normalisation integral,

$$\int...\int\Phi_T^*\Phi_T d^3r_1 d^3r_2, \qquad ...13.39$$

always cancels out and does not need to be evaluated.

The mean of the values of the "local energy",

$$\frac{1}{\Phi_T}\hat{H}\Phi_T, \qquad ...13.40$$

at the various points along the Monte Carlo random walk then gives an estimate of the energy expectation value. By the variational principle, the exact energy expectation value is always greater than or equal to the true ground state energy; but the Monte Carlo estimate has statistical errors and may lie below the true ground state energy if these are large enough. Anyway, the better the trial wavefunction, the closer to the true ground state energy the variational estimate should be.

In this project you are given a possible trial wavefunction,

$$\Phi = e^{-2r_1}e^{-2r_2}e^{r_{12}/2}. \qquad ...13.41$$

Use the variational Monte Carlo technique to calculate variational estimates of the true ground state energy.

Before you start programming, you will need the analytic expressions for the local energy. This involves some nasty algebra, but the answer is,

$$\frac{1}{\Phi}\hat{H}\Phi = -\frac{17}{4}-\frac{r_1\cdot r_{12}}{r_1 r_{12}}+\frac{r_2\cdot r_{12}}{r_2 r_{12}}. \qquad ...13.42$$

where r_1 and r_2 are the coordinates of the 2 atoms relative to the nucleus and $r_{12}=r_2-r_1$.

The Monte Carlo moves can be made by generating random numbers (use a library routine to do this) and adding them to the electron coordinates. I suggest that you update all six electron position coordinates $(x_1,y_1,z_1,x_2,y_2,z_2)$ each move, and so you will need six random numbers each time.

The accepted lore is that the Metropolis algorithm is most efficient when the step size is chosen to keep the acceptance probability close to 0.5. However, the method should work in principle no matter what the step size and you should try a few different step sizes to confirm that this is indeed the case. The starting positions of the two electrons can be chosen randomly, but remember that the Metropolis algorithm only samples the probability distribution exactly in the limit as the number of moves tends to infinity. You will therefore have to throw away the results from the moves near the beginning of the run and only start accumulating the values of the local energy once things have settled down. You should experiment to find out how many moves you need to throw away.

The statistical errors in Monte Carlo calculations decrease like $1/\sqrt{N}$, where N is the total number of moves after the initial equilibration period. The errors therefore improve only slowly as the length of the run is increased. You will not be able (and should not attempt) to attain great accuracy. However, you should think hard about the magnitude of the statistical errors involved. Calculating the variance of the values in the list of energies accumulated during the random walk is easy and you should certainly do it.

Numerical methods of Monte-Carlo Methods

Numerical methods which make use of random numbers are called *Monte-Carlo methods*— after the famous casino. The obvious applications of such methods are in *stochastic physics*: e.g., statistical thermodynamics. However, there are other, less obvious, applications: e.g., the evaluation of multi-dimensional integrals.

Random Numbers

No numerical algorithm can generate a truly random sequence of numbers, However, there exist algorithms which generate repeating sequences of M (say) integers which are, to a fairly good approximation, randomly distributed in the range 0 to M–1. Here, M is a (hopefully) large integer. This type of sequence is termed *psuedo-random*.

The most well-known algorithm for generating psuedo-random sequences of integers is the so-called *linear congruental* method. The formula linking the nth and $(n+1)$th integers in the sequence is

$$I_{n+1} = (AI_n + C) \bmod M, \quad \text{...13.43}$$

where A, C, and M are positive integer constants. The first number in the sequence, the so-called "seed" value, is selected by the user.

Consider an example case in which $A = 7$, $C = 0$, and $M = 10$. A typical sequence of numbers generated by formula (13.43) is

$$I = \{3,1,7,9,3,1,\cdots\}. \quad \text{...13.44}$$

Evidently, the above choice of values for A, C, and M is not a particularly good one, since the sequence repeats after only four iterations. However, if A, C, and M are properly chosen then the sequence is of maximal length (*i.e.*, of length M), and approximately randomly distributed in the range 0 to M–1.

The function listed below is an implementation of the linear congruental method.

```
// random.cpp
// Linear congruential psuedo-random number generator.
// Generates psuedo-random sequence of integers in
//   range 0 .. RANDMAX.
#define RANDMAX 6074    // RANDMAX = M - 1
int random (int seed = 0)
{
   static int next = 1;
   static int A = 106;
   static int C = 1283;
   static int M = 6075;
   if (seed) next = seed;
   next = next * A + C;
   return next
  M;
}
```

The keyword static in front of a local variable declaration indicates that the programme should preserve the value of that variable between function calls. In other words, if the static variable next has the value 999 on exit from function random then the next time this function is called next will have exactly the same value. Note that the values of non-static local variables are not preserved between function calls. The = 0 in the first line of function random is a default value for the argument seed. In fact, random can be called in one of two ways. Firstly, random can be called with no argument: i.e., random (): in which case, seed is given the default value 0. Secondly, random can be called with an integer argument: i.e., random (n): in which case, the value of seed is set to n. The first way of calling random just returns the next integer in the psuedo-random sequence. The second way seeds the sequence with the value n (*i.e.*, I_1 is set to n), and then returns the next integer in the sequence (*i.e.*, I_2). Note that the function prototype for random takes the form int random (int = 0): the = 0 indicates that the argument is optional.

The above function returns a pseudo-random integer in the range 0 to RANDMAX (where RANDMAX takes the value $M-1$). In order to obtain a random variable x, uniformly distributed in the range 0 to 1, we would write

```
x = double (random ()) / double (RANDMAX);
```

Now if x is truly random then there should be no correlation between successive values of x. Thus, a good way of testing our random number generator is to plot x_j versus x_{j+1} (where x_j corresponds to the j th number in the psuedo-random sequence) for many different values of j. For a good random number generator, the plotted points should densely fill the unit square. Moreover, there should be no discernible pattern in the distribution of points.

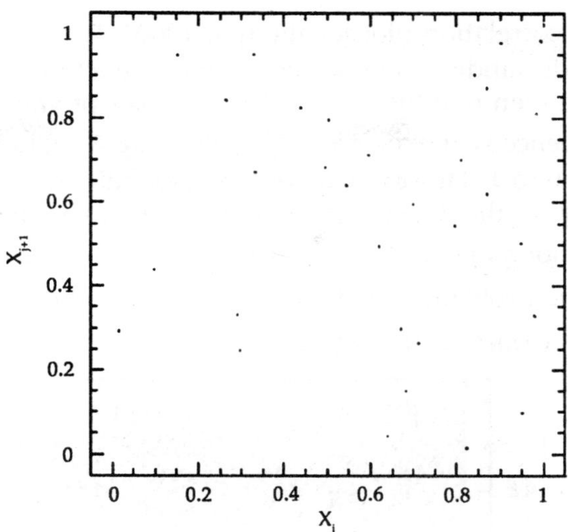

Figure 13.1: *Plot of x_j versus x_{j+1} for $j = 1{,}10000$. Here, the x_j are random values, uniformly distributed in the range 0 to 1, generated using a linear congruental psuedo-random number generator characterised by $A = 106, C = 1283$, and $M = 6075$*

Figure 13.1 shows a correlation plot for the first 10000 $x_j - x_{j+1}$ pairs generated using a linear congruental psuedo-random number generator characterised by $A = 106$, $C = 1283$, and $M = 6075$. It can be seen that this is a poor choice of values for $A, C,$ and M, since the pseudo-random sequence repeats after a few iterations, yielding x_j values which do not densely fill the interval 0 to 1.

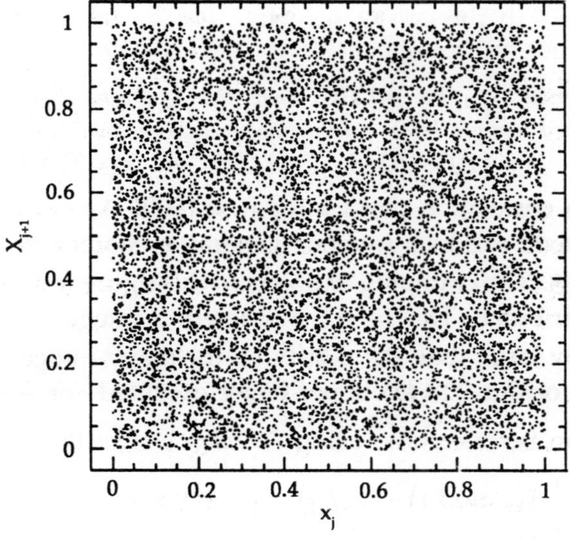

Figure 13.2: *Plot of x_j versus x_{j+1} for j=1,10000. Here, the x_j are random values, uniformly distributed in the range 0 to 1, generated using a linear congruental psuedo-random number generator characterised by $A = 107$, $C = 1283$, and $M = 6075$*

Figure 13.2 shows a correlation plot for the first 10000 $x_j - x_{j+1}$ pairs generated using a linear congruential psuedo-random number generator characterised by $A = 107$, $C = 1283$, and $M = 6075$. It can be seen that this is a far better choice of values for *A*, *C*, and *M*, since the pseudo-random sequence is of maximal length, yielding x_j values which are fairly evenly distributed in the range 0 to 1. However, if we look carefully at figure 13.2, we can see that there is a slight tendency for the dots to line up in the horizontal and vertical directions. This indicates that the x_j are not quite randomly distributed: i.e., there is some correlation between successive x_j values. The problem is that *M* is too low: i.e., there is not a sufficiently wide selection of different x_j values in the interval 0 to 1.

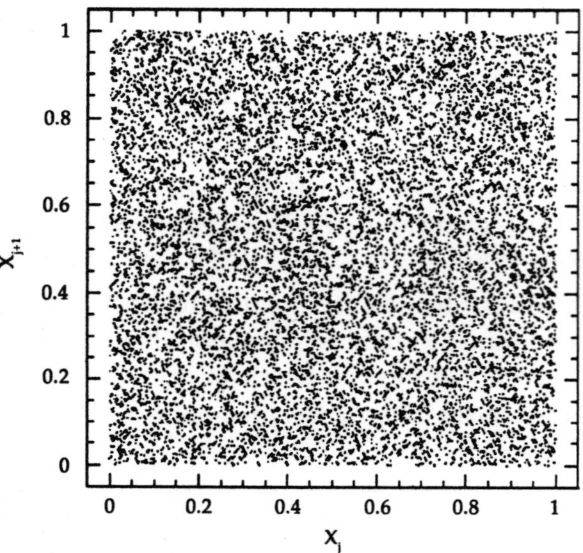

***Figure 13.3:** Plot of x_j versus x_{j+1} for j=1,10000. Here, the x_j are random values, uniformly distributed in the range 0 to 1, generated using a linear congruential psuedo-random number generator characterised by* $A = 1103515245$, $C = 12345$, *and* $M = 32768$

Figure 13.3 shows a correlation plot for the first 10000 $x_j - x_{j+1}$ pairs generated using a linear congruential psuedo-random number generator characterised by $A = 1103515245$, $C = 12345$, and $M = 32768$. The clumping of points in this figure indicates that the x_j are again not quite randomly distributed. This time the problem is integer overflow: i.e., the values of *A* and *M* are sufficiently large that $AI_n > 10^{32} - 1$ for many integers in the pseudo-random sequence. Thus, the algorithm (13.43) is not being executed correctly.

Integer overflow can be overcome using *Schrange's algorithm.* If $y = (Az) \bmod M$ then

$$y = \begin{cases} A(z \bmod q) - r(z/q) & \text{if } y > 0 \\ A(z \bmod q) - r(z/q) + M & \text{otherwise} \end{cases}, \qquad \text{...13.45}$$

where $q = M / A$ and $r = M\%A$. The so-called *Park and Miller* method for generating a pseudo-random sequence corresponds to a linear congruential method characterised by the values $A = 16807$, $C = 0$, and $M = 2147483647$. The function listed below implements this method, using Schrange's algorithm to avoid integer overflow.

```
// random.cpp
// Park and Miller's psuedo-random number generator.
#define RANDMAX 2147483646 // RANDMAX = M - 1
int random (int seed = 0)
{
    static int next = 1;
    static int A = 16807;
    static int M = 2147483647;      // 2^31 - 1
    static int q = 127773;          // M / A
    static int r = 2836;            // M % A
    if (seed) next = seed;
    next = A * (next % q) - r * (next / q);
    if (next < 0) next += M;
    return next;
}
```

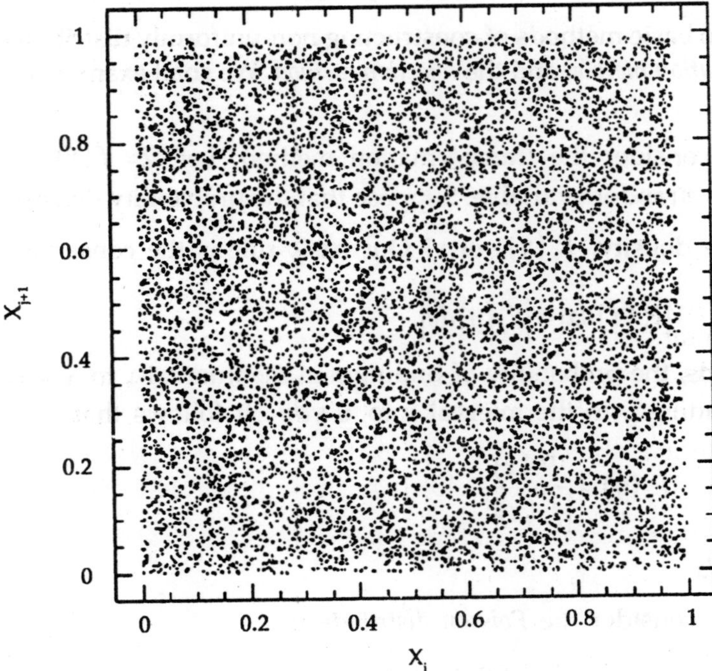

***Figure 13.4:** Plot of x_j versus x_{j+1} for $j = 1,10000$. Here, the x_j are random values, uniformly distributed in the range 0 to 1, generated using Park & Miller's psuedo-random number generator*

Figure 13.4 shows a correlation plot for the first 10000 $x_j - x_{j+1}$ pairs generated using Park & Miller's method. We can now see no pattern whatsoever in the plotted points. This indicates that the x_j are indeed randomly distributed in the range 0 to 1. From now on, we shall use Park & Miller's method to generate all the psuedo-random numbers needed in our investigation of Monte-Carlo methods.

Distribution Functions

Let $P(x)dx$ represent the probability of finding the random variable *x* in the interval *x* to $x+dx$. Here, $P(x)$ is termed a *probability density*. Note that $P=0$ corresponds to no chance, whereas $P=1$ corresponds to certainty. Since it is certain that the value of *x* lies in the range $-\infty$ to $+\infty$, probability densities are subject to the normalising constraint

$$\int_{-\infty}^{+\infty} P(x)dx = 1. \qquad ...13.46$$

Suppose that we wish to construct a random variable which is uniformly distributed in the range x_1 to x_2. In other words, the probability density of *x* is

$$P(x) = \begin{cases} 1/(x_2 - x_1) & \text{if } x_1 \le x \le x_2 \\ 0 & \text{otherwise} \end{cases}. \qquad ...13.47$$

Such a variable is constructed as follows

```
x = x1 + (x2 - x1) * double (random ()) / double (RANDMAX);
```

There are two basic methods of constructing non-uniformly distributed random variables: i.e., the *transformation method* and the *rejection method*. We shall examine each of these methods in turn.

Let us first consider the transformation method. Let $y = f(x)$, where *f* is a known function, and *x* is a random variable. Suppose that the probability density of *x* is $P_x(x)$. What is the probability density, $P_y(y)$, of *y*? Our basic rule is the conservation of probability:

$$|P_x(x)dx| = |P_y(y)dy|. \qquad ...13.48$$

In other words, the probability of finding *x* in the interval *x* to $x+dx$ is the same as the probability of finding *y* in the interval *y* to $y+dy$. It follows that

$$P_y(y) = \frac{P_x(x)}{|f'(x)|}, \qquad ...13.49$$

where $f' = df/dx$.

For example, consider the *Poisson distribution*:

$$P_y(y) \begin{cases} e^{-y} & \text{if } 0 \le y \le \infty \\ 0 & \text{otherwise} \end{cases}. \qquad ...13.50$$

Let $y = f(x) = -\ln x$, so that $|f'| = 1/x$. Suppose that

$$P_x(x) = \begin{cases} 1 & \text{if } 0 \le x \le 1 \\ 0 & \text{otherwise} \end{cases}. \quad ...13.51$$

It follows that

$$P_y(y) = \frac{1}{|f'|} = x = e^{-y}, \quad ...13.52$$

with $x = 0$ corresponding to $y = \infty$, and $x = 1$ corresponding to $y = 0$. We conclude that if

```
x = double (random ()) / double (RANDMAX);
y = - log (x);
```

then y is distributed according to the Poisson distribution.

The transformation method requires a differentiable probability distribution function. This is not always practical. In such cases, we can use the rejection method instead.

Suppose that we desire a random variable y distributed with density $P_y(y)$ in the range y_{min} to y_{max}. Let P_{ymax} be the maximum value of $P(y)$ in this range (see Figure 13.5). The rejection method is as follows. The variable y is sampled randomly in the range y_{min} to y_{max}. For each value of y we first evaluate $P_y(y)$. We next generate a random number x which is uniformly distributed in the range 0 to P_{ymax}. Finally, if $P_y(y) < x$ then we reject the y value; otherwise, we keep it. If this prescription is followed then y will be distributed according to $P_y(y)$.

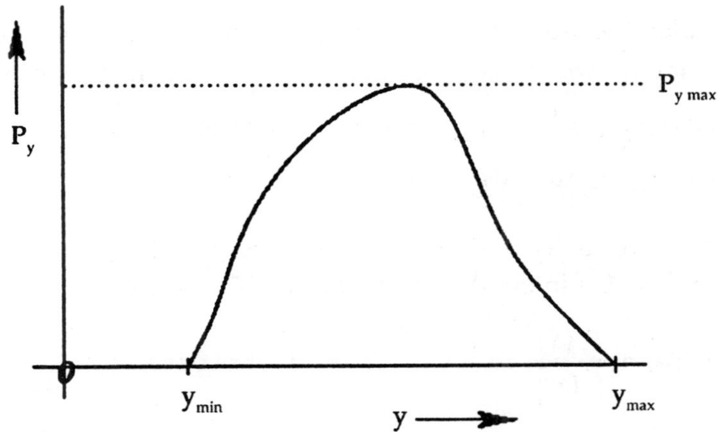

***Figure 13.5:** The rejection method*

As an example, consider the Gaussian distribution:

$$P_y(y) = \frac{\exp\left[(y-\bar{y})^2/2\sigma^2\right]}{\sqrt{2\pi}\sigma}, \quad ...13.53$$

where $\overline{y}$ is the mean value of y, and σ is the standard deviation. Let

$$y_{\min} = \overline{y} - 4\sigma, \qquad ...13.54$$

$$y_{\max} = \overline{y} + 4\sigma, \qquad ...13.55$$

since there is a negligible chance that y lies more than 4 standard deviations from its mean value. It follows that

$$P_{y\max} = \frac{1}{\sqrt{2\pi}\sigma}, \qquad ...13.56$$

with the maximum occurring at $y = \overline{y}$. The function listed below employs the rejection method to return a random variable distributed according to a Gaussian distribution with mean and standard deviation sigma:

```
// gaussian.cpp
// Function to return random variable distributed
// according to Gaussian distribution with mean
// and standard deviation sigma.
#define RANDMAX 2147483646
int random (int = 0);
double gaussian (double mean, double sigma)
{
   double ymin = mean - 4. * sigma;
   double ymax = mean + 4. * sigma;
   double Pymax = 1. / sqrt (2. * M_PI) / sigma;
   // Calculate random value uniformly distributed
   //   in range ymin to ymax
  double y = ymin + (ymax - ymin) * double (random ()) / double (RANDMAX);
   // Calculate Py
   double Py = exp (- (y - mean) * (y - mean) / 2. / sigma / sigma) /
      sqrt (2. * M_PI) / sigma;
  // Calculate random value uniformly distributed in range 0 to Pymax
   double x = Pymax * double (random ()) / double (RANDMAX);
   // If x > Py reject value and recalculate
   if (x > Py) return gaussian (mean, sigma);
   else return y;
}
```

Figure 13.6 illustrates the performance of the above function. It can be seen that the function successfully returns a random value distributed according to the Gaussian distribution.

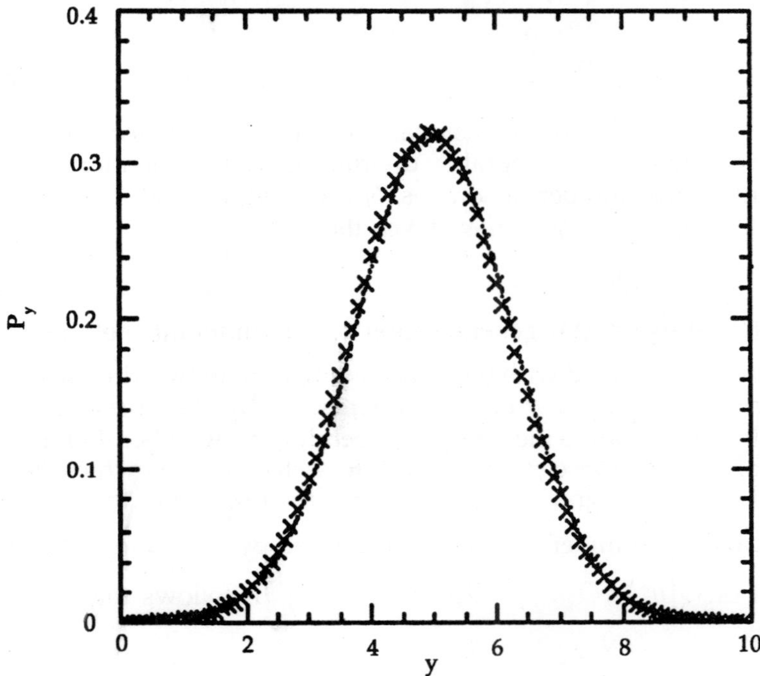

***Figure 13.6:** A million values returned by function* gaussian *with* mean = 5. *and* sigma = 1.25. *The values are binned in 100 bins of width 0.1. The figure shows the number of points in each bin divided by a suitable normalisation factor. A Gaussian curve is shown for comparison*

Monte-Carlo Integration

Consider a one-dimensional integral: $\int_{x_l}^{x_h} f(x)\,dx$. We can evaluate this integral numerically by dividing the interval x_l to x_h into N identical subdivisions of width

$$h = \frac{x_h - x_l}{N}. \qquad \text{...13.57}$$

Let x_i be the *midpoint* of the *i*th subdivision, and let $f_i = f(x_i)$. Our approximation to the integral takes the form

$$\int_{x_i}^{x_h} f(x)\,dx \approx \sum_{i=1}^{N} f_i h \qquad \text{...13.58}$$

This integration method—which is known as the *midpoint method*—is not particularly accurate, but is very easy to generalise to multi-dimensional integrals.

What is the error associated with the midpoint method? Well, the error is the product of the error per subdivision, which is $O(h^2)$, and the number of subdivisions, which is $O(h^{-1})$.

The error per subdivision follows from the linear variation of $f(x)$ within each subdivision. Thus, the overall error is $O(h^2)\times O(h^{-1})=O(h)$. Since, $h\propto N^{-1}$, we can write

$$\int_{x_l}^{x_h} f(x)\,dx \simeq \sum_{i=1}^{N} f_i h + O(N^{-1}). \qquad ...13.59$$

Let us now consider a two-dimensional integral. For instance, the area enclosed by a curve. We can evaluate such an integral by dividing space into identical squares of dimension *h*, and then counting the number of squares, *N* (say), whose midpoints lie within the curve. Our approximation to the integral then takes the form

$$A \simeq N h^2. \qquad ...13.60$$

This is the two-dimensional generalisation of the midpoint method.

What is the error associated with the midpoint method in two-dimensions? Well, the error is generated by those squares which are intersected by the curve. These squares either contribute wholly or not at all to the integral, depending on whether their midpoints lie within the curve. In reality, only those parts of the intersected squares which lie within the curve should contribute to the integral. Thus, the error is the product of the area of a given square, which is $O(h^2)$, and the number of squares intersected by the curve, which is $O(h^{-1})$. Hence, the overall error is $O(h^2)\times O(h^{-1})=O(h)=O(N^{-1/2})$. It follows that we can write

$$A = N h^2 + O(N^{-1/2}). \qquad ...13.61$$

Let us now consider a three-dimensional integral. For instance, the volume enclosed by a surface. We can evaluate such an integral by dividing space into identical cubes of dimension *h*, and then counting the number of cubes. *N* (say), whose midpoints lie within the surface. Our approximation to the integral then takes the form

$$V \simeq N h^3. \qquad ...13.62$$

This is the three-dimensional generalisation of the midpoint method.

What is the error associated with the midpoint method in three-dimensions? Well, the error is generated by those cubes which are intersected by the surface. These cubes either contribute wholly or not at all to the integral, depending on whether their midpoints lie within the surface. In reality, only those parts of the intersected cubes which lie within the surface should contribute to the integral. Thus, the error is the product of the volume of a given cube, which is $O(h^3)$, and the number of cubes intersected by the surface, which is $O(h^{-2})$. Hence, the overall error is $O(h^3)\times O(h^{-2})=O(h)=O(N^{-1/3})$. It follows that we can write

$$V = N h^3 + O(N^{-1/3}). \qquad ...13.63$$

Let us, finally, consider using the midpoint method to evaluate the volume, *V*, of a *d*-dimensional hypervolume enclosed by a $(d-1)$-dimensional hypersurface. It is clear that

$$V = N h^d + O(N^{-1/d}), \qquad ...13.64$$

where N is the number of identical hypercubes into which the hypervolume is divided. Note the increasingly slow fall-off of the error with N as the dimensionality, d, becomes greater. The explanation for this phenomenon is quite simple.

Suppose that $N = 10^6$. With $N = 10^6$ we can divide a unit line into (identical) subdivisions whose linear extent is 10^{-6}, but we can only divide a unit area into subdivisions whose linear extent is 10^{-3}, and a unit volume into subdivisions whose linear extent is 10^{-2}.

Thus, for a fixed number of subdivisions the grid spacing (and, hence, the integration error) increases dramatically with increasing dimension.

Let us now consider the so-called *Monte-Carlo method* for evaluating multi-dimensional integrals. Consider, for example, the evaluation of the area, A, enclosed by a curve, C. Suppose that the curve lies wholly within some simple domain of area A', as illustrated in figure 13.7.

Let us generate N' points which are *randomly* distributed throughout A'. Suppose that N of these points lie within curve C. Our estimate for the area enclosed by the curve is simply

$$A = \frac{N}{N'} A'. \qquad \text{...13.65}$$

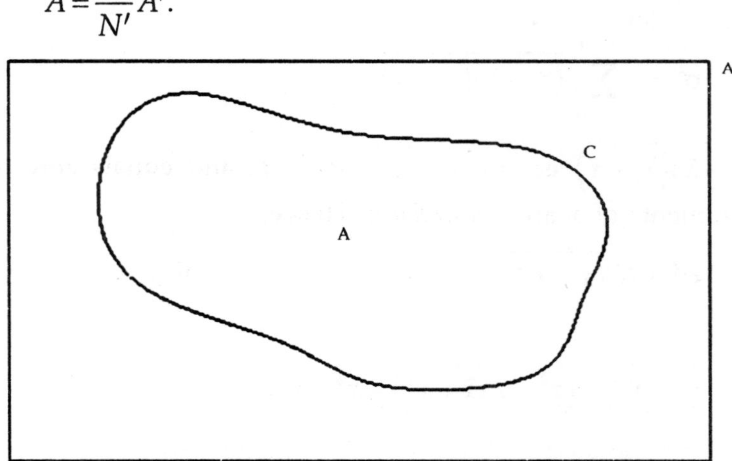

***Figure 13.7:** The Monte-Carlo integration method*

What is the error associated with the Monte-Carlo integration method? Well, each point has a probability $p = A/A'$ of lying within the curve.

Hence, the determination of whether a given point lies within the curve is like the measurement of a random variable x which has two possible values: 1 (corresponding to the point being inside the curve) with probability p, and 0 (corresponding to the point being outside the curve) with probability $1-p$.

If we make N' measurements of x (*i.e.*, if we scatter N' points throughout A') then the number of points lying within the curve is

$$N = \sum_{i=1,N'} x_i, \qquad \text{...13.66}$$

where x_i denotes the ith measurement of x. Now, the mean value of N is

$$\bar{N} = \sum_{i=1,N'} \bar{x} = N'\bar{x}, \qquad ...13.67$$

where

$$\bar{x} = 1 \times p + 0 \times (1 - p) = p. \qquad ...13.68$$

Hence,

$$\bar{N} = N'p = N'\frac{A}{A'}, \qquad ...13.69$$

which is consistent with equation (13.65). We conclude that, on average, a measurement of N leads to the correct answer. But, what is the scatter in such a measurement? Well, if σ represents the *standard deviation* of N then we have

$$\sigma^2 = \overline{\left(N - \bar{N}\right)^2}, \qquad ...13.70$$

which can also be written

$$\sigma^2 = \sum_{i,j,=1,N'} \overline{(x_i - \bar{x})(x_j - \bar{x})}. \qquad ...13.71$$

However, $\overline{(x_i - \bar{x})(x_j - \bar{x})}$ equals $\overline{(x - \bar{x})^2}$ if $i = j$, and equals zero, otherwise, since successive measurements of x are *uncorrelated*. Hence,

$$\sigma^2 = N'\overline{(x - x)^2}. \qquad ...13.72$$

Now,

$$\overline{(x - \bar{x})^2} = \overline{\left(x^2 - 2x\bar{x} + \bar{x}^2\right)} = \overline{x^2} - \bar{x}^2, \qquad ...13.73$$

and

$$\overline{x^2} = 1^2 \times p + 0^2 \times (1 - p) = p. \qquad ...13.74$$

Thus,

$$\overline{(x - \bar{x})^2} = p - p^2 = p(1 - p), \qquad ...13.75$$

giving

$$\sigma = \sqrt{N'p(1 - p)}. \qquad ...13.76$$

Finally, since the likely values of N lie in the range $N = \bar{N} \pm \sigma$, we can write

$$N = N'\frac{A}{A'} \pm \sqrt{N'\frac{A}{A'}\left(1 - \frac{A}{A'}\right)}. \qquad ...13.77$$

It follows from equation (13.65) that

$$A = A'\frac{N}{N'} \pm \frac{\sqrt{A(A'-A)}}{\sqrt{N'}}. \quad \text{...13.78}$$

In other words, the error scales like $(N')^{-1/2}$.

The Monte-Carlo method generalises immediately to *d*-dimensions. For instance, consider a *d*-dimensional hypervolume *V* enclosed by a $(d-1)$-dimensional hypersurface *A*. Suppose that *A* lies wholly within some simple hypervolume V'.

We can generate N' points randomly distributed throughout V'. Let *N* be the number of these points which lie within *A*. It follows that our estimate for *V* is simply

$$V = \frac{N}{N'}V'. \quad \text{...13.79}$$

Now, there is nothing in our derivation of equation (13.78) which depends on the fact that the integral in question is two-dimensional. Hence, we can generalise this equation to give

$$V = V'\frac{N}{N'} \pm \frac{\sqrt{V(V'-V)}}{\sqrt{N'}}. \quad \text{...13.80}$$

We conclude that the error associated with Monte-Carlo integration *always* scales like $(N')^{-1/2}$, irrespective of the dimensionality of the integral.

We are now in a position to compare and contrast the midpoint and Monte-Carlo methods for evaluating multi-dimensional integrals. In the midpoint method, we fill space with an *evenly spaced* mesh of *N* (say) points (*i.e.*, the midpoints of the subdivisions), and the overall error scales like $N^{-1/d}$, where *d* is the dimensionality of the integral.

In the Monte-Carlo method, we fill space with *N* (say) *randomly distributed* points, and the overall error scales like $N^{-1/2}$, irrespective of the dimensionality of the integral. For a one-dimensional integral (*d*=1), the midpoint method is *more efficient* than the Monte-Carlo method, since in the former case the error scales like N^{-1}, whereas in the latter the error scales like $N^{-1/2}$.

For a two-dimensional integral (*d*=2), the midpoint and Monte-Carlo methods are both *equally efficient*, since in both cases the error scales like $N^{-1/2}$. Finally, for a three-dimensional integral (*d*=3), the midpoint method is *less efficient* than the Monte-Carlo method, since in the former case the error scales like $N^{-1/3}$, whereas in the latter the error scales like $N^{-1/2}$.

We conclude that for a sufficiently high dimension integral the Monte-Carlo method is always going to be more efficient than an integration method (such as the midpoint method) which relies on a uniform grid.

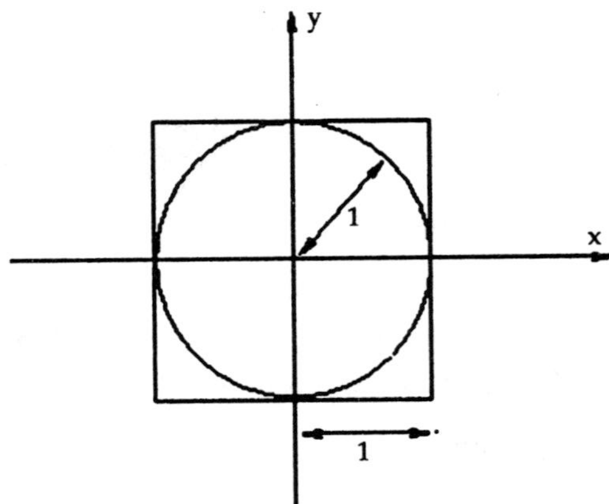

Figure 13.8: Example calculation: volume of unit-radius 2-dimensional sphere enclosed in a close-fitting 2-dimensional cube

Up to now, we have only considered how the Monte-Carlo method can be employed to evaluate a rather special class of integrals in which the integrand function can only take the values 0 or 1. However, the Monte-Carlo method can easily be adapted to evaluate more general integrals. Suppose that we wish to evaluate $\int f\, dV$, where f is a general function and the domain of integration is of arbitrary dimension. We proceed by randomly scattering N points throughout the integration domain and calculating f at each point. Let x_i denote the ith point. The Monte-Carlo approximation to the integral is simply

$$\int f\, dv = \frac{1}{N} \sum_{i=1,N} f(x_i) + O\left(\frac{1}{\sqrt{N}}\right). \quad \text{...13.81}$$

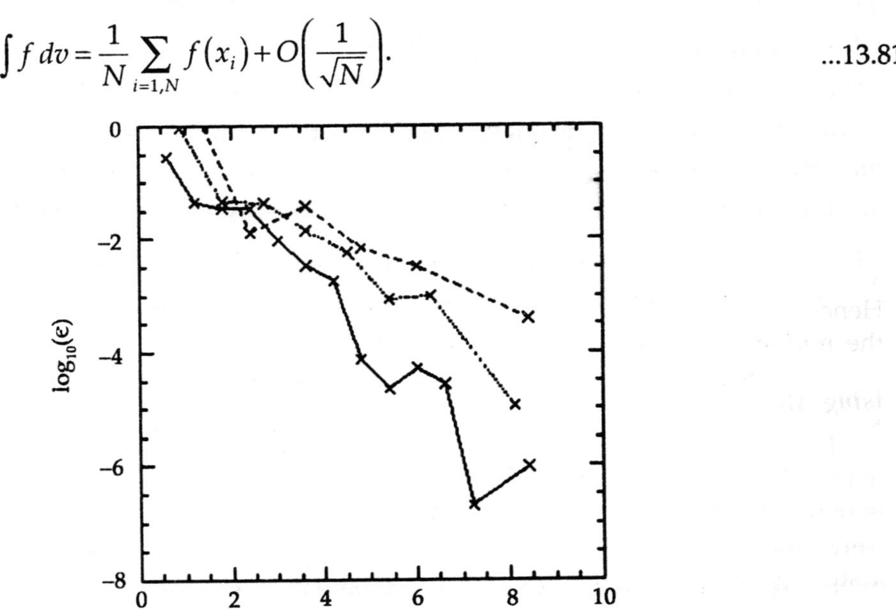

Figure 13.9: The integration error, ϵ, versus the number of grid-points, N, for three integrals evaluated using the midpoint method. The integrals are the area of a unit-radius circle (solid curve), the volume of a unit-radius sphere (dotted curve), and the volume of a unit-radius 4-sphere (dashed curve)

We end this section with an example calculation. Let us evaluate the volume of a unit-radius d-dimensional sphere, where d runs from 2 to 4, using both the midpoint and Monte-Carlo methods. For both methods, the domain of integration is a cube, centred on the sphere, which is such that the sphere just touches each face of the cube, as illustrated in figure 13.8.

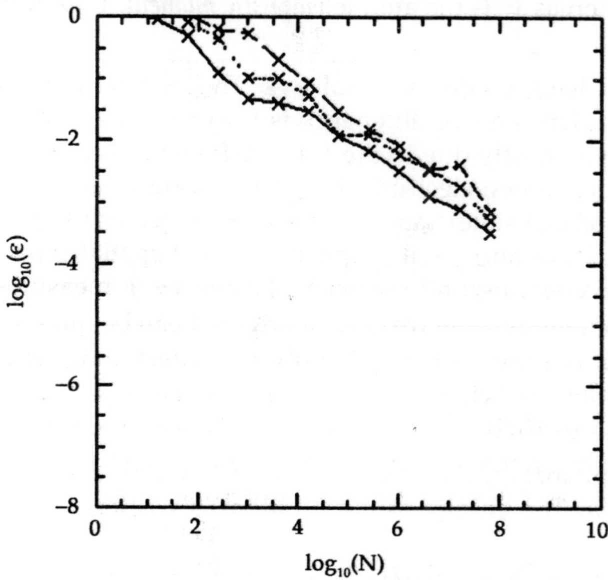

Figure 13.10: *The integration error, ε, versus the number of points, N, for three integrals evaluated using the Monte-Carlo method. The integrals are the area of a unit-radius circle (solid curve), the volume of a unit-radius sphere (dotted curve), and the volume of a unit-radius 4-sphere (dashed curve)*

Figure 13.9 shows the integration error associated with the midpoint method as a function of the number of grid-points, N. It can be seen that as the dimensionality of the integral increases the error falls off much less rapidly as N increases.

Figure 13.10 shows the integration error associated with the Monte-Carlo method as a function of the number of points, N. It can be seen that there is very little change in the rate at which the error falls off with increasing N as the dimensionality of the integral varies. Hence, as the dimensionality, d, increases the Monte-Carlo method eventually wins out over the midpoint method.

Ising Model

Ferromagnetism arises when a collection of atomic spins align such that their associated magnetic moments all point in the same direction, yielding a net magnetic moment which is macroscopic in size. The simplest theoretical description of ferromagnetism is called the *Ising model*. This model was invented by Wilhelm Lenz in 1920: it is named after Ernst Ising, a student of Lenz who chose the model as the subject of his doctoral dissertation in 1925.

Consider N atoms in the presence of a z-directed magnetic field of strength H. Suppose that all atoms are identical spin-1/2 systems. It follows that either $s_i = +1$ (spin up) or $s_i = -1$ (spin down), where s_i is (twice) the z-component of the ith atomic spin. The total energy of the system is written:

$$E = -J\sum_{<ij>} s_i s_j - \mu H \sum_{i=1,N} s_i. \qquad ...13.82$$

Here, $< ij >$ refers to a sum over *nearest neighbour* pairs of atoms. Furthermore, J is called the *exchange energy*, whereas μ is the atomic *magnetic moment*. Equation (13.82) is the essence of the Ising model.

The physics of the Ising model is as follows. The first term on the right-hand side of equation (13.82) shows that the overall energy is lowered when neighbouring atomic spins are aligned. This effect is mostly due to the *Pauli exclusion principle*. Electrons cannot occupy the same quantum state, so two electrons on neighbouring atoms which have parallel spins (*i.e.*, occupy the same orbital state) cannot come close together in space. No such restriction applies if the electrons have anti-parallel spins. Different spatial separations imply different electrostatic interaction energies, and the exchange energy, J, measures this difference. Note that since the exchange energy is *electrostatic* in origin, it can be quite large: i.e., $J \sim 1eV$. This is far larger than the energy associated with the direct magnetic interaction between neighbouring atomic spins, which is only about 10^{-4} eV. However, the exchange effect is very short-range; hence, the restriction to nearest neighbour interaction is quite realistic.

Our first attempt to analyse the Ising model will employ a simplification known as the *mean field approximation*. The energy of the ith atom is written

$$e_i = -\frac{J}{2}\sum_{k=1,z} s_k s_i - \mu H s_i, \qquad ...13.83$$

where the sum is over the z nearest neighbours of atom i. The factor 1/2 is needed to ensure that when we sum to obtain the total energy,

$$E = \sum_{i=1,N} e_i, \qquad ...13.84$$

we do not count each pair of neighbouring atoms twice.

We can write

$$e_i = -\mu H_{\text{eff}} s_i, \qquad ...13.85$$

where

$$H_{\text{eff}} = H + \frac{J}{2\mu}\sum_{k=1,z} s_k. \qquad ...13.86$$

Here, H_{eff} is the *effective magnetic field*, which is made up of two components: the external field, H, and the internal field generated by neighbouring atoms.

Consider a single atom in a magnetic field H_m. Suppose that the atom is in thermal equilibrium with a heat bath of temperature T. According to the well-known Boltzmann distribution, the mean spin of the atom is

$$\bar{s} = \frac{e^{+\beta\mu H_m} - e^{-\beta\mu H_m}}{e^{+\beta\mu H_m} + e^{-\beta\mu H_m}}, \qquad ...13.87$$

where $\beta = 1/kT$, and k is the Boltzmann constant. The above expression follows because the energy of the "spin up" state ($s = +1$) is $-\mu H_m$, whereas the energy of the "spin down" state ($s = -1$) is $+\mu H_m$. Hence,

$$\bar{s} = \tanh(\beta \mu H_m). \qquad ...13.88$$

Let us assume that all atoms have *identical* spins: i.e., $s_i = \bar{s}$. This assumption is known as the "mean field approximation". We can write

$$H_{eff} = H + \frac{zJ\bar{s}}{2\mu}. \qquad ...13.89$$

Finally, we can combine equations (13.88) and (13.89) (identifying H_m and H_{eff}) to obtain

$$\bar{s} = \tanh\{\beta\mu H + \beta zJ\bar{s}/2\}. \qquad ...13.90$$

Note that the heat bath in which a given atom is immersed is simply the rest of the atoms. Hence, T is the temperature of the atomic array. It is helpful to define the *critical temperature,*

$$T_c = \frac{zJ}{2k}, \qquad ...13.91$$

and the *critical magnetic field,*

$$H_c = \frac{kT_c}{\mu} = \frac{zJ}{2\mu}. \qquad ...13.92$$

Equation (13.90) reduces to

$$\bar{s} = \tanh\left\{\frac{T_c}{T}\left(\frac{H}{H_c} + \bar{s}\right)\right\}. \qquad ...13.93$$

The above equation cannot be solved analytically. However, it is fairly easily to solve numerically using the following iteration scheme:

$$\bar{s}_{i+1} = \tanh\left\{\frac{T_c}{T}\left(\frac{H}{H_c} + \bar{s}_i\right)\right\}. \qquad ...13.94$$

The above formula is iterated until $\bar{s}_{i+1} \to \bar{s}_i$.

It is helpful to define the *net magnetisation,*

$$M = \mu \sum_{i=1,N} s_i = \mu N \bar{s}, \qquad ...13.95$$

the net energy,

$$E = \sum_{i=1,N} e_i = -NkT_c\left(\frac{H}{H_c} + \bar{s}\right)\bar{s}, \qquad ...13.96$$

and the *heat capacity,*

$$C = \frac{dE}{dT}. \qquad ...13.97$$

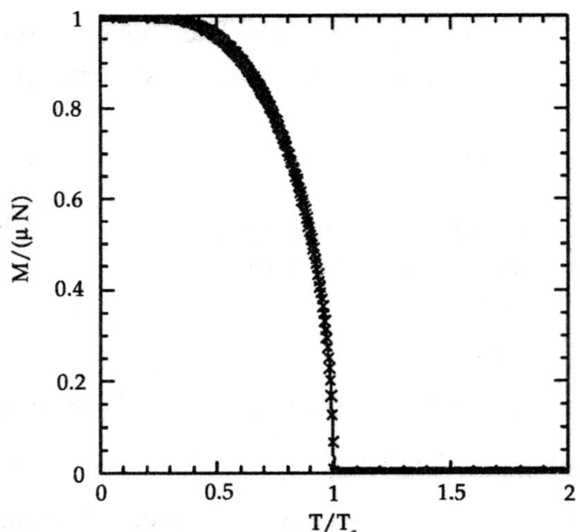

Figure 13.11: The net magnetisation, M, of a collection of N ferromagnetic atoms as a function of the temperature, T, in the absence of an external magnetic field. Calculation performed using the mean field approximation

Figures 13.12, 13.11, and 13.13 show the net magnetisation, net energy, and heat capacity calculated from the iteration formula (13.94) in the absence of an external magnetic field (*i.e.*, with $H = 0$). It can be seen that below the critical (or "Curie") temperature, T_c, there is *spontaneous magnetisation*: i.e., the exchange effect is sufficiently large to cause neighbouring atomic spins to spontaneously align. On the other hand, thermal fluctuations completely eliminate any alignment above the critical temperature. Moreover, at the critical temperature there is a *discontinuity* in the first derivative of the energy, E, with respect to the temperature, T. This discontinuity generates a downward jump in the heat capacity, C, at $T = T_c$. The sudden loss of spontaneous magnetisation as the temperature exceeds the critical temperature is a type of *phase transition*.

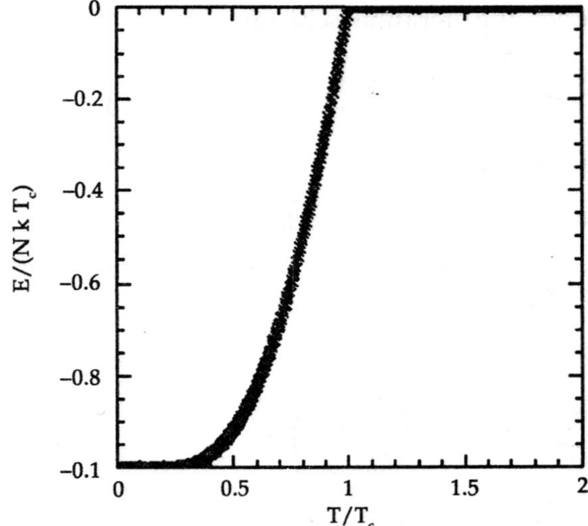

Figure 13.12: The net energy, E, of a collection of N ferromagnetic atoms as a function of the temperature, T, in the absence of an external magnetic field. Calculation performed using the mean field approximation

Now, according to the conventional classification of *phase transitions*, a transition is *first-order* if the energy is discontinuous with respect to the order parameter (*i.e.*, in this case, the temperature), and *second-order* if the energy is continuous, but its first derivative with respect to the order parameter is discontinuous, etc. We conclude that the loss of spontaneous magnetisation in a ferromagnetic material as the temperature exceeds the critical temperature is a second-order phase transition.

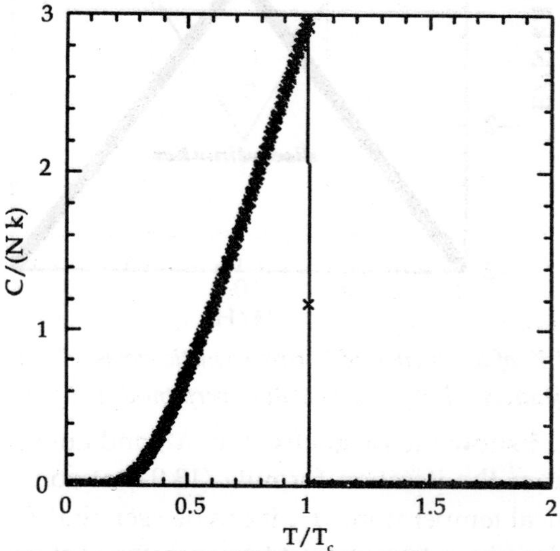

Figure 13.13: *The heat capacity, C, of a collection of N ferromagnetic atoms as a function of the temperature, T, in the absence of an external magnetic field. Calculation performed using the mean field approximation*

In order to see an example of a first-order phase transition, let us examine the behaviour of the magnetisation, M, as the external field, H, is varied at constant temperature, T.

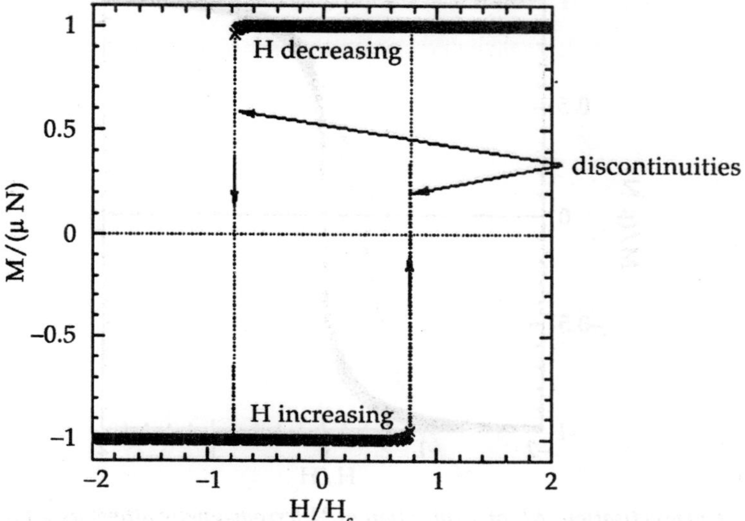

Figure 13.14: *The net magnetisation, M, of a collection of N ferromagnetic atoms as a function of the external magnetic field, H, at constant temperature, $T < T_c$. Calculation performed using the mean field approximation*

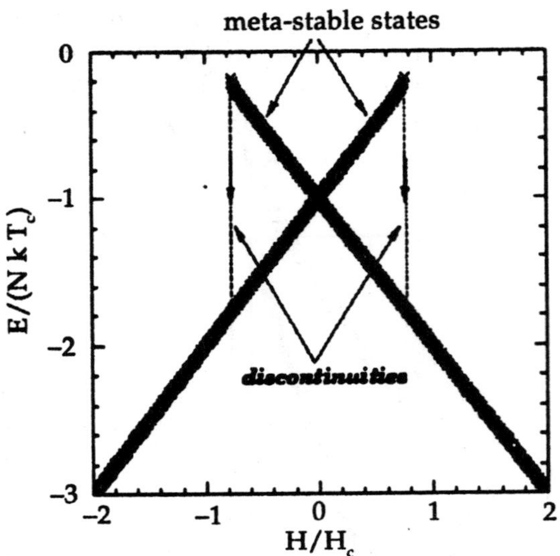

Figure 13.15: The net energy, E, of a collection of N ferromagnetic atoms as a function of the external magnetic field, H, at constant temperature, $T < T_c$. Calculation performed using the mean field approximation

Figures 13.14 and 13.15 show the magnetisation, M, and energy, E, versus external field-strength, H, calculated from the iteration formula (13.94) at some constant temperature, T, which is less than the critical temperature, T_c. It can be seen that E is *discontinuous*, indicating the presence of a first-order phase transition. Moreover, the system exhibits *hysteresis*—meta-stable states exist within a certain range of H values, and the magnetisation of the system at fixed T and H (within the aforementioned range) depends on its *past history*: i.e., on whether H was increasing or decreasing when it entered the meta-stable range.

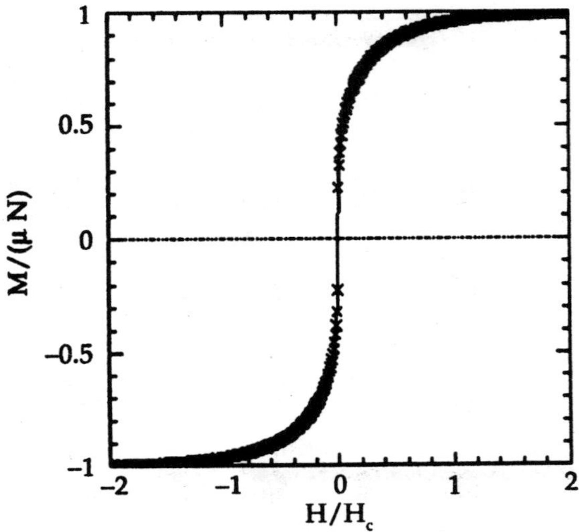

Figure 13.16: The net magnetisation, M, of a collection of N ferromagnetic atoms as a function of the external magnetic field, H, at constant temperature, $T = T_c$. Calculation performed using the mean field approximation

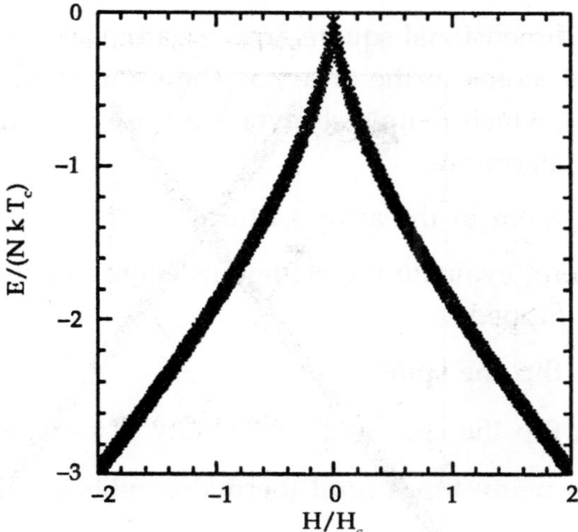

Figure 13.17: *The net energy, E, of a collection of N ferromagnetic atoms as a function of the external magnetic field, H, at constant temperature,* $T = T_c$. *Calculation performed using the mean field approximation*

Figures 13.16 and 13.17 show the magnetisation, *M*, and energy, *E*, versus external field-strength, *H*, calculated from the iteration formula (13.94) at a constant temperature, *T*, which is equal to the critical temperature, T_c. It can be seen that *E* is now *continuous*, and there are no meta-stable states. We conclude that first-order phase transitions and hysteresis only occur, as the external field-strength is varied, when the temperature lies below the critical temperature: i.e., when the ferromagnetic material in question is capable of spontaneous magnetisation.

The above calculations, which are based on the mean field approximation, correctly predict the existence of first- and second-order phase transitions when $H \neq 0$ and $H = 0$, respectively. However, these calculations get some of the details of the second-order phase transition wrong. In order to do a better job, we must abandon the mean field approximation and adopt a Monte-Carlo approach.

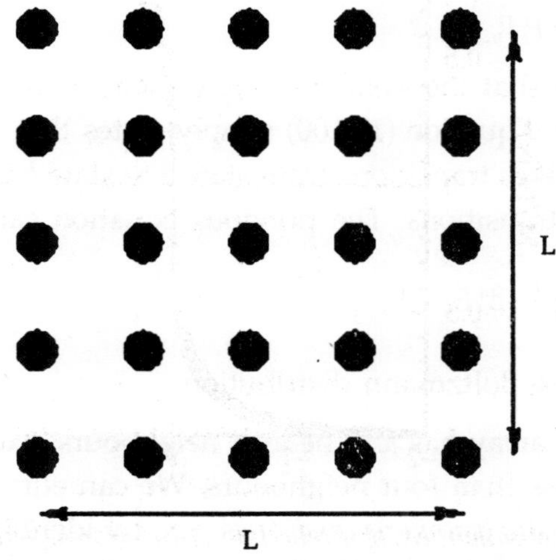

Figure 13.18: *A two-dimensional array of atoms*

Let us consider a two-dimensional square array of atoms. Let L be the size of the array, and $N = L^2$ the number of atoms in the array, as shown in figure 13.18. The Monte-Carlo approach to the Ising model, which completely avoids the use of the mean field approximation, is based on the following algorithm:

- Step through each atom in the array in turn:
  - For a given atom, evaluate the change in energy of the system, ΔE, when the atomic spin is flipped.
  - If $\Delta E < 0$ then flip the spin.
  - If $\Delta E > 0$ then flip the spin with probability $P = \exp(-\beta\Delta E)$.
- Repeat the process many times until thermal equilibrium is achieved.

The purpose of the algorithm is to shuffle through all possible states of the system, and to ensure that the system occupies a given state with the Boltzmann probability: i.e., with a probability proportional to $\exp(-\beta E)$, where E is the energy of the state.

In order to demonstrate that the above algorithm is correct, let us consider flipping the spin of the *i*th atom. Suppose that this operation causes the system to make a transition from state *a* (energy, E_a) to state *b* (energy, E_b). Suppose, further, that $E_a < E_b$. According to the above algorithm, the probability of a transition from state *a* to state *b* is

$$P_{a\to b} = \exp[-\beta(E_b - E_a)], \qquad \text{...13.98}$$

whereas the probability of a transition from state *b* to *a* state is

$$P_{a\to b} = 1. \qquad \text{...13.99}$$

In thermal equilibrium, the well-known *principal of detailed balance* implies that

$$P_a P_{a\to b} = P_b P_{b\to a}, \qquad \text{...13.100}$$

where P_a is the probability that the system occupies state *a*, and P_b is the probability that the system occupies state *b*. Equation (13.100) simply states that in thermal equilibrium the rate at which the system makes transitions from state *a* to state *b* is equal to the rate at which the system makes reverse transitions. The previous equation can be rearranged to give

$$\frac{P_b}{P_a} = \exp[-\beta(E_b - E_a)], \qquad \text{...13.101}$$

which is consistent with the Boltzmann distribution.

Now, each atom in our array has *four* nearest neighbours, except for atoms on the edge of the array, which have less than four neighbours. We can eliminate this annoying special behaviour by adopting *periodic boundary conditions*: i.e., by identifying opposite edges of the array. Indeed, we can think of the array as existing on the surface of a torus.

It is helpful to define

$$T_0 = \frac{J}{k}. \qquad \text{...13.102}$$

Now, according to mean field theory,

$$T_c = \frac{zJ}{2k} = 2T_0. \qquad \text{...13.103}$$

The evaluation of

$$C = \lim_{\Delta T \to 0} \frac{\Delta E}{\Delta T} \qquad \text{...13.104}$$

via the direct method is difficult due to statistical noise in the energy, E. Instead, we can make use of a standard result in equilibrium statistical thermodynamics:

$$C = \frac{\sigma_E^2}{kT^2}, \qquad \text{...13.105}$$

where σ_E is the standard deviation of fluctuations in E. Fortunately, it is fairly easy to evaluate σ_E: we can simply employ the standard deviation in E from step to step in our Monte-Carlo iteration scheme.

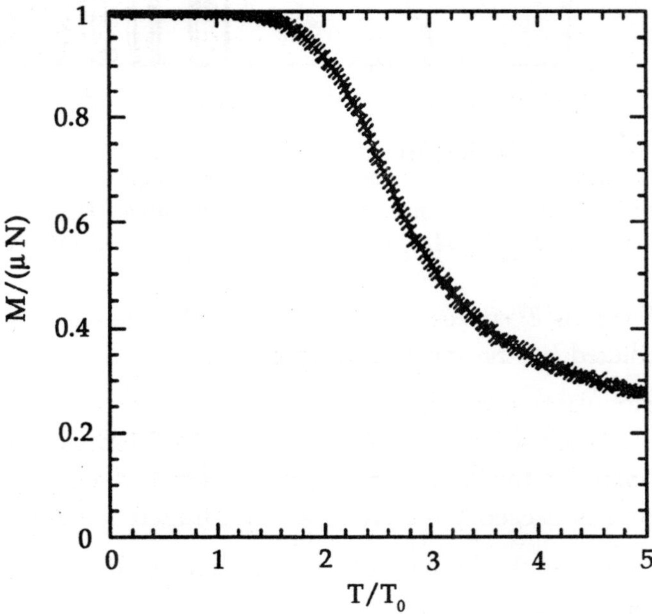

Figure 13.19: *The net magnetisation, M, of a 5×5 array of ferromagnetic atoms as a function of the temperature, T, in the absence of an external magnetic field. Monte-Carlo simulation*

Figures 13.19-13.26 show magnetisation and heat capacity versus temperature curves for $L = 5$, 10, 20, and 40 in the absence of an external magnetic field. In all cases, the Monte-Carlo simulation is iterated 5000 times, and the first 1000 iterations are discarded when evaluating σ_E (in order to allow the system to attain thermal equilibrium). The two-dimensional array of atoms is initialised in a fully aligned state for each different value of the temperature. Since there is no external magnetic field, it is irrelevant whether the magnetisation, M, is positive or negative. Hence, M is replaced by $|M|$ in all plots.

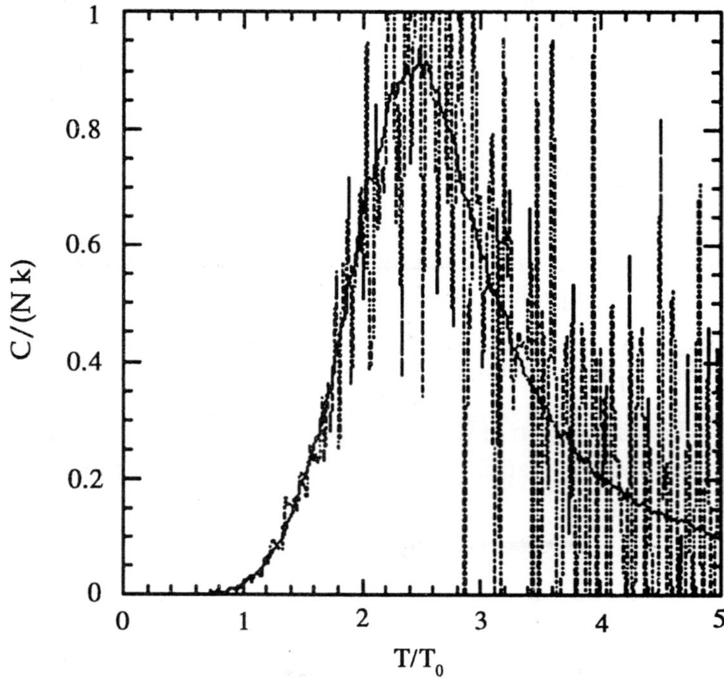

Figure 13.20: *The heat capacity, C, of a 5×5 array of ferromagnetic atoms as a function of the temperature, T, in the absence of an external magnetic field. Monte-Carlo simulation. The solid curve shows the heat capacity calculated from equation (13.104), whereas the dotted curve shows the heat capacity calculated from equation (13.105)*

Note that the M versus T curves generated by the Monte-Carlo simulations look very much like those predicted by the mean field model.

The resemblance increases as the size, L, of the atomic array increases. The major difference is the presence of a magnetisation "tail" for $T > T_c$ in the Monte-Carlo simulations: i.e., in the Monte-Carlo simulations the spontaneous magnetisation does not collapse to zero once the critical temperature is exceeded—there is a small lingering magnetisation for $T > T_c$.

The C versus T curves show the heat capacity calculated directly (*i.e.*, $C = \Delta E / \Delta T$), and via the identity $C = \sigma_E^2 / kT^2$.

The latter method of calculation is clearly far superior, since it generates significantly less statistical noise. Note that the heat capacity *peaks* at the critical temperature: i.e., unlike the mean field model, C is not zero for $T > T_c$. This effect is due to the residual magnetisation present when $T > T_c$.

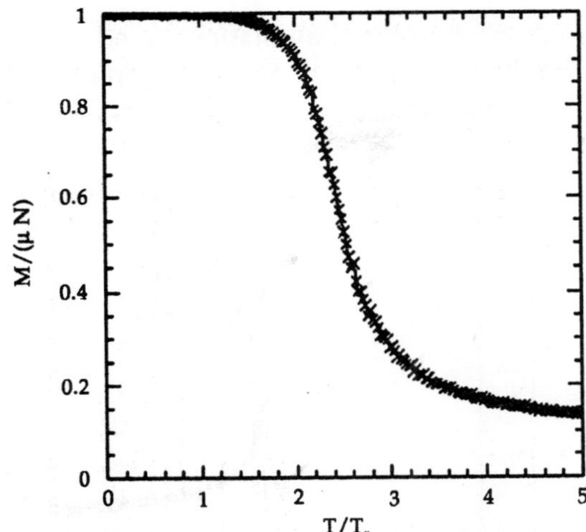

Figure 13.21: The net magnetisation, M, of a 10×10 array of ferromagnetic atoms as a function of the temperature, T, in the absence of an external magnetic field. Monte-Carlo simulation

Our best estimate for T_c is obtained from the location of the peak in the C versus T curve in figure 13.26. We obtain $T_c = 2.27T_0$. Recall that the mean field model yields $T_c = 2T_0$. The exact answer for a two-dimensional array of ferromagnetic atoms is

$$T_c = \frac{2T_0}{\ln\left(1+\sqrt{2}\right)} = 2.27T_0, \qquad \text{...13.106}$$

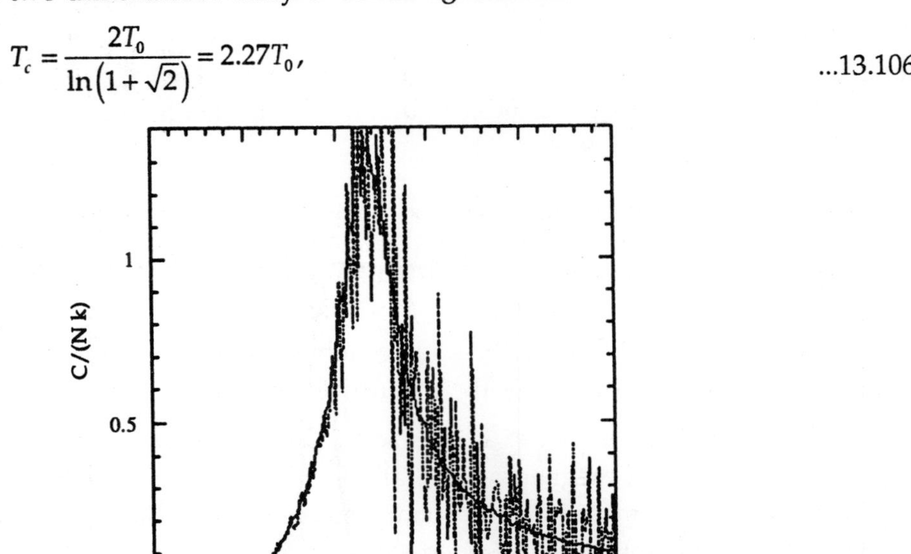

Figure 13.22: The heat capacity, C, of a 10×10 array of ferromagnetic atoms as a function of the temperature, T, in the absence of an external magnetic field. Monte-Carlo simulation. The solid curve shows the heat capacity calculated from equation (13.104), whereas the dotted curve shows the heat capacity calculated from equation (13.105)

which is consistent with our Monte-Carlo calculations. The above analytic result was first obtained by Onsager in 1944. Incidentally, Onsager's analytic solution of the 2-D Ising model is one of the most complicated and involved calculations in all of theoretical physics. Needless to say, no one has ever been able to find an analytic solution of the Ising model in more than two dimensions.

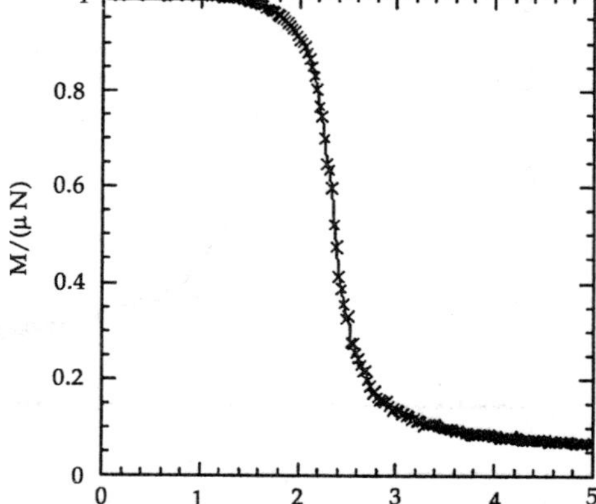

Figure 13.23: *The net magnetisation, M, of a 20×20 array of ferromagnetic atoms as a function of the temperature, T, in the absence of an external magnetic field. Monte-Carlo simulation*

Note, from figure 13.20, 13.22, 13.24, and 13.26, that the height of the peak in the heat capacity curve at $T = T_c$ increases with increasing array size, L. Indeed, a close examination of these figures yields $C_{max}/Nk = 0.95$, for $L = 5$, $C_{max}/Nk = 1.34$ for $L = 10$, $C_{max}/Nk = 1.77$ for $L = 20$ and $C_{max}/Nk = 2.16$ for $L = 40$. Figure 13.27 shows C_{max}/Nk plotted against $\ln L$ for $L = 5$, 10, 20, and 40. It can be seen that the points lie on a very convincing straight-line, which strongly suggests that

$$\frac{C_{max}}{kN} \propto \ln L. \qquad \text{...13.107}$$

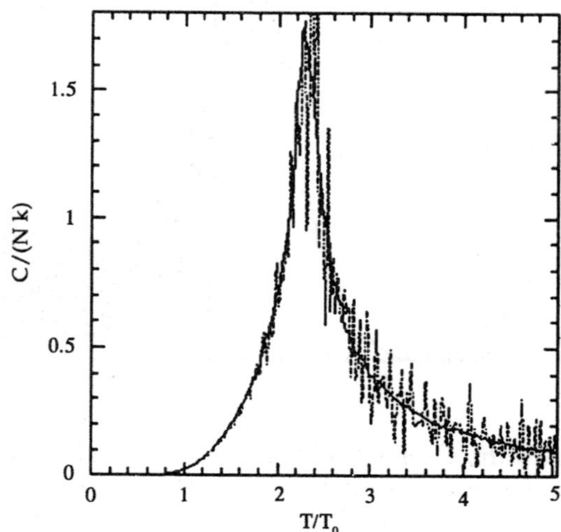

Figure 13.24: *The heat capacity, C, of a 20×20 array of ferromagnetic atoms as a function of the temperature, T, in the absence of an external magnetic field. Monte-Carlo simulation. The solid curve shows the heat capacity calculated from equation (13.104), whereas the dotted curve shows the heat capacity calculated from equation (13.105)*

Of course, for physical systems,

$$L \sim \sqrt{N_A} \sim 10^{12},$$

where N_A is Avogadro's number. Hence, C is effectively *singular* at the critical temperature (since $\ln N_A \gg 1$), as sketched in figure 13.28. This observation leads us to revise our definition of a second-order phase transition. It turns out that actual discontinuities in the heat capacity almost never occur. Instead, second-order phase transitions are characterised by a *local quasi-singularity* in the heat capacity.

Recall, from equation (13.105), that the typical amplitude of energy fluctuations is proportional to the square-root of the heat capacity (*i.e.*, $\sigma_E \propto \sqrt{C}$). It follows that the amplitude of energy fluctuations becomes *extremely large* in the vicinity of a second-order phase transition.

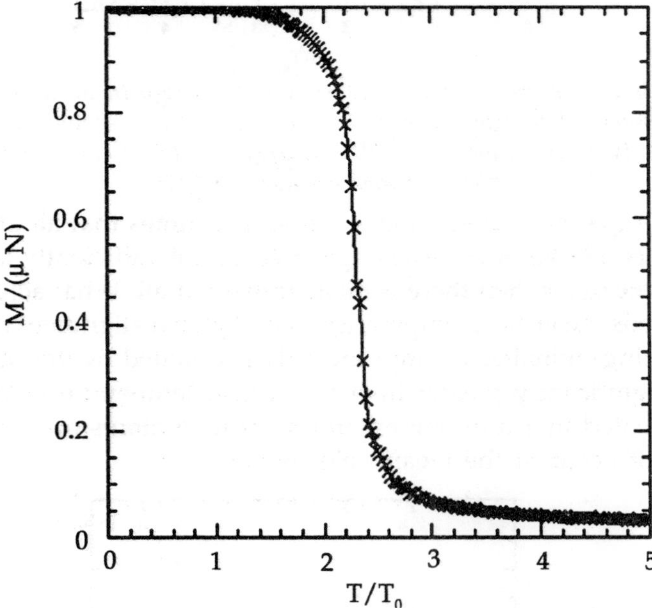

Figure 13.25: *The net magnetisation, M, of a 40×40 array of ferromagnetic atoms as a function of the temperature, T, in the absence of an external magnetic field. Monte-Carlo simulation*

Now, the main difference between our mean field and Monte-Carlo calculations is the existence of residual magnetisation for $T > T_c$ in the latter case. Figures 13.29-13.33 show the magnetisation pattern of a 40×40 array of ferromagnetic atoms, in thermal equilibrium and in the absence of an external magnetic field, calculated at various temperatures. It can be seen that for $T = 20T_0$ the pattern is essentially random. However, for $T = 5T_0$, small clumps appear in the pattern. For $T = 3T_0$, the clumps are somewhat bigger. For $T = 2.32T_0$, which is just above the critical temperature, the clumps are global in extent. Finally, for $T = 1.8T_0$, which is a little below the critical temperature, there is almost complete alignment of the atomic spins.

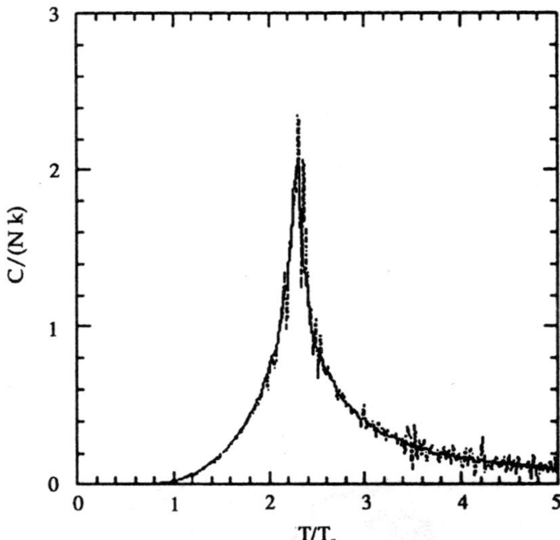

Figure 13.26: The heat capacity, C, of a 40×40 array of ferromagnetic atoms as a function of the temperature, T, in the absence of an external magnetic field. Monte-Carlo simulation. The solid curve shows the heat capacity calculated from equation (13.104), whereas the dotted curve shows the heat capacity calculated from equation (13.105)

The problem with the mean field model is that it assumes that all atoms are situated in identical environments. Hence, if the exchange effect is not sufficiently large to cause global alignment of the atomic spins then there is no alignment at all. What actually happens when the temperature exceeds the critical temperature is that global alignment disappears, but local alignment (*i.e.*, clumping) remains. Clumps are only eliminated by thermal fluctuations once the temperature is significantly greater than the critical temperature. Atoms in the middle of the clumps are situated in a different environment than atoms on the clump boundaries. Hence, clumps cannot occur in the mean field model.

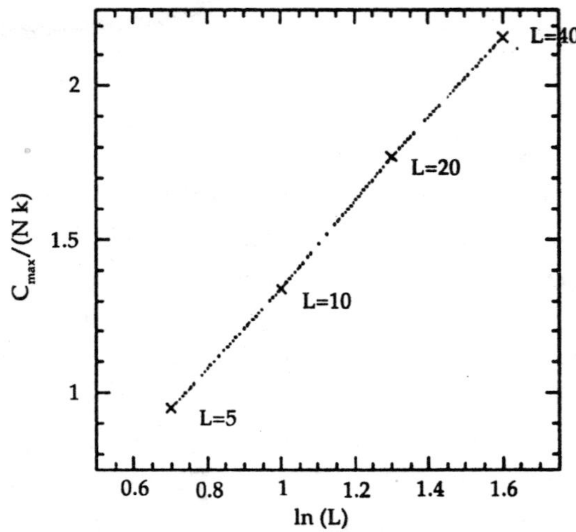

Figure 13.27: The peak value of the heat capacity (normalised by N k) versus the logarithm of the array size for a two-dimensional array of ferromagnetic atoms in the absence of an external magnetic field. Monte-Carlo simulation

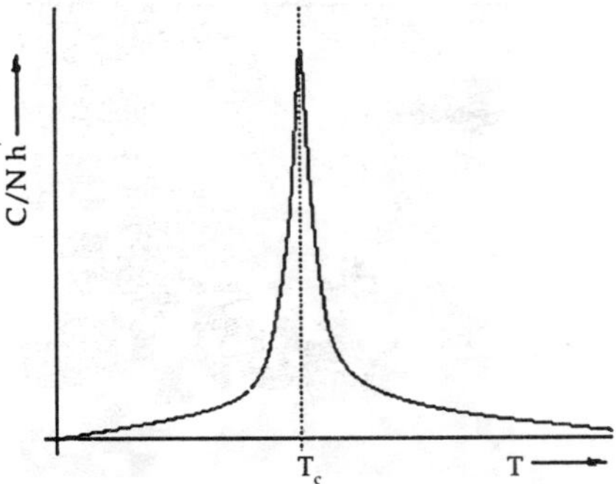

Figure 13.28: *A sketch of the expected variation of the heat capacity versus the temperature for a physical two-dimensional ferromagnetic system*

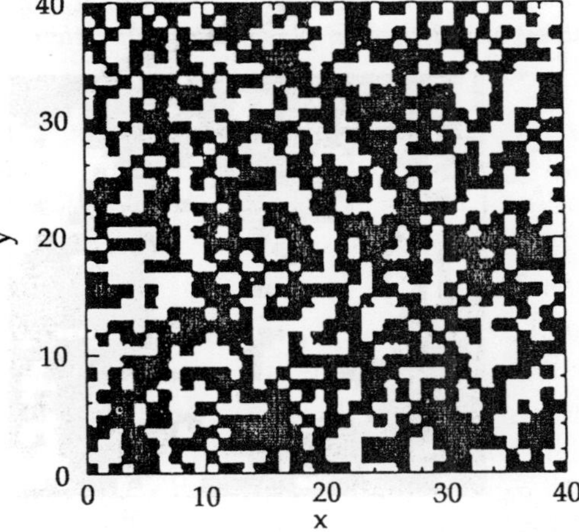

Figure 13.29: *Magnetisation pattern of 40×40 a array of ferromagnetic atoms in thermal equilibrium and in the absence of an external magnetic field. Monte-Carlo calculation with* $T = 20T_0$. *Black/white squares indicate atoms magnetised in plus/minus z-direction, respectively*

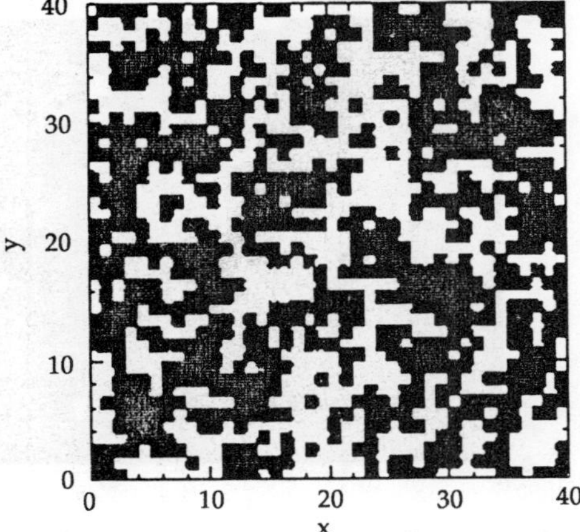

Figure 13.30: *Magnetisation pattern of a 40×40 array of ferromagnetic atoms in thermal equilibrium and in the absence of an external magnetic field. Monte-Carlo calculation with* $T = 5T_0$. *Black/white squares indicate atoms magnetised in plus/minus z-direction, respectively*

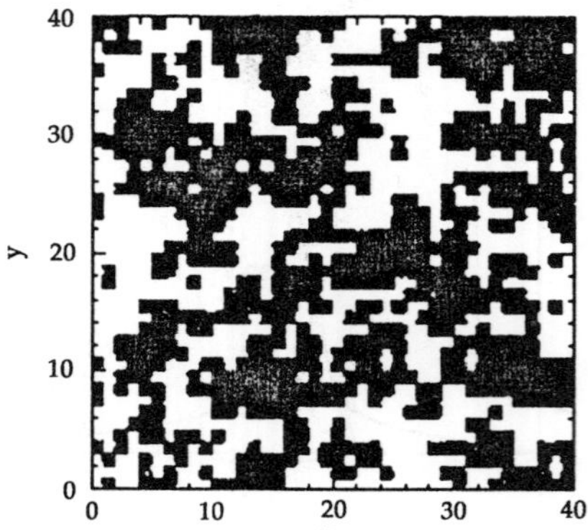

Figure 13.31: Magnetisation pattern of a 40×40 array of ferromagnetic atoms in thermal equilibrium and in the absence of an external magnetic field. Monte-Carlo calculation with $T = 3T_0$. *Black/white squares indicate atoms magnetised in plus/minus z-direction, respectively*

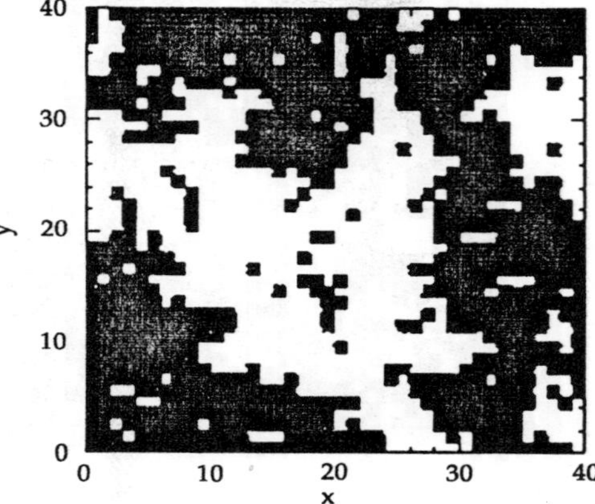

Figure 13.32: Magnetisation pattern of a 40×40 array of ferromagnetic atoms in thermal equilibrium and in the absence of an external magnetic field. Monte-Carlo calculation with $T = 2.32T_0$. *Black/white squares indicate atoms magnetised in plus/minus z-direction, respectively*

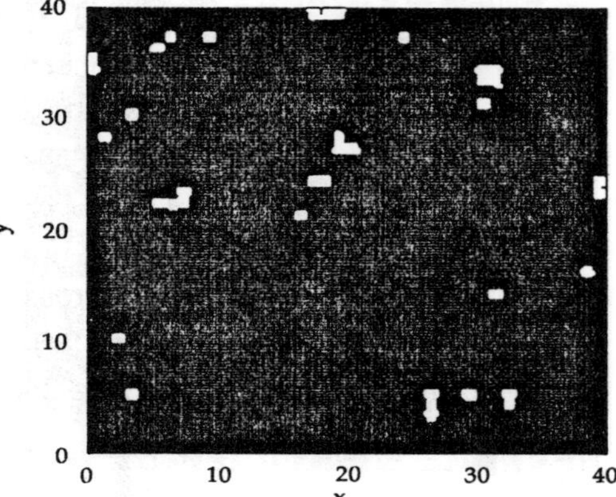

Figure 13.33: Magnetisation pattern of a 40×40 array of ferromagnetic atoms in thermal equilibrium and in the absence of an external magnetic field. Monte-Carlo calculation with $T = 1.8T_0$. *Black/white squares indicate atoms magnetised in plus/minus z-direction, respectively*

14

Temperature of Ising Spin Models

The behaviour of real magnetic systems depends on temperature, so obviously there has to be some aspect of the model which corresponds to physical temperature.

In physics energy is any measurable quantity with dimension *mass.length²/time²* (e.g., kinetic energy = $\frac{1}{2}mv^2$ for an object of mass m moving with speed v). In molecular dynamics, temperature is given by the average translational molecular kinetic energy of the molecules. In thermodynamics, temperature is the integrating factor of the differential equation referred to as *the first law of thermodynamics*, $du = c_v dT$ for a perfect gas, where du is the internal energy proportional to the temperature change, c_v is the specific heat at constant volume and T is the Kelvin temperature.

However in the definition of a spin model, there is nothing corresponding to the physical property of temperature. But we do have a quantity of energy, namely, the interaction energy between spins, denoted by J. It is convenient to assign to J the value 1.380658 x 10^{-23} joules, since this is the same numerical value as Boltzmann's constant, $k_B = 1.380658 \times 10^{-23}$ J/K, implying that $J/k_B = 1$ (this ratio has units of degrees Kelvin).

Temperature enters when we consider the dynamics of a spin system, specifically, when we consider the probability that a spin will change its spin value. The parameter T which enters these calculations is the analogue of temperature. The units of ΔE_{ji} are joules, the units of k_B are joules per degree Kelvin, and the units of T are degrees K, so the quantity $\Delta E_{ji}/(k_B T)$ is dimensionless, as it should be if it is to be used as an exponent of e in the definitions of $P(S_i \rightarrow S_j)$.

Cluster Flip Dynamics Algorithms

The Metropolis and the Glauber algorithms may (or may not) be close to reality in the way they represent the dynamics of real systems, but in computer simulations they are slow to achieve a condition of equilibrium when the temperature is close to the critical temperature. For this reason algorithms were developed which are more efficient (in the sense that equilibrium is attained more quickly).

The Swendsen-Wang Algorithm

One of these is the Swendsen-Wang algorithm, which is as follows: In a particular state of a spin system there are clusters of spins. *A spin cluster* consists of a set of spins, each of which is a nearest neighbour to at least one other spin in the cluster, with all spins having the same spin value. The Swendsen-Wang process consists of two parts: The first part is the creation of "virtual" spin clusters, as follows: If two spins are connected by an open lattice bond and those spins have the same value then a "virtual" bond is created between them with probability $1 - e^{-2/T}$, where T is the temperature. Thereby each spin cluster is partitioned into (smaller) virtual spin clusters. The second part of the Swendsen-Wang process is as follows: For each virtual spin cluster, select a spin value at random (from among all possible spin values) and assign that spin value to all spins in the virtual cluster. This process is then repeated.

One application of this process results in each spin being considered for change once and once only, and so is equivalent to a single sweep through the lattice in the Metropolis and the Glauber algorithms.

The Wolff Algorithm

The Wolff algorithm is as follows: A spin is chosen at random as the seed spin for the growth of a virtual spin cluster. The cluster is grown by adding spins as follows: For each spin added (starting with the seed spin) each of its nearest neighbour spin with the same spin value is added to the virtual cluster with a probability of $1 - e^{-2/T}$ (the same as in the Swendsen-Wang algorithm). This process is repeated until no new spin is added. Then, in the case of Ising spins, all spins in the virtual cluster are flipped (to assume their opposite values), and in the case of the q-state Potts model all spins receive a spin value which is randomly chosen except that it is different from their current value. The process is then repeated.

Time

A model of a physically real system must have an analogue of time. In the Metropolis, Glauber and Swendsen-Wang algorithms there is a natural unit of (analogous) time, namely, a single sweep through the lattice, during which each spin is considered once and once only for change.

For the Wolff algorithm the situation is less straightforward. We might take a cluster flip as a unit of time, but since the size of the Wolff cluster may vary over two orders of magnitude between low and high temperature, and the size of a Wolff cluster is inversely proportional to temperature, we would then have to say that many more spins are flipped per unit of time at low temperature than at high temperature, which is contrary to what we find in real systems.

An alternative approach to defining a unit of time for use with the Wolff algorithm is as follows: In a Swendsen-Wang time unit (assuming the Ising model) approximately half of the spins get flipped (since approximately half of the Swendsen-Wang clusters get flipped). Every Wolff cluster gets flipped, so in the Wolff process we can take as a time unit a sequence of a variable number of Wolff cluster flips such that the sum of the cluster sizes (which equals the number of spins flipped) equals (on average) one-half the number of spins. (At low temperatures most Wolff clusters are large, so we have to adjust continually the target number of Wolff clusters making up a time unit in order to maintain this relation between the number of spins flipped and the number of units of time elapsed.) Then the number of spin flips (on average) is the same in one unit of time for the Wolff algorithm as for the Swendsen-Wang algorithm.

For the q-state Potts model, in a Swendsen-Wang time unit approximately (q-1)/q of the spins get flipped, since for each virtual spin cluster the probability of the new spin value being different from the old is (q-1)/q. Every Wolff cluster gets flipped, so in the Wolff process with the q-state Potts model we can take as a time unit a sequence of a variable number of Wolff cluster flips such that the sum of the cluster sizes equals (q-1)/q times the number of spins.

In the simulations done in this research using the Wolff algorithm time is measured in terms of the units defined in the previous two paragraphs.

Boundary Conditions

Every computational model is finite, and no singularities or discontinuities are ever observed in a finite system. The discontinuities and singularities which are of interest in the study of critical phenomena occur only in ideal infinite systems, and cannot be observed in any model which can be realised physically (and thus not in a computer model). The behaviour of actual models is only an approximation to ideal behaviour.

The difficulty of using a finite lattice to simulate an infinite lattice can be eased somewhat by the use of so-called *periodic boundary conditions*. We wish to embed the finite lattice in a virtual infinite lattice consisting of multiple copies of the finite lattice. This is accomplished by arranging for sites on one edge (or face, etc.) of the finite lattice to have as nearest neighbours sites on the opposite edge (or face, etc.), so that all sites have the same number of nearest neighbours whether or not they occur on an edge (or face) of the lattice. All simulations in this study are performed using periodic boundary conditions.

If sites on an edge (or face) have only the usual set of nearest neighbours (less in number than sites in the interior of the lattice) then the model is said to possess *free boundary conditions*. Sometimes mixed boundary conditions are used, whereby the lattice "wraps around" only at some edges, e.g. to obtain (in the case of a square lattice) a cylindrical topology (compared to a toroidal topology in the case of periodic boundary conditions on a square lattice).

Other kinds of boundary condition are possible, such as the *heliacal*, obtained by taking a chain of sites and placing it on a helix so that n sites form one loop (thus creating a square lattice geometry), resulting in site m having as nearest neighbours sites n+m (above) and n-m.

Finite-Size Effects

As noted above, the discontinuities and singularities which are of interest in the study of critical phenomena occur only in ideal infinite systems. E.g., in the infinite square Ising model there is a discontinuity at T_c in the rate of change of magnetisation M with respect to temperature T. For $T > T_c$, $M=0$, whereas for $T < T_c$ there is a spontaneous non-zero magnetisation. At T_c (moving from higher temperature to lower) there is a change from a constant zero M to an increasing M. But in a finite system the mean absolute value of M is non-zero even above T_c and there is no discontinuity in the rate of change at T_c. Only as the size of the model increases is there a better approximation to the ideal behaviour. Unfortunately larger systems require much longer computation time.

Estimates for various properties of the infinite system (such as critical exponents) may be made by assuming the validity of so-called *finite-size scaling*. Measurements of a quantity are made with a series of increasing lattice sizes, *L1, L2, L3*, ..., and the quantity is then plotted against the (decreasing) reciprocal of the size. If a linear fit to the data points is possible then we can extrapolate to zero to obtain an estimate of the value for a lattice of infinite size.

It is important to note that in an infinite system the non-zero spontaneous magnetisation is stable, but in a finite system it is "metastable", i.e., it will occasionally (due to unlikely sequences of spin flips mostly to the opposite of the dominant value) flip to the same value of opposite sign, where it stays for awhile before flipping back again.

Thus in spin model studies, with lattices of finite size, below T_c it is often advisable (especially when using cluster flip algorithms) to monitor the absolute value of the magnetisation, rather than the magnetisation itself, since if results are obtained by averaging (at each timepoint) over many runs the metastable positive magnetisation and the metastable negative magnetisation will average out to zero, and no non-zero stable magnetisation will be observed.

Critical Temperatures of Pure Ising Spin Models

Binder Cumulant

Binder and Heerman (1988) define the *reduced fourth-order cumulant* (a.k.a. the *Binder cumulant*) for a lattice of linear size L as:

$$U_L = 1 - \left\{ M^{(4)}{}_L / \left[3.\left(M^{(2)}{}_L \right)^2 \right] \right\}$$

where $M^{(2)}{}_L$ is the mean of the squares of the magnetisation (with lattice size L) and $M^{(4)}{}_L$ is the mean of the fourth powers of the magnetisation (with averages taken over systems at equilibrium at a constant temperature). They state:

For $T > T_c$ and $L >> \xi$ [the correlation length], one can show that U_L decreases towards zero as $U_L \propto L^{-d}$. For $T < T_c$ and $L >> \xi$ ... U_L tends to $U_\infty = 2/3$. For $L << \xi$, on the other hand, U_L varies only weakly with temperature and linear dimension ... This behaviour of the cumulant makes it very useful for obtaining estimates of T_c ... One may plot U_L versus T for various L's and estimate T_c from the common intersection point of these curves."

Thus suppose we plot U_L against T for some lattice of size L. Then for another lattice of size $L' > L$ and any T, if $T < T_c$ we shall obtain $U_{L'} > U_L$, and if $T > T_c$ then $U_{L'} < U_L$. Thus when graphs of U_L against T are plotted for a number of lattice sizes they should intersect at a point (or at least their pairwise intersections should be fairly close), the T-value of the intersection point giving an estimate of T_c.

Looked at from another perspective, for a suitable scaling function U

$$U_L = U[(T/T_c - 1).L^{1/v}]$$

where v is the critical exponent of the correlation length .

Thus when $T = T_c$ $U_L = U(0.L^{1/v}) = U(0)$, so at T_c U_L has a value independent of L, and so the graphs of U_L against T for various L should all pass through a single point at T_c.

The results of the simulations described on the following pages are compared with values for the critical temperatures for the various lattice types given by Fisher, which were obtained by series expansion. Occasionally a result will be compared with a more recently published result obtained by Monte Carlo studies.

Some simulations were done using Swendsen-Wang dynamics and some using Wolff dynamics.

Thermal Equilibrium

When measuring the properties of a system it is usual to ensure that the system is in equilibrium. By definition a system is in equilibrium when its bulk properties remain constant (or at least fluctuate closely around a constant mean value) over time, or more exactly, over a time period long enough in the context of the study.

Suppose we boil water, place it in a beaker, and proceed to observe its temperature. Under normal conditions the water will cool and after perhaps an hour it will be at the same temperature as the surrounding air. If the temperature does not change (except for minute fluctuations) over, say, an hour, we can say that the water in the cup has reached thermal equilibrium. Yet if we wait twelve hours we shall probably find that the temperature of the water has changed, because the air temperature has changed, so had the water really reached thermal equilibrium?

Since the properties of spin models do not depend on properties of the systems used to simulate them (such as the temperature of the computer on which the simulation is performed) the question of when thermal equilibrium is attained is somewhat simpler in this case. But basically the idea is the same: A spin system is in thermal equilibrium, with respect to a set of measurable properties (e.g., magnetisation and autocorrelation), if those properties have remained constant (or at least have fluctuated closely around a constant mean) over a period of time (or rather, the analogue of time in the model) judged by the investigator to be sufficiently long that no change is likely in the absence of external influences.

Thus if we wish to measure, e.g., the magnetisation of a spin system at equilibrium, we must simply allow the system to evolve until the magnetisation appears stable. Or rather, the average magnetisation, since normally we take averages of measurements at a particular timepoint over many runs.

If we use the Metropolis or the Glauber algorithms to drive the spin system then many timesteps may be required before the magnetisation becomes stable, especially if the temperature is close to the critical temperature of the system. For this reason dynamics algorithms were developed which cause the spin system to reach equilibrium much more quickly. In this study the Swendsen-Wang and the Wolff algorithms have usually been used, because with these the system typically attains an equilibrium state after less than 40 timesteps.

When measuring some quantity such as the Binder cumulant, it is thus necessary to ascertain first, by experiment, how long it takes for this to stabilise under the algorithm in use, for the lattice sizes and the temperatures at which one wishes to measure the quantity. Then one must decide for how many further timesteps the simulation will be run, since the final measurement of the quantity is the average of the quantity as measured over all timepoints following that at which equilibrium is deemed to have been reached.

For example, suppose we wish to measure the critical temperature, using measurement of the Binder cumulant, of the 2d Ising model on the triangular lattice, and that we plan to use lattice sizes of 20, 30, 40 and 60, temperatures in the range 3.5 through 3.8, and to average over 1000 sample runs in each case. How many timesteps are required for the system to reach equilibrium, and for how long should we run the system further?

Four simulations were done, using the Swendsen-Wang algorithm, for the four extreme cases: Lattice sizes 20 and 60, and temperatures 3.5 and 3.8. The measurements of the absolute magnetisation and of the Binder cumulant over timepoints 0 through 80 are shown in Figures 14.1 and 14.2. This shows that for $T = 3.5$ these quantities stabilise by timepoint 20 and fluctuate hardly at all. For $T = 3.8$ they appear to fluctuate about a mean from timepoint 30 onwards, and in the case of the Binder cumulant the fluctuations are larger for lattice size 60.

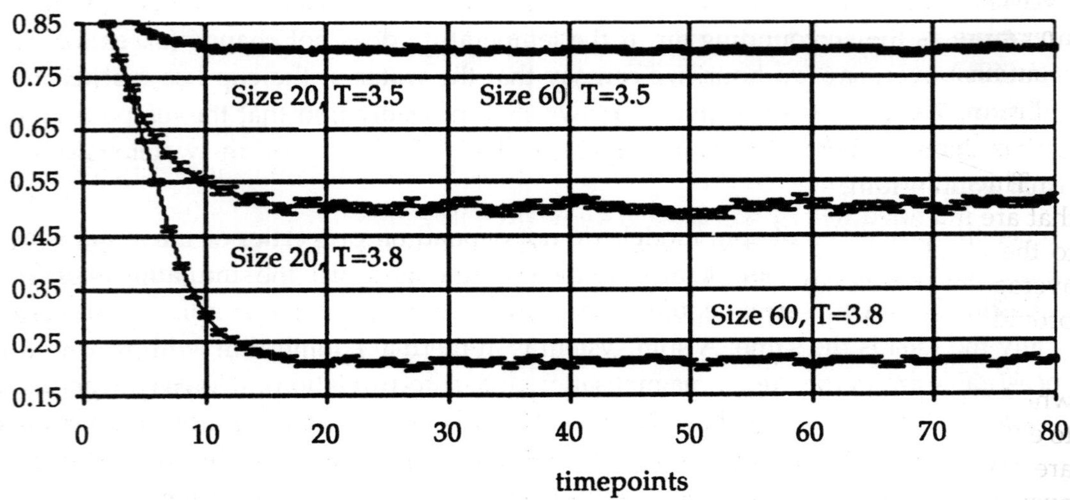

***Figure 14.1:** Absolute magnetisation, Ising triangular, 1000 samples*

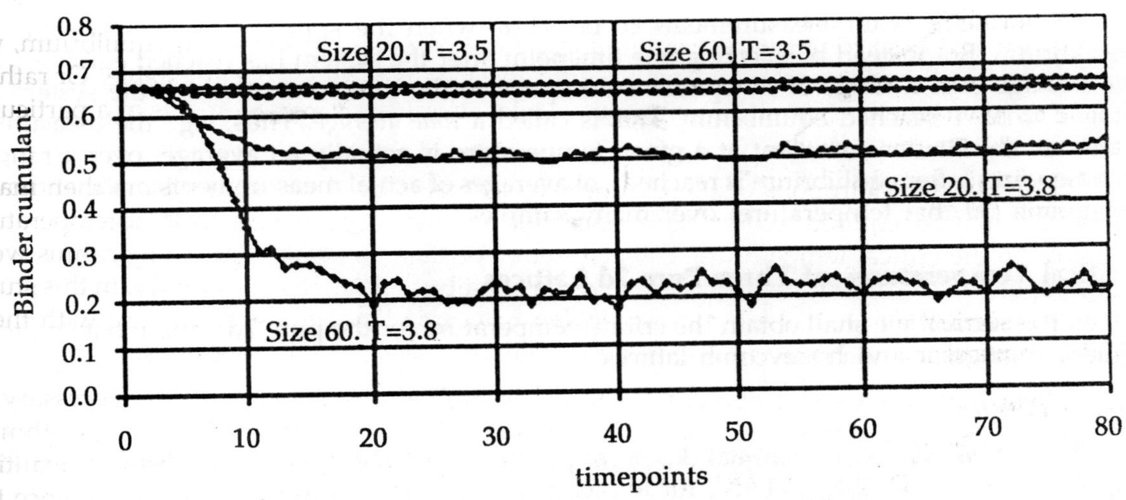

Figure 14.2: Binder cumulant, Ising triangular, 1000 samples

Figure 14.3 shows the magnified fluctuations of the Binder cumulant over the time period 40 through 80 (together with the absolute magnetisation). The linear trendline for the Binder cumulant data is almost horizontal.

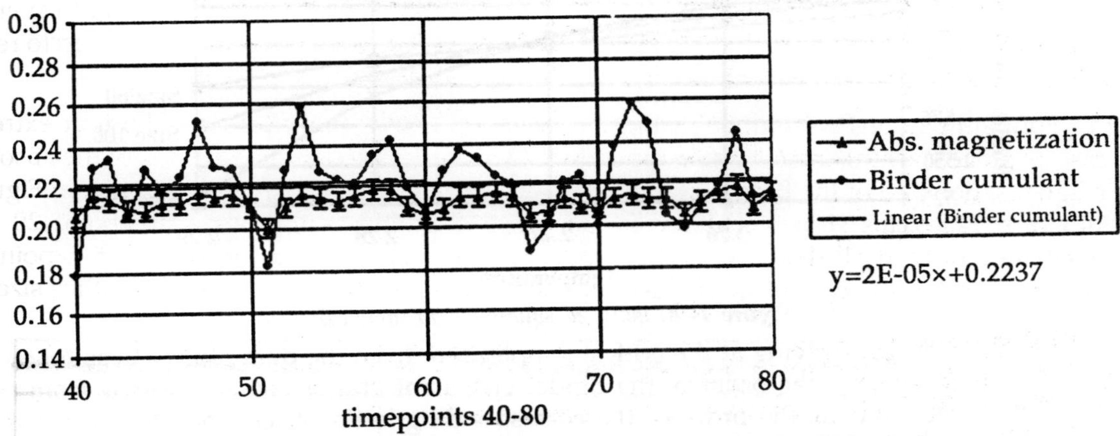

Figure 14.3: Ising triangular, lattice size 60×60, T=3.8, 1000 samples

Thus it is reasonable to conclude that with the Swendsen-Wang algorithm the simulations that are intended to be performed are such that the system reaches equilibrium with regard to the absolute magnetisation and the Binder cumulant by timepoint 30, and that a time average over timesteps 40 through 80 of the sample averages of the Binder cumulant will provide a satisfactory measurement of this quantity.

To clarify: A simulation actually consists of many "runs" (a.k.a. "samples"), in each of which the spin system is started from some given initial state (e.g., all spins up) and is allowed to evolve over a certain number of timepoints. At each timepoint in a single run measurements are made (e.g., magnetisation). The final value for the measurement at each timepoint is the average at that timepoint over all samples.

We normally want measurements to be made when the spin system is in a state of equilibrium. But instead of selecting one timepoint after the system has reached equilibrium we average over a range of timepoints subsequent to the timepoint at which the system is judged to have reached equilibrium. This is called a *time average*. Thus, e.g., the measured value for the Binder cumulant at a given temperature is actually an average, over a range of timepoints (after equilibrium is reached), of averages of actual measurements made at those timepoints (at that temperature) over many samples.

Critical Temperatures of Three Pure 2d Lattices

In this section we shall obtain the critical temperature for the pure 2d Ising model on the square, triangular and honeycomb lattices.

Square Lattice

Using Wolff dynamics simulations were performed for the pure square Ising model on six lattices of size 60, 100 and 150, for ten temperatures in the range 2.2 through 2.35 (1000 samples each), and the Binder cumulant was measured. The plots of the Binder cumulant against T, for each size and for five temperatures in the range 2.25 - 2.29, are given in Figure 14.4. At T = 2.27 (close to the critical temperature) the largest error is 0.0022.

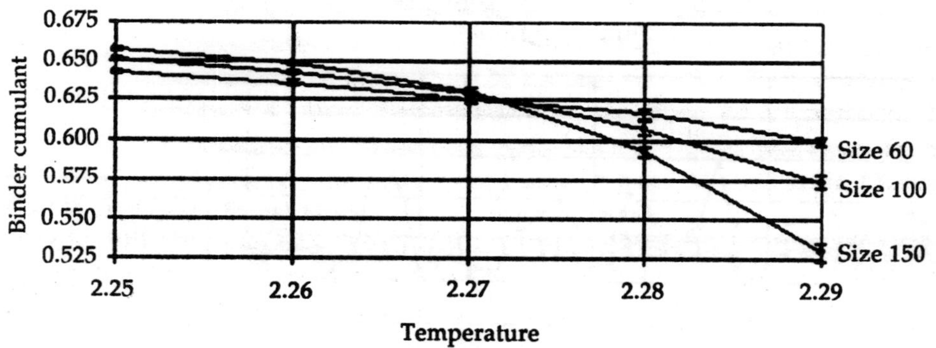

***Figure 14.4:** Ising 2d square (1000 samples)*

The procedure for arriving at the critical temperature from the data plotted is as follows: For higher temperatures the order of the Binder cumulant graphs (when viewing from top to bottom) is the same as the order of the corresponding lattice sizes, and for lower T the reverse is the case. One looks for the highest T below the intersection point(s) at which the error bars do not overlap, and for the lowest T above the intersection point(s) at which they also do not overlap. T_c is then the mean of these two, and their difference (divided by 2) gives the error bar for T_c.

Thus from the data plotted in figure 14.4 we can conclude that T_c for the square lattice is 2.27(1). Although the precision is not high, this compares well with the actual value of 2.269185 obtained from Onsager's analytical solution.

Triangular Lattice

Using Swendsen-Wang dynamics simulations were performed for the pure triangular Ising model on lattices of size 20, 40 and 60, for twelve temperatures in the range 3.5 through 3.8 (in each case 1000 samples were used).

Figure 14.5 shows the plots of the Binder cumulant against temperature for each lattice size for six temperatures in the range 3.62 through 3.66. By examination of the error bars, we may conclude that the critical temperature for the triangular lattice is 3.64(2). This compares well with the value given by Fisher of 3.6410.

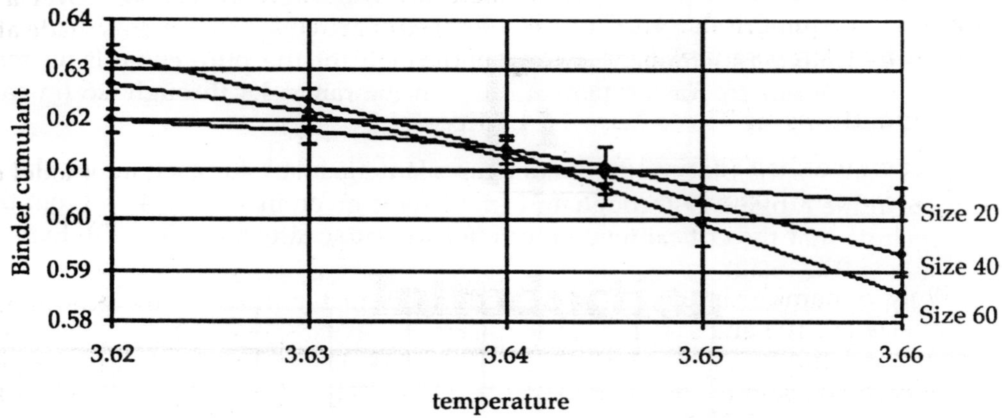

***Figure 14.5:** Ising 2d triangular (1000 samples)*

Honeycomb Lattice

Using Swendsen-Wang dynamics simulations were performed for the pure honeycomb Ising model on lattices of size 20, 40 and 60, for seven temperatures in the range 1.50 through 1.53 (in each case 1000 samples were used).

Figure 14.6 shows the plots of the Binder cumulant against temperature for each lattice size for five temperatures in the range 1.51 through 1.53. By examining the error bars in the plots we may conclude that the critical temperature for the triangular lattice is 1.52(1) This compares well with the value given by Fisher of 1.5187.

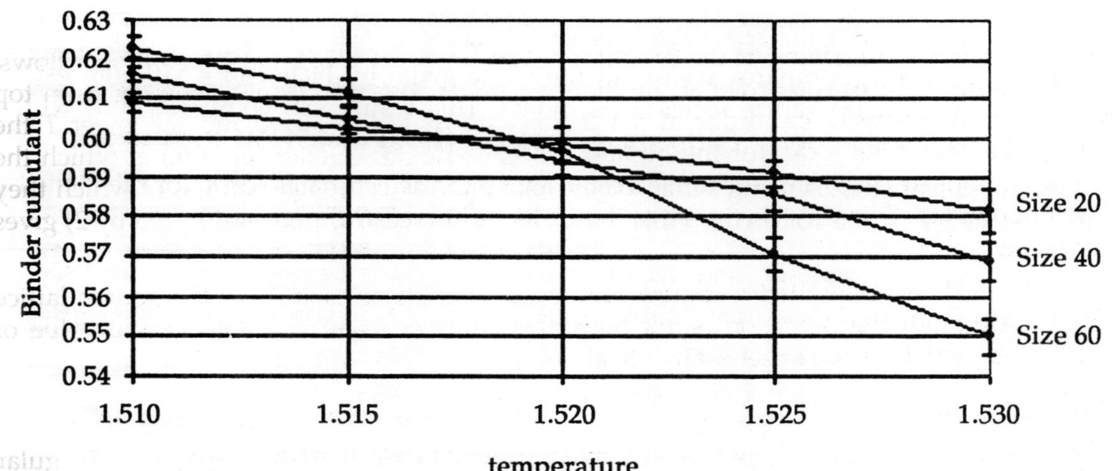

***Figure 14.6:** Binder cumulant, Ising honeycomb, 1000 samples*

Critical Temperatures of Two Pure 3d Lattices

In this section we shall obtain the critical temperatures for the pure 3d Ising model on the cubic and the diamond lattices.

Cubic Lattice

Using Wolff dynamics simulations were performed for the pure cubic Ising model on lattices of size 8, 12 and 16, for ten temperatures in the range 4.4 through 4.6 (in each case 1000 samples were used).

Figure 14.7 shows the plots of the Binder cumulant against T for each lattice size for five temperatures in the range 4.49 through 4.54 (the data is given in Table 3.4.5). From this data we may conclude that the critical temperature for the cubic lattice is 4.515(25). Expressed as $K_c = 1/T_c$ this is 0.2215(12).

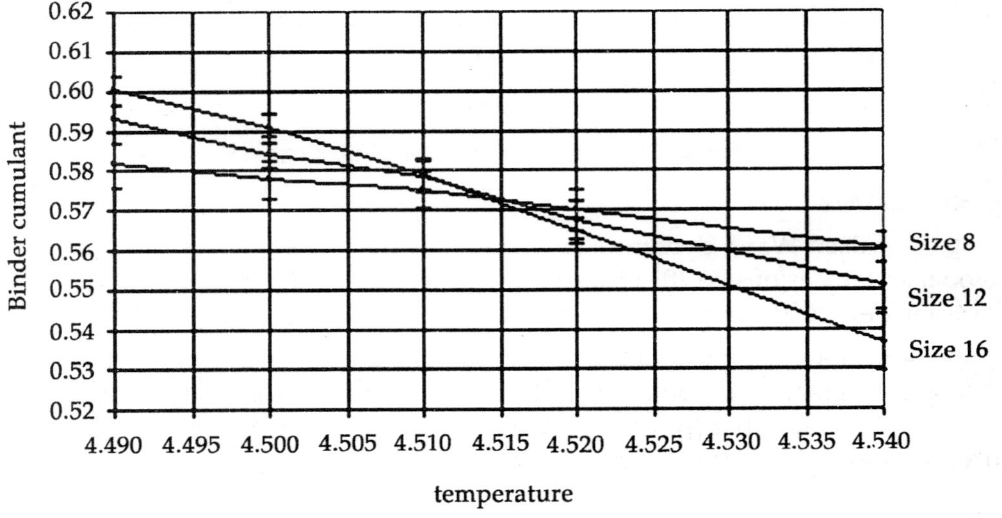

Figure 14.7: *Binder cumulant, Ising cubic, 1000 samples*

This result for T_c compares well with the value given by Fisher of 4.5103 (an estimate obtained by high-temperature series expansions). This result also compares well with the value obtained by the Monte Carlo study of Heuer (1993) of 4.5115(1).

Several recent studies using series expansions provide estimates for K_c for the cubic Ising model, as stated in the following table (in order of increasing precision):

0.221 67 (2)	Salman and Adler (1991)	Low temperature series expansion
0.221 658 (5)	Salman and Adler (1991)	High temperature series expansion
0.221 665 (5)	Adler (1983)	Series expansion
0.221 649 (4)	Blote and Kamieniarz (1994)	Monte Carlo study
0.221 654 4 (10)	Livet (1991)	Monte Carlo study
0.221 654 6 (10)	Blote, Luijten and Heringa (1995)	Series expansion
0.221 654 4 (3)	Talapov and Blote (1996)	Series expansion

Although the value for K_c obtained from this simulation software (namely, 0.2215(12)) is much less precise than the results stated above, the value is accurate; the lack of precision is due to the small number of samples used (1000 for each temperature).

At the critical temperature 1000 samples produce an error of c. 0.0041 in the measurement of the Binder cumulant. Using 6000 samples reduces the error to c. 0.0016, consistent with the fact that reducing the error by a factor of n requires increasing the sample size by n^2 (since $0.0041/0.0016 = 2.56$ and $\sqrt{6 = 2.45}$). To reduce the error further to 0.0005 it would be necessary to use c. 60,000 samples, which would require c. 230 hours of computing time on a 330 MHz Intel PC.

Diamond Lattice

Using Wolff dynamics simulations were performed for the pure diamond Ising model on lattices of size 8, 12 and 16, for seven temperatures in the range 2.6 through 2.8 (in each case 1000 samples were used).

Figure 14.8 shows the plots of the Binder cumulant against temperature for each lattice size for five temperatures in the range 2.65 through 2.75 . We may conclude that the critical temperature for the cubic lattice is 2.700(25). This compares well with the value given by Fisher of 2.7040.

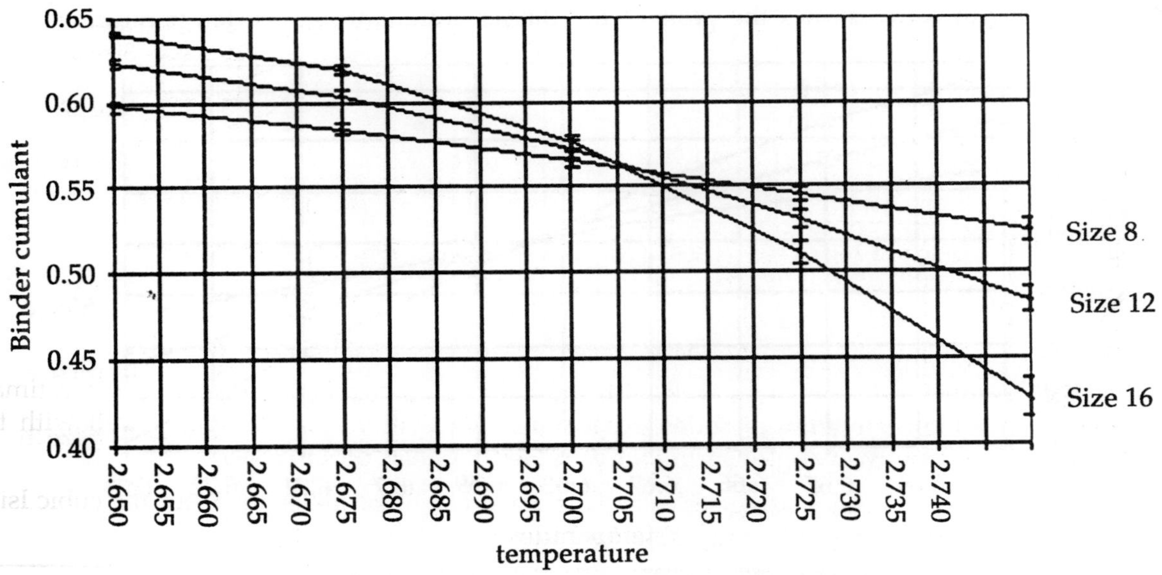

Figure 14.8: Binder cumulant, Ising diamond, 1000 samples

Critical Temperature of the Pure 4d Hypercubic Lattice

Using Wolff dynamics simulations were performed for the pure 4d hypercubic Ising model on lattices of size 6, 8 and 10, for ten temperatures in the range 6.3 through 7.0 (in each case 1000 samples were used). Figure 14.9 shows the plots of the Binder cumulant against temperature for each lattice size for seven temperatures in the range 6.50 through 6.85. The plots all intersect at 6.68 and we may conclude that the critical temperature for the 4d hypercubic lattice is 6.68(7).

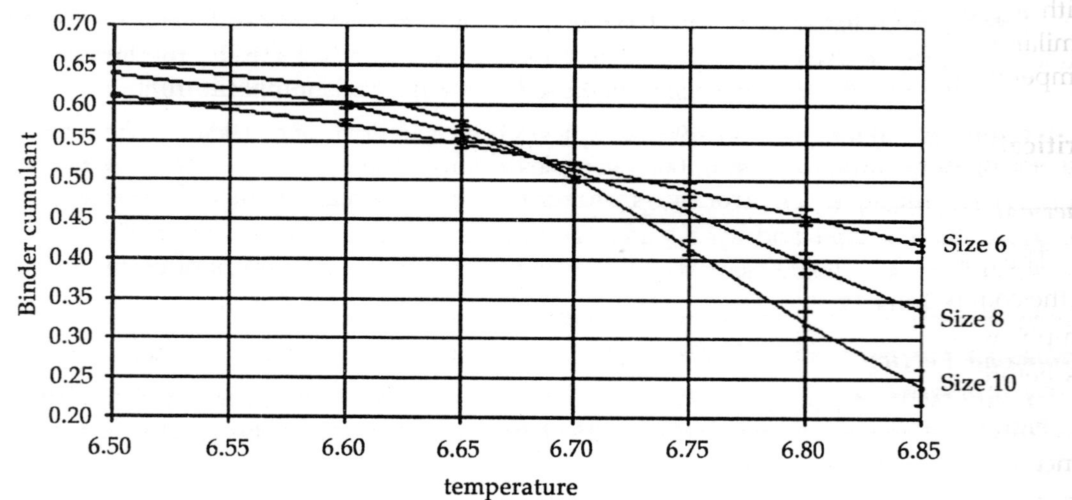

***Figure 14.9:** Binder cumulant, Ising 4d hypercubic, 1000 samples*

The experiment was repeated with modifications in order to reduce the error in T_c. Simulations were performed for six temperatures in the range 6.63 through 6.73, using 4000 samples. Figure 14.10 shows the plots of the Binder cumulant against temperature.

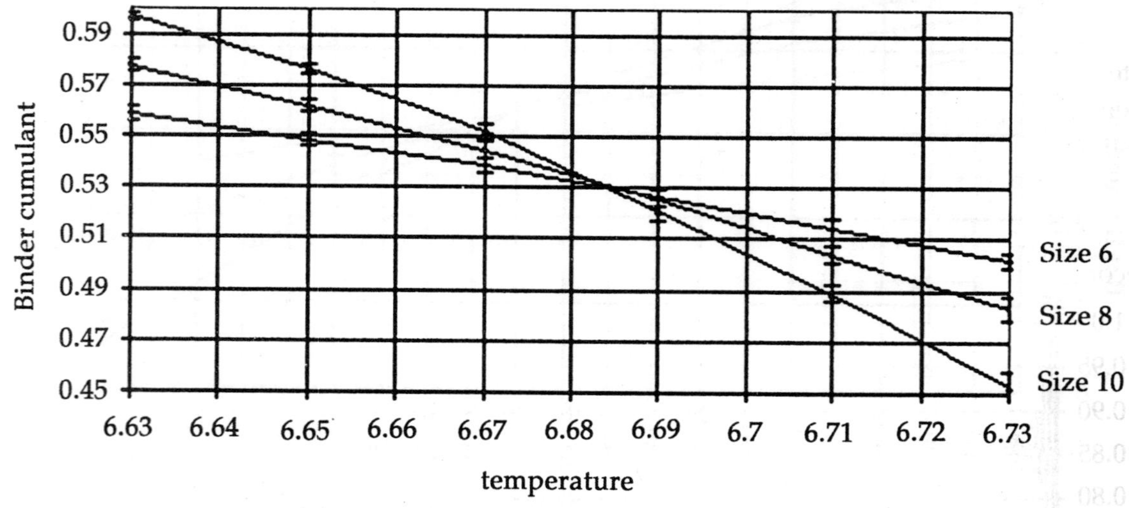

***Figure 14.10:** Binder cumulant, Ising 4d hpercubic, 4000 samples*

From this we may conclude that the critical temperature for the Ising 4d hypercubic lattice is 6.68(3).

Gaunt, Sykes and McKenzie (1979) conducted high-temperature series expansion studies of susceptibility in the 4*d* Ising hypercubic model, and obtained an estimate for $1/v_c$ of 6.732(2), where $v = \tanh(J/k_B T)$. Taking $J/k_B = 1$ this gives an estimate for T_c of 6.682(2).

Thus the Monte Carlo result produced by the simulation software compares well with the more precise estimate of Gaunt *et al.* obtained from series expansions. To obtain a result with a precision $1/10^{th}$ of that obtained by the second experiment (i.e., ±0.003), and thus similar to that of Gaunt's result, it would be necessary to perform the experiment with temperatures in the range 6.680 - 6.684 using 100 x 4,000 = 400,000 samples.

Critical Temperatures of Dilute Ising Spin Models

Thermal Equilibrium in Dilute Systems

The question of deciding when a spin system has reached equilibrium has been discussed in the context of pure spin systems. When we come to deal with dilute spin systems we cannot simply assume that a dilute system will equilibrate in the same time that a pure system will. As before, we need to conduct trials.

Figures 14.11 through 14.14 show results for an Ising square lattice, size 40x40, with site concentration 0.8 at the low temperature of 0.3 (which is about 20 per cent of the critical temperature for this degree of site dilution) with Wolff dynamics. 5000 samples were used to obtain averages at each timepoint.

Figure 14.11 shows both the absolute magnetisation and the Binder cumulant. It seems that the absolute magnetisation is stable from timepoint 100 onwards, and this is confirmed by closer inspection of the plot for timepoints 100 through 200 as given in figure 14.12.

On the other hand, the Binder cumulant continues to rise, although very slowly, even after timepoint 100. However, from timepoint 160 onwards the increase is very small. It thus seems that to obtain an equilibrium value for the Binder cumulant in this case we would be justified in averaging over timepoints 160 through 200.

For other values of the site concentration and the temperature similar trials were done to ascertain the earliest timepoint at which the magnetisation and the Binder cumulant become stable.

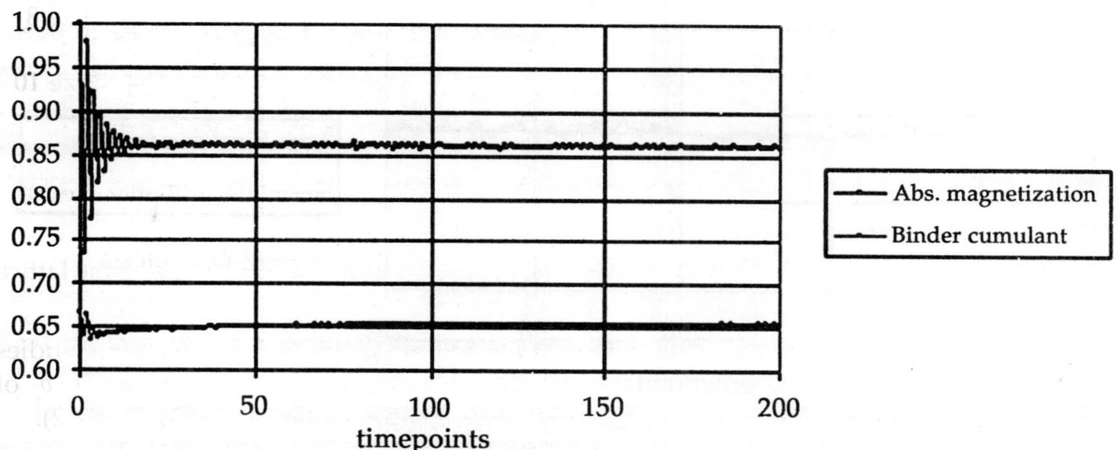

Figure 14.11: *Ising square, 40×40, site-diluted (p=0.8) T=0.3, Wolff, 5000 samples*

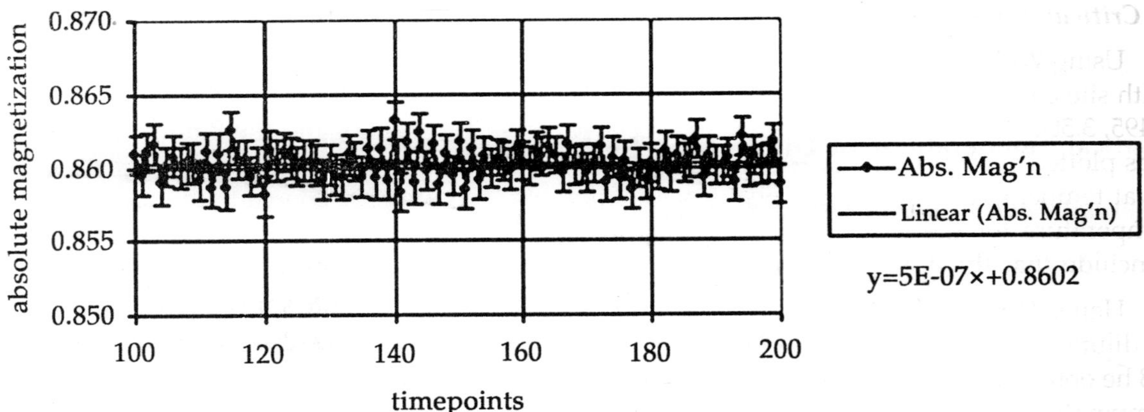

Figure 14.12: Ising square, 40×40, site-diluted (p=0.8) T=0.3, Wolff, 5000 samples

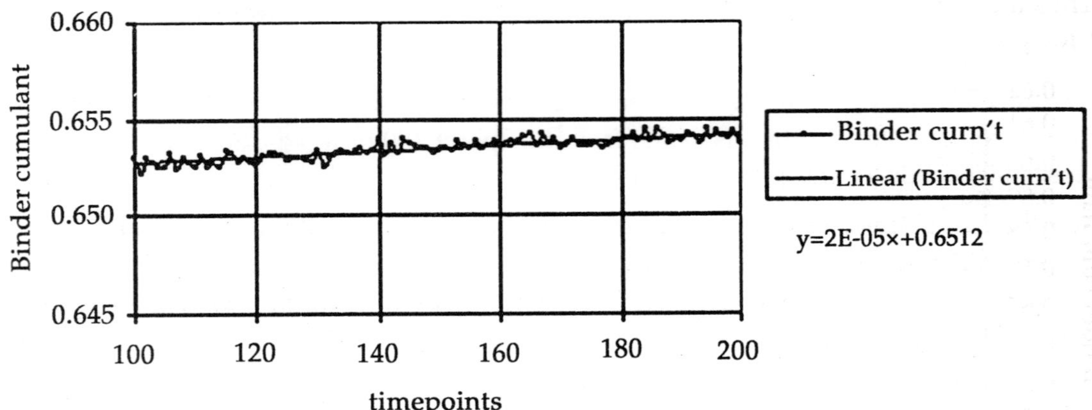

Figure 14.13: Ising square, 40×40, site-diluted (p=0.8) T=0.3, Wolff, 5000 samples

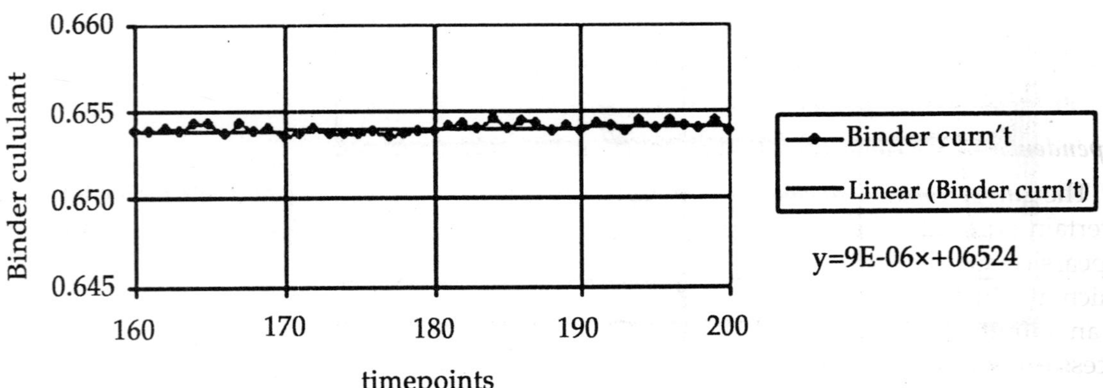

Figure 14.14: Ising square, 40×40, site-diluted (p=0.8) T=0.3, Wolff, 5000 samples

A Critical Temperature of the Ising Cubic Lattice with Site Dilution

Using Wolff dynamics simulations were performed for the site-diluted cubic Ising model, with site concentration 0.8, on lattices of size 10, 20 and 30, for six temperatures, 3.485, 3.490, 3.495, 3.500, 3.505 and 3.510 (in each case 2000 samples were used), and the Binder cumulant was plotted against temperature. There is slight overlap of the error bars for size 10 and size 20 at temperatures 3.495 and 3.505. If we ignore this then we may conclude that the critical temperature for the cubic lattice with site concentration 0.8 is 3.500(5). Otherwise we can conclude that the critical temperature is 3.500(10), i.e. 3.50(1).

Heuer (1993), using Monte Carlo methods, studied the cubic Ising model over a range of dilutions and obtained critical temperatures using the Binder cumulant method. For $p = 0.8$ he obtained a value of 3.4992(5), and says that his result "agrees very well with the critical temperature $T_c(p = 0.8) = 3.4991(5)$ of Wang *et al.* who have determined T_c from the extrapolated susceptibility maximum."

Thus the result given by the simulation software provides confirmation of its validity not only for pure lattices but also for site-diluted lattices.

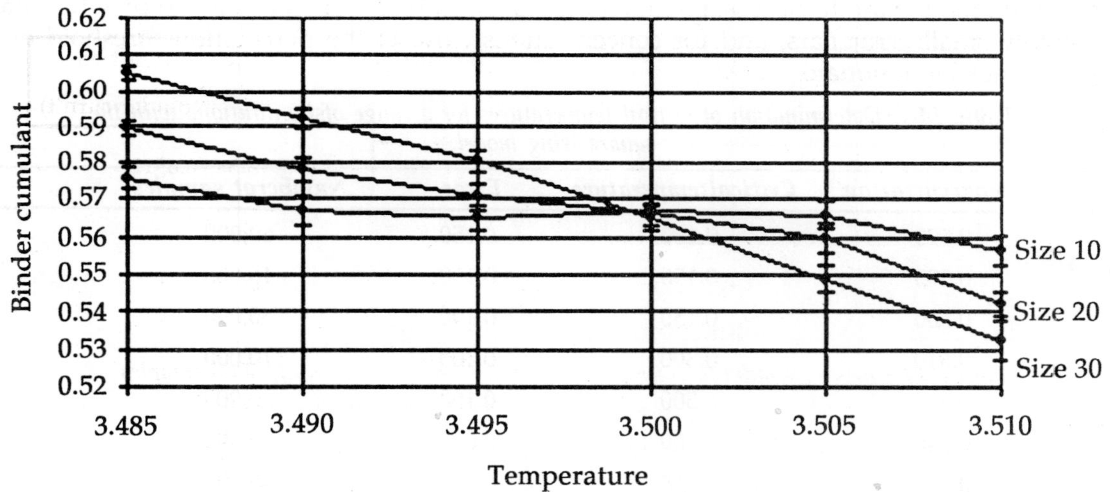

Figure 14.15: *Ising cubic, site-diluted (p=0.8), Binder cumulant, Wolff, 2000 samples*

Dependence of Critical Temperature on Site Dilution in the Ising Square Lattice

The *site percolation threshold* is the lowest concentration of occupied sites in a lattice of a certain type (all of whose bonds are open) at which an infinite cluster of occupied sites appears (when the lattice sites are occupied with a certain concentration of sites), and below which all clusters of occupied sites are finite. This, of course, applies only to the ideal case of an infinite lattice. When estimating the site percolation threshold by experiment one necessarily uses finite lattices and looks for the concentration at which a *percolating cluster* of occupied sites first appears (i.e., a cluster of sites which extends from one side of the lattice to the opposite side). For the square lattice the site percolation threshold is 0.592746, a result deduced by Sykes and Essam.

As the site concentration of a spin model on a lattice is decreased the critical temperature decreases. (This is because a smaller site concentration implies fewer connections among spins, thus it is easier for thermal fluctuations to overcome the tendency to long-range order.) In the ideal case the critical temperature becomes zero when the site concentration is decreased to the site percolation threshold.

The way in which the critical temperature of a spin model depends upon site concentration can be ascertained by measuring T_c for a series of site concentrations from the pure case (p=1.0) down to the percolation threshold.

When measuring the critical temperature of a site-diluted lattice the main difference from the pure case is that averages must also be obtained over various configurations of occupied sites. In the pure case, since all sites are occupied, there is only one possible site configuration. When some sites are unoccupied there are many possible site configurations, and Monte Carlo studies of site-diluted lattices must also average over a random sampling of site configurations.

As the site concentration decreases it is necessary to average over a larger number of samples in order to reduce the size of the error bars. Whereas a few thousand samples is sufficient for small error bars in the case of site concentrations of 0.9 and above, for lower concentrations it may be necessary to average over as many as 100,000 samples to obtain acceptably small error bars, and for concentrations close to the percolation threshold even this may not be adequate.

Table 14.1: *Determination of critical temperatures for a range ofsite dilutions with the square Ising model.*

Siteoncentration	*Criticaltemperature*	*Error*	*Numberof samples*
0.590	0.150	0.150	60,000
0.593	0.150	0.150	16,000
0.600	0.150	0.100	30,000
0.610	0.300	0.100	10,000
0.625	0.500	0.100	30,000
0.640	0.700	0.050	10,000
0.645	0.755	0.035	40,000
0.650	0.790	0.040	40,000
0.660	0.850	0.020	40,000
0.680	0.940	0.050	20,000
0.700	1.040	0.040	12,000
0.725	1.160	0.040	8,000
0.756	1.300	0.050	1,000
0.800	1.500	0.060	2,000
0.837	1.660	0.030	500
0.900	1.915	0.055	1,000
1.000	2.269	0.000	

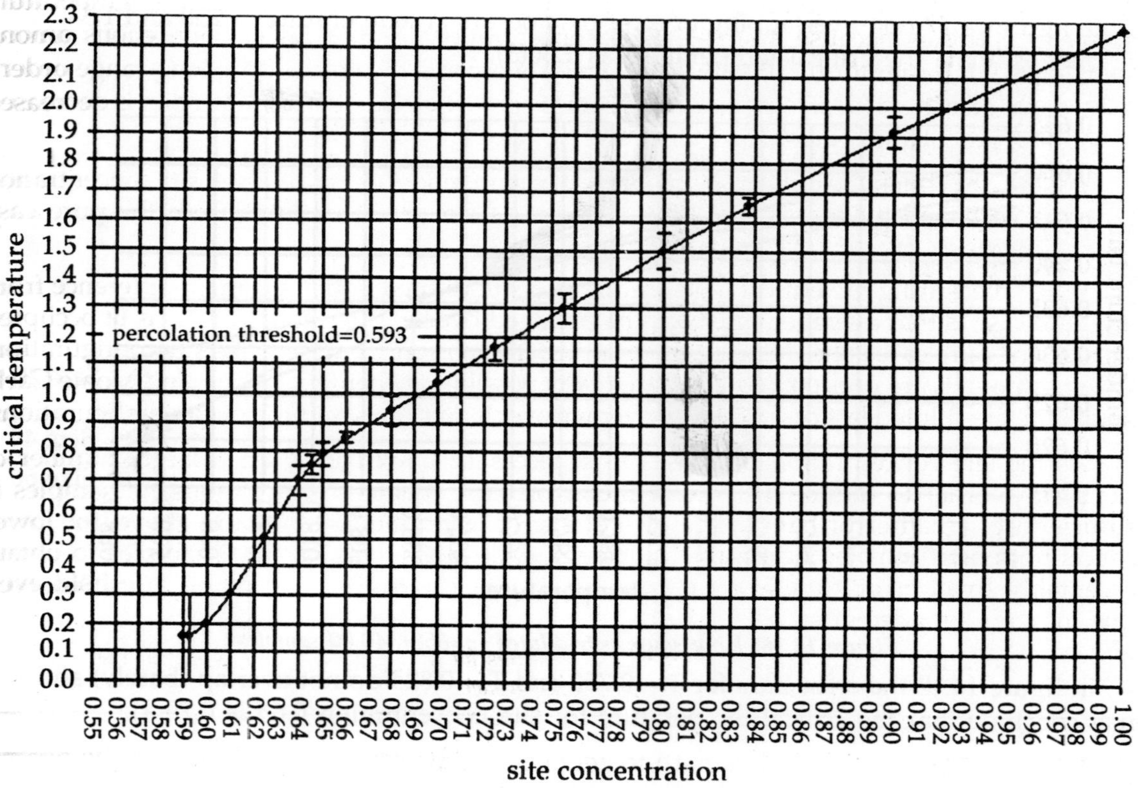

critical temperature

percolation threshold=0.593

site concentration

Figure 14.16: *Critical temperature vs. site concentration, Ising square model*

With the lattice sizes (20 - 60) and the number of samples (up to 60,000) used here the Binder cumulant method for determining T_c works well only for site-diluted lattices with site concentrations higher than about 0.65. figures 14.17 and 14.18 show the plots for the Binder cumulant in the cases of p = 0.756 (1000 samples) and p = 0.65 (40,000 samples) respectively.

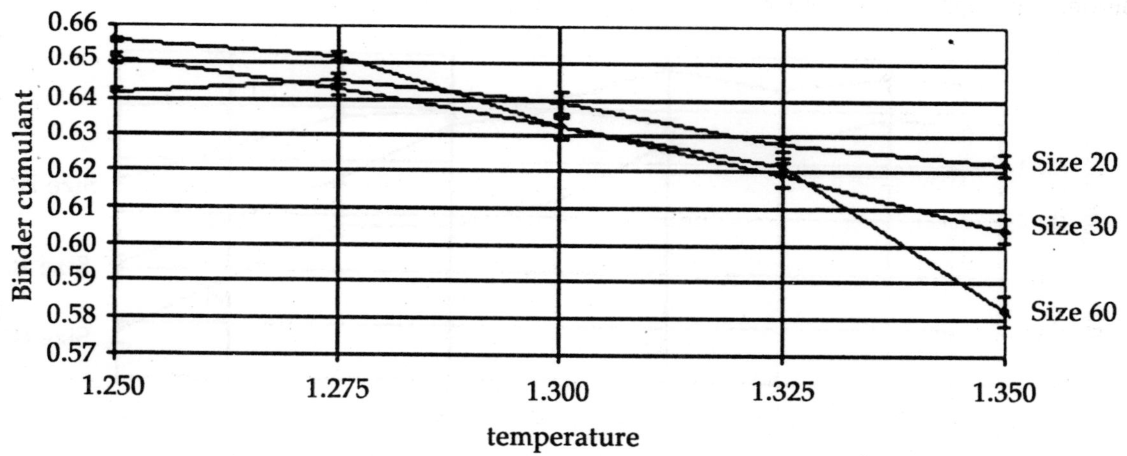

Figure 14.17: *Ising square, site-diluted (p=0.756), 1,000 samples*

In figure 14.17 non-overlapping of error bars occurs at temperatures 1.250 and 1.350 and we may estimate the critical temperature as 1.30(5).

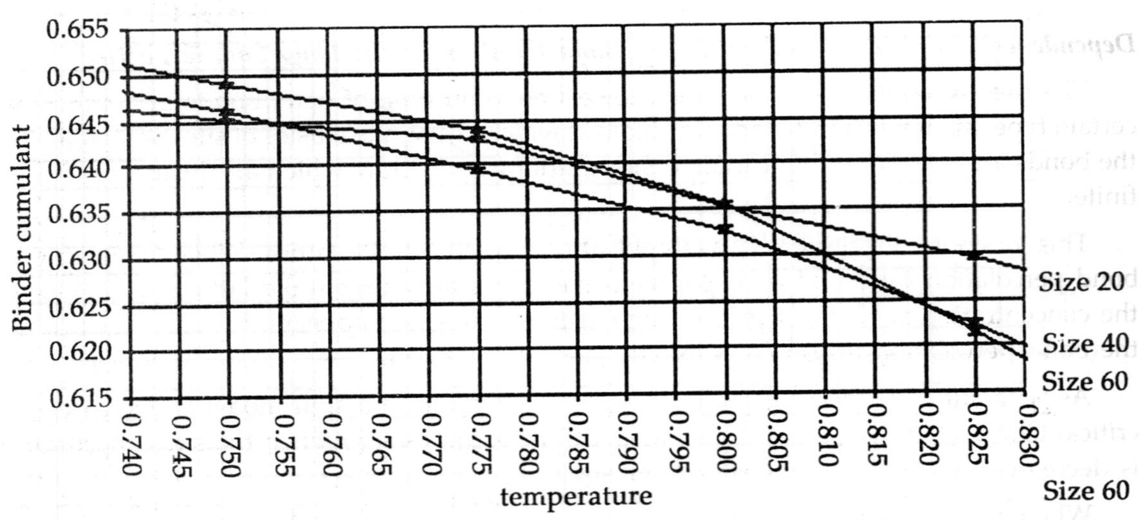

Figure 14.18: Ising squre, site-diluted (p=650), 40,00 samples

In figure 14.18 the error bars for T = 0.750 and T = 0.825 do not overlap (barely) and we may conclude that the critical temperature for site concentration 0.65 is 0.79(4).

Below p = 0.65, however, it is hard to get good estimates. figure 14.19 shows the Binder cumulant data for the case of p = 0.600 for the temperature range of 0.1 through 0.5 using 30,000 samples. There is little on which to base an estimate of T_c. Something like 0.2(1) is about the best one can do from this data.

In fact it appears that the Binder cumulant method for the determination of the critical temperature breaks down below a site concentration of about 0.65, because the plots for the Binder cumulant for different lattice sizes do not intersect as expected.

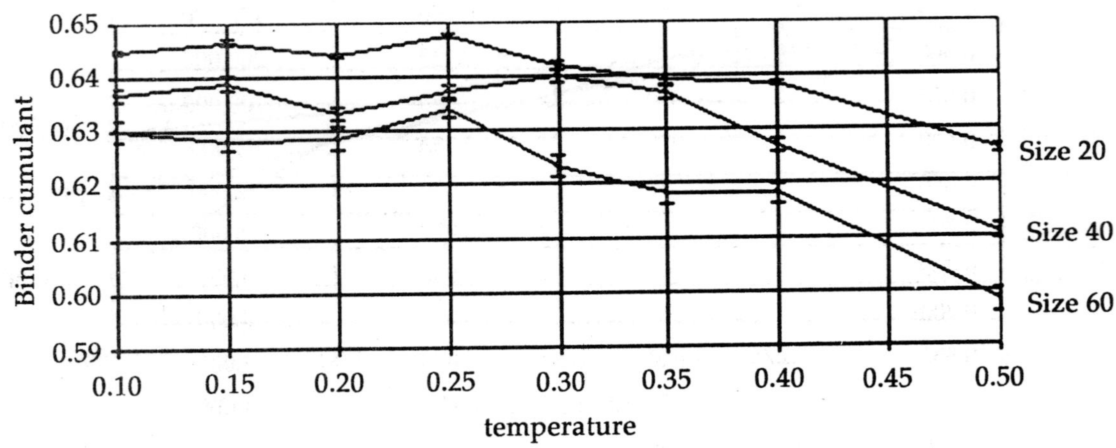

Figure 14.19: Isling square, site-diluted (p=0.600), 30,000 samples

It could be that the expression for U_L given, $U_L = U[(T/T_c - 1).L^{1/\nu}]$, upon which the Binder cumulant metnod is based, is an oversimplification, and that some correction for strongly site-diluted lattices is required.

Dependence of Critical Temperature on Bond Dilution in the Ising Square Lattice

The *bond percolation threshold* is the lowest concentration of open bonds in a lattice of a certain type (all of whose sites are occupied) at which an infinite cluster of sites appears (when the bonds are opened with a certain concentration), and below which all clusters of sites are finite.

This, of course, applies only to the ideal case of an infinite lattice. When estimating the bond percolation threshold by experiment one necessarily uses finite lattices and looks for the concentration at which a percolating cluster of sites first appears. For the square lattice the bond percolation threshold is exactly 0.5.

As with site concentration, as the bond concentration of a spin model is decreased, the critical temperature decreases. In the ideal case it becomes zero when the site concentration is decreased to the bond percolation threshold.

When measuring the critical temperature of a bond-diluted lattice the main difference from the pure case is that averages must also be obtained over various configurations of open bonds.

In the pure case, since all bonds are open, there is only one possible bond configuration. When some bonds are closed there are many possible bond configurations, and Monte Carlo studies of bond-diluted lattices must average over a random sampling of bond configurations.

Table 14.2: Determination of critical temperatures for a range ofbond dilutions with the square Ising model.

Bond concentration	*Critical temperature*	*Error*	*Number of samples*
0.525	0.200	0.200	200,000
0.550	0.550	0.200	128,000
0.560	0.675	0.125	200,000
0.575	0.830	0.050	100,000
0.600	0.950	0.050	64,000
0.650	1.150	0.050	32,000
0.700	1.320	0.070	16,000
0.750	1.500	0.050	8,000
0.800	1.650	0.050	4,000
0.850	1.820	0.070	2,000
0.900	1.970	0.070	1,000
1.000	2.269		

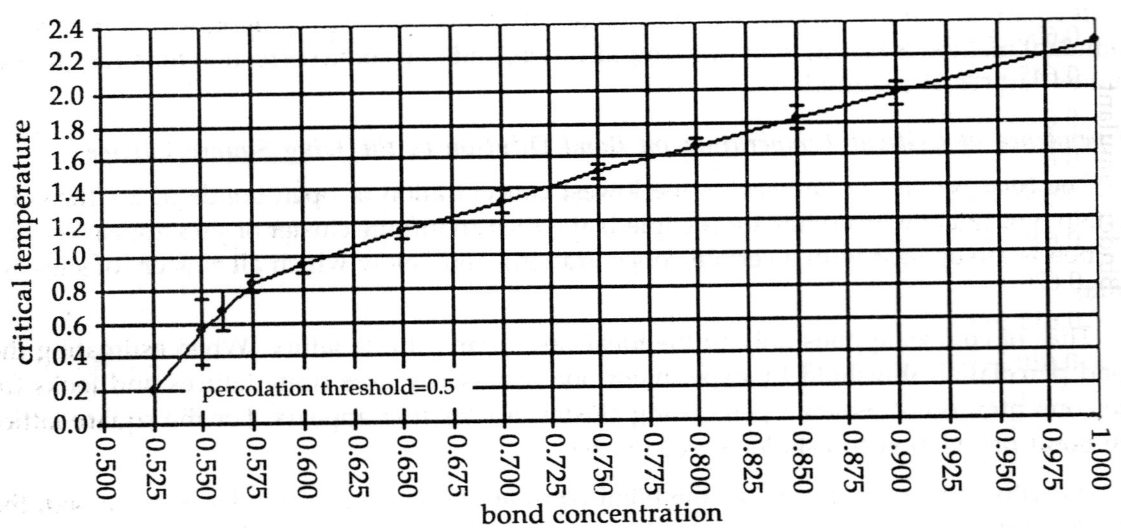

Figure 14.20: Critical temperature vs. bond concentration, Ising 2d square model

As the bond concentration decreases it is necessary to average over a larger number of samples in order to reduce the size of the error bars.

Whereas a few thousand samples is sufficient for small error bars in the case of bond concentrations of 0.9 and above, for lower concentrations it may be necessary to average over as many as 100,000 samples to obtain acceptable error bars, e.g. for bond concentration 0.575 as shown in figure 14.21.

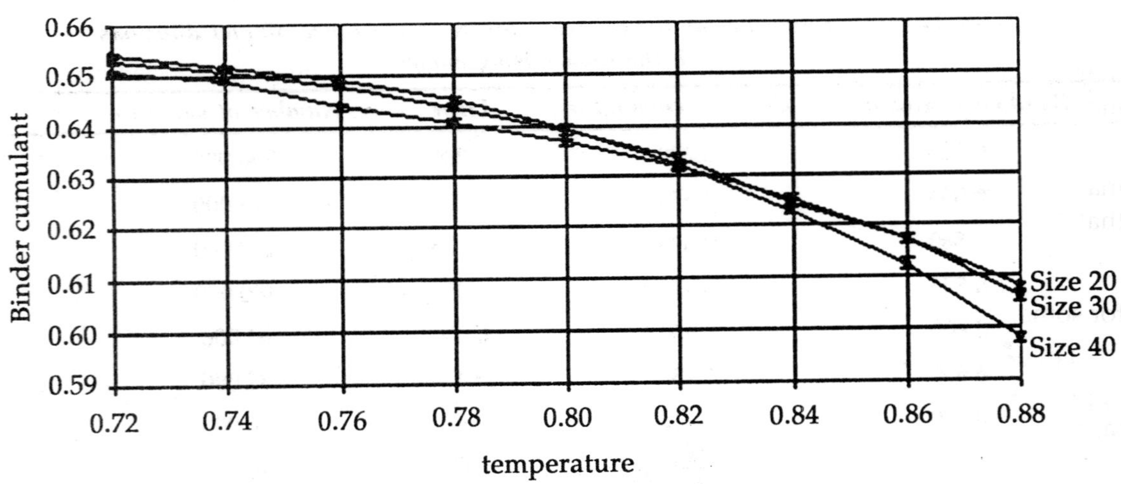

Figure 14.21: Ising square, bond-diluted (p=0.575), 100,000 samples

For bond concentrations lower than 0.55 (just above the bond percolation threshold of 0.5) it is difficult to obtain an estimate of the critical temperature, because the plots for lattice sizes 20, 30 and 40 do not intersect, as shown in figure 14.22.

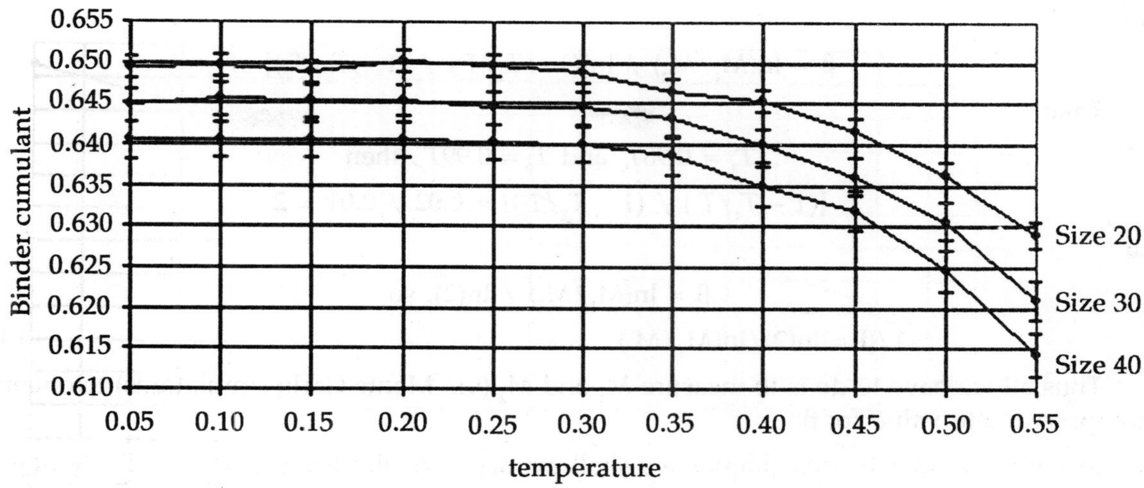

Figure 14.22: *Ising square, bond-diluted (p=0.525), 200,000 samples*

As in the case of site-dilution, it appears that the Binder cumulant method of determining critical temperatures breaks down as the bond concentration approaches the bond percolation threshold, because the plots do not intersect (despite the use of a large number of samples).

Extraction of the Critical Exponent of the Magnetisation

Suppose $A(\varepsilon)$ is a function describing the behaviour of a quantity, such as magnetisation, in terms of $\varepsilon = 1 - T/T_c$, where T is temperature and T_c is the critical temperature. Then the limit λ as $\varepsilon \to 0$ of $\ln[A(\varepsilon)]/\ln(\varepsilon)$, if it exists, is called the *critical exponent* associated with A. In this case we write $A \sim \varepsilon^{\lambda}$ ("~" is read as "is asymptotically equal to").

Among the critical exponents for magnetic systems are those for specific heat, α, magnetisation, β, isothermal susceptibility, γ, and correlation length, ν. These are not all independent, and it is possible to derive inequalities such as $\alpha + 2\beta + \gamma \geq 2$.

From the above definition of a critical exponent we have that in the ideal case the magnetisation M just below the critical temperature T_c varies as $(1 - T/T_c)^{\beta}$. This does not imply that $M(T) = a.(1 - T/T_c)^{\beta}$, for some constant a. Rather we have

$$M(T) = a.(1 - T/T_c)^{\beta}.[1 + a'.(1 - T/T_c)^{\beta'} + ...]$$

for some constants a, β, a', β', ...

However, near the critical point the leading term is dominant, in which case $M(T)$ is approximated by the expression $a.(1 - T/T_c)^{\beta}$. This approximation will now be used to obtain (approximate) estimates of β for several kinds of lattices.

Consider two temperature values, T_1 and T_2, just below T_c, with $T_1 < T_2$, then

$$M_1 = M(T_1) = a.(1 - T_1/T_c)^{\beta} \text{ and } M_2 = M(T_2) = a.(1 - T_2/T_c)^{\beta},$$

so

$$M_1/M_2 = (1 - T_1/T_c)^{\beta} \;/\; (1 - T_2/T_c)^{\beta} = [(1 - T_1/T_c) \;/\; (1 - T_2/T_c)]^{\beta}$$

so

$$\beta = \ln(M_1/M_2) \ / \ \ln[(1 - T_1/T_c) \ / \ (1 - T_2/T_c)]$$

Take

$$T_1 = 0.98T_c \text{ and } T_2 = 0.99T_c, \text{ then}$$

$$[(1 - T_1/T_c) \ / \ (1 - T_2/T_c)] = 0.02 \ / \ 0.01 = 2$$

so

$$\beta = \ln(M_1/M_2) \ / \ \ln(2), \text{ so}$$

$$1/\beta = \ln(2)/\ln(M_1/M_2) \qquad \text{...14.1}$$

Thus all we have to do is to measure M_1 and M_2 (via Monte Carlo simulations) to obtain an approximate value for β.

In the remainder of this chapter we shall extract b for the Ising model on the square, triangular and honeycomb lattices, and then we shall consider the 3-state Potts model.

Ising Model on the Square Lattice

The simulation programme was used to extract the value of β for the pure square Ising model using the method described above to obtain values for M_1 and M_2.

For lattices of size 100, 200 and 400, values of the absolute magnetisation were obtained for temperatures 2.223801 and 2.246493 (resp. 98% and 99% of the critical temperature of 2.269185), using 1000 samples for lattice sizes 200 and 400 and 4000 samples for lattice size 100, with simulations over timepoints 0 through 50, averaging over timepoints 31 through 50. Wolff dynamics was used (which produces a stable value for the absolute magnetisation after 20 timesteps). The results are shown in figure 14.23.

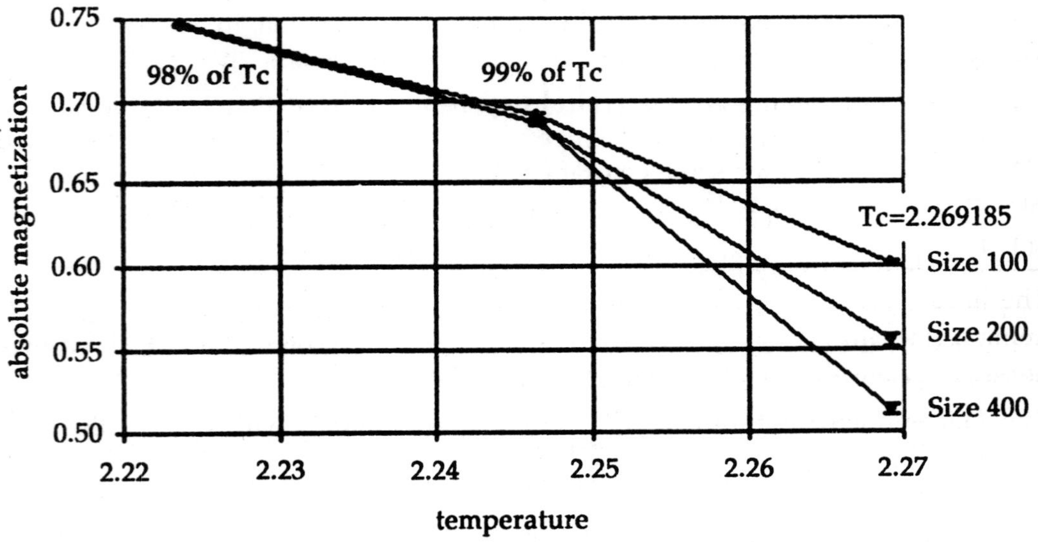

Figure 14.23: *Ising square lattice*

To obtain the "ideal" values of of M_1 and M_2, i.e., the values for the infinite lattice, we plot the values obtained by measuring these quantities in the three finite lattices against the reciprocal of the lattice size, and then perform a linear fit to the data, as shown in figure 14.24 for M_1, which is based on the data in table 14.3 (below). The intersection of the straight line with the vertical axis gives the ideal ("infinite") value.

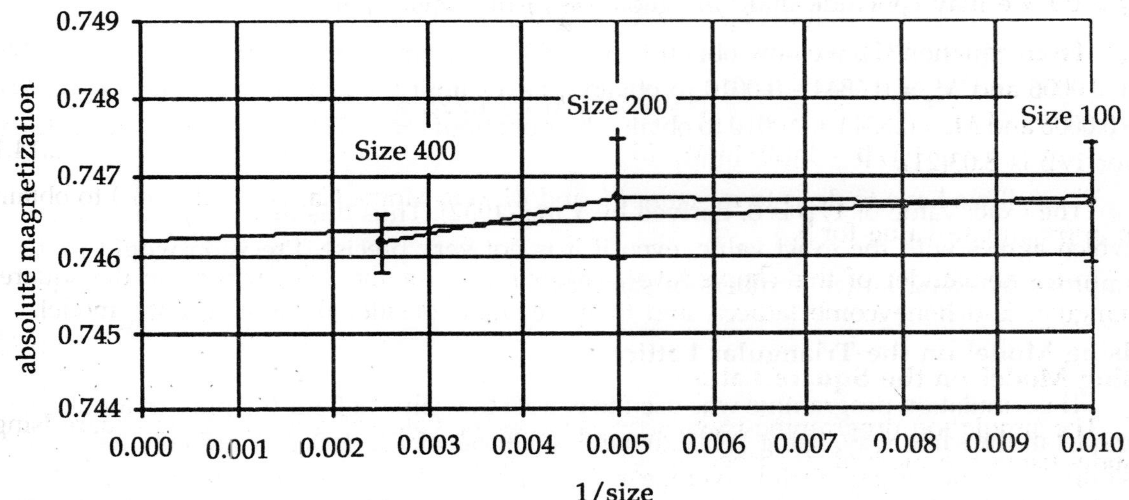

Figure 14.24: *M1*

Table 14.3

Size	*100*	*200*	*400*
Reciprocal of size	0.0100	0.0050	0.0025
M1			
Abs. mag.	0.7467	0.7467	0.7462
Std. dev.	0.0008	0.0008	0.0004
M2			
Abs. mag.	0.6909	0.6869	0.6861
Std. dev.	0.0012	0.0014	0.0007

The linear fit shown in figure 14.24 was done in Excel. A "least squares" method of obtaining a linear fit, $y = a + bx$, to data with error bars (and also without), which has the advantage of providing an error estimate for the intersection a with the vertical axis, is given in Press *et al.*, 1992. The present writer has written a programme which implements this method, and this programme is described, "The FIT Programmes". This will be used in this chapter and in the next to obtain error estimates.

For the data in table 14.3 the FIT programme gives, for M_1 and for M_2 respectively, the data fitted to $y(x) = a + bx$:

```
a = 0.7461, siga = 0.0006          a = 0.6844, siga = 0.0010
b = 0.0727, sigb = 0.1182          b = 0.6330, sigb = 0.1847
chi^2 = 0.1420, Q = 0.7063         chi^2 = 0.2738, Q = 0.6008
```

Thus the ideal values of M_1 and M_2 respectively are 0.7461(6) and 0.6844(10). Since $q > 0.1$ we may conclude that "the goodness-of-fit is believable".

From equation (1), we now obtain $1/\beta = 8.030$. To estimate the error we take $M_1 = 0.7461 + 0.0006$ and $M_2 = 0.6844 - 0.0010$ to obtain a lower limit for $1/\beta$ of 7.825, and $M_1 = 0.7461 - 0.0006$ and $M_2 = 0.6844 + 0.0010$ to obtain an upper limit for $1/\beta$ of 8.247, so our final estimate for $1/\beta$ is 8.03(21).

The exact value of $1/\beta$ is 8, derived by Yang (1952). Thus this method produces a result which agrees with the exact value, even if it is not very precise. Precision would probably improve by using larger numbers of samples.

Ising Model on the Triangular Lattice

The simulation programme was used to extract the value of β for the pure triangular Ising model in the same way, using Wolff dynamics, lattices of size 100, 200 and 400, 1000 or 4000 samples, and obtaining the time average in the same way. Values of the absolute magnetisation were obtained for temperatures 3.56818 and 3.60459.

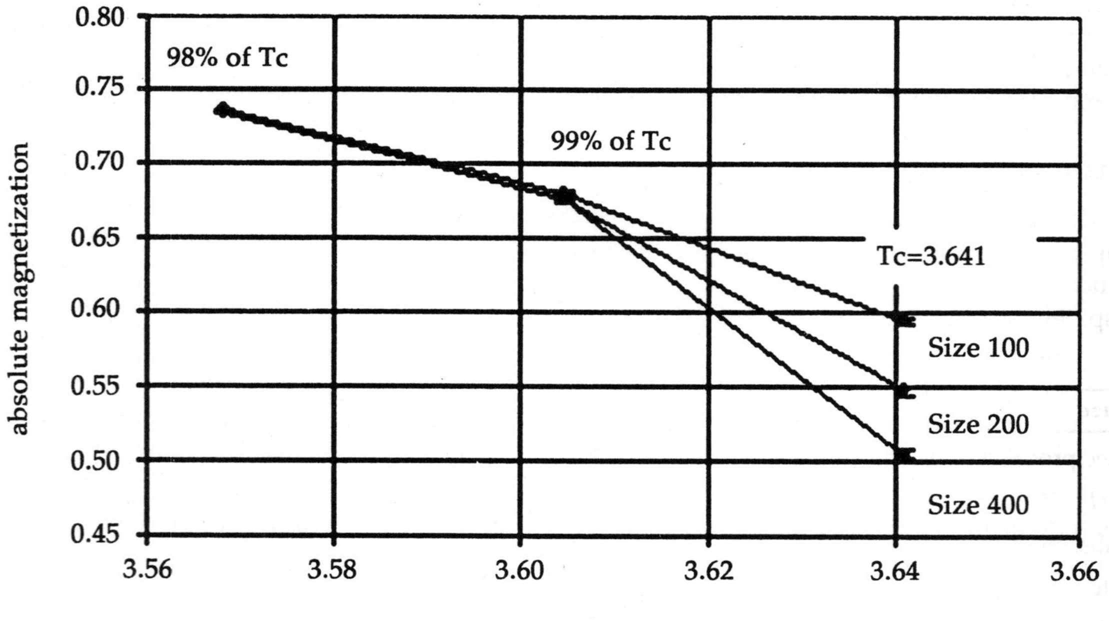

Figure 14.25: Ising triangular lattice

From Figure 14.25 we obtain Table 14.4:

Table 14.4

Size	*100*	*200*	*400*
Reciprocal of size	0.0100	0.0050	0.0025
M1			
Abs. mag.	0.7351	0.7360	0.7350
Std. dev.	0.0007	0.0007	0.0004
M2			
Abs. mag.	0.6804	0.6766	0.6755
Std. dev.	0.0012	0.0014	0.0007

For the data in Table 14.4 the FIT programme gives, for M_1 and for M_2 respectively, the data fitted to $y(x) = a + bx$:

```
a = 0.7351, siga = 0.0006          a = 0.6738, siga = 0.0010
b = 0.0252, sigb = 0.1071          b = 0.6487, sigb = 0.1847
chi^2 = 1.5181, Q = 0.2179         chi^2 = 0.1217, Q = 0.7272
```

Thus we may conclude that the ideal values of M_1 and M_2 respectively are 0.7351(6) and 0.6738(10).

From equation (14.1) we obtain $1/\beta = 7.961$. We use the same procedure (adding and subtracting errors to M_1 and M_2) to obtain the error for $1/\beta$, namely, $(8.176-7.756)/2 = 0.210$, so our final estimate for $1/\beta$ is 7.96(21). This is close both to the estimate for the square lattice and to the exact value of 8.

Ising Model on the Honeycomb Lattice

The simulation programme was used to extract the value of β for the pure honeycomb Ising model in the same way, except that measurements were also made with lattice size 800. Values of the absolute magnetisation were obtained for temperatures 1.488326 and 1.503513 (resp. 98% and 99% of the critical temperature of 1.5187) but not for the critical temperature.

Table 14.5

Size	*100*	*200*	*400*	*800*
Reciprocal of size	0.01000	0.00500	0.00250	0.00125
M1				
Abs. mag.	0.76258	0.76362	0.76372	0.76373
Std. dev.	0.00062	0.00040	0.00028	0.00020
M2				
Abs. mag.	0.69991	0.70209	0.70242	0.70234
Std. dev.	0.00111	0.00079	0.00051	0.00037

For the data in table 14.5 the FIT programme gives, for M_1 and for M_2 respectively, the data fitted to $y(x) = a + bx$:

M1:

```
     a  =  +0.7639,  siga  =  0.0002
     b  =  -0.1064,  sigb  =  0.0668
chi^2  =  0.6709,       Q  =  0.7150
```

M2:

```
     a  =  +0.7028,  siga  =  0.0004
     b  =  -0.2335,  sigb  =  0.1222
chi^2  =  0.9495,       Q  =  0.6220
```

Thus we may conclude that the ideal values of M_1 and M_2 respectively are 0.7639(2) and 0.7028(4).

From equation (1) we now obtain $1/\beta = 8.311$. We use the same procedure (adding and subtracting errors to M_1 and M_2) to obtain the error for $1/\beta$, namely, (8.401-8.222)/2 = 0.090, so the final estimate for $1/\beta$ is 8.31(9). The exact value is 8, so this measurement gives a result higher than expected. This is probably due to the use of the approximation to $M(t)$ mentioned.

3-state Potts Model on the Square Lattice

Before proceeding to extract the value of β for the 3-state Potts model the behaviour of the absolute magnetisation using Wolff dynamics was studied. Simulations were performed using the 3-state Potts model on a 200x200 square lattice at 99 per cent of the critical temperature of 0.994973. The absolute magnetisation was found to stabilise after 40 timesteps, as shown in figure 14.26. Accordingly the absolute magnetisation was measured by taking a time average over timesteps 50 through 100.

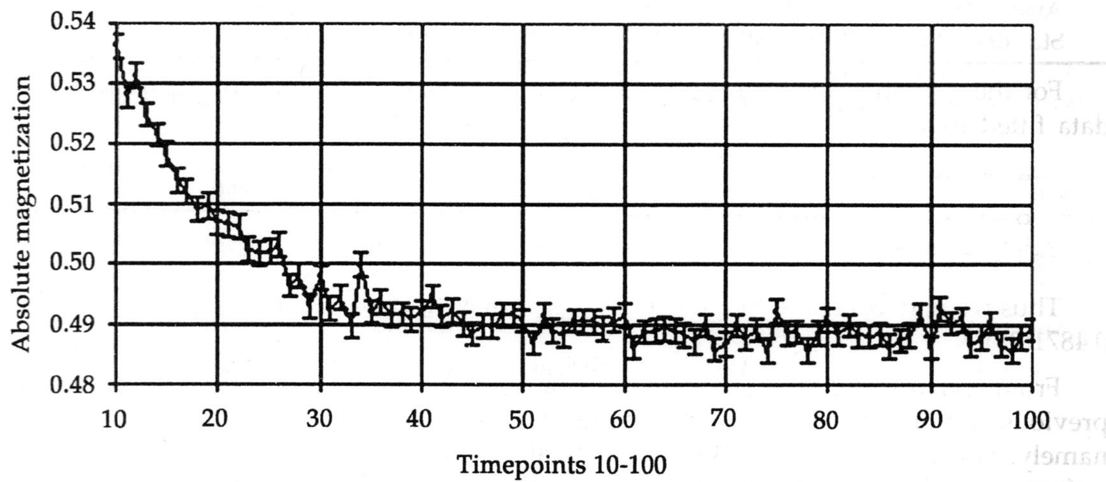

Figure 14.26: *3-state Potts model, square lattice, 200×200 T=0.985023=99% of Tc, 8000 samples*

The simulation programme was used to extract the value of β for the 3-state Potts model on the square lattice in the same way as for the Ising models. Values of the absolute magnetisation were obtained for temperatures 0.975073 and 0.985023 (resp. 98% and 99% of the critical temperature of 0.994973).

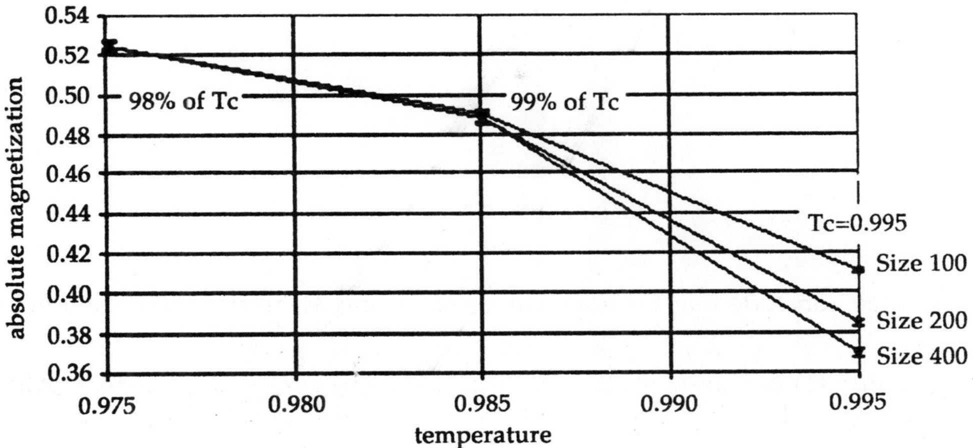

Figure 14.27: *3-state Potts, square model*

From figure 14.27 we obtain Table 14.6:

Table 14.6

Size	*100*	*200*	*400*
Reciprocal of size	0.0100	0.0050	0.0025
M1			
Abs. mag.	0.5250	0.5247	0.5248
Std. dev.	0.0015	0.0021	0.0017
M2			
Abs. mag.	0.4907	0.4887	0.4882
Std. dev.	0.0014	0.0019	0.0027

For the data in Table 14.6 the FIT programme gives, for M_1 and for M_2 respectively, the data fitted to $y(x) = a + bx$:

```
    a  =  0.5247,  siga  =  0.0021                    a  =  0.4871,  siga  =  0.0028
    b  =  0.0311,  sigb  =  0.2954                    b  =  0.3598,  sigb  =  0.3463
chi^2  =  0.0047,       Q  =  0.9455             chi^2  =  0.0157,        Q  =  0.9002
```

Thus we may conclude that the ideal values of M_1 and M_2 respectively are 0.5247(21) and 0.4871(28).

From equation (14.1) we now obtain $1/\beta = 9.322$. We use the same procedure as in the previous sections (adding and subtracting errors to M_1 and M_2) to obtain the error for $1/\beta$, namely, $(10.727\text{-}8.240)/2 = 1.243$, so our final estimate for $1/\beta$ is 9.32 ± 1.24. The error is unfortunately large but this result is at least consistent with the conjectured exact value of 9.

Bibliography

Anil, Date: *Introduction to Computational Fluid Dynamics,* Cambridge University Press, 2006.

Arakawa, A. and V. R. Lamb: *Methods of Computational Physics,* Academic Press, 1977.

B. Alder, S. Fernbach and M. Rotenberg: *Methods in Computational Physics,* Academic Press, New York, 1963.

B. Thiagarajan and Pa Rajalakshmi: *Computational Biology,* MJP Pub., 2009.

Bennett, C.H.: *Efficient Estimation of Free Energy Differences from Monte Carlo Data,* 1976.

————: *Exact Defect Calculations in Model Substances, in Diffusion in Solids: Recent Developments,* A.S. Nowick and J.J. Burton, Academic Press, New York, 1975.

————: *Molecular Dynamics and the Theory of Point Defect Diffusion,* France, 1976.

————: *Structural Changes at the Lennard-Jones Glass Transition,* N.J. Grant and B.C. Giessen, M.I.T. 1976.

————: *Dissipation-Error Tradeoff in Proofreading, BioSystems,* 1979.

C. H. Bennett, David P. DiVincenzo, Christopher A. Fuchs, Tal Mor, Eric Rains, Peter W. Shor, and John A. Smolin: *Quantum Nonlocality without Entanglement,* 1999.

C.H. Bennett and G. Brassard: *An Update on Quantum Cryptography in Advances in Cryptology,* Springer, Berlin, 1985.

————: *Quantum Public Key Distribution,* in IBM Technical Disclosure Bulletin, 1985.

————: *On the Role of Dissipation in Stabilizing Complex and Nonergodic Behavior in Locally Interacting Discrete Systems,* 1985.

Cees Oomens, Marcel Brekelmans and Frank Baaijens: *Biomechanics: Concepts and Computation,* Cambridge University Press, Cambridge, 1998.

Charles, F. Van Loan: *Introduction to Scientific Computing: A Matrix-Vector Approach Using MATLAB,* Prentice-Hall, London, 2000.

D Jolly: *Computational Chemistry,* Ivy Pub, 2008.

D. E. Stevenson: *Science, Computational Science and Computer Science: At a Crossroads,* Communications of the ACM, 1994.

D. Manocha and J. Canny: *Multipolynomial Resultant Algorithms: Computer Science Division Report*, University of California, Berkeley, 1993.

D.M. Etter: *Engineering Problem Solving with MATLAB*, Prentice-Hall, New Jersey, 1997.

Dan, Gusfield: *Algorithms on Strings, Trees, and Sequences: Computer Science and Computational Biology*, Cambridge University Press, Cambridge, 1997.

David, M. Hirst: *A Computational Approach to Chemistry*, Blackwell Sci., 1990.

David, Poole: *Computational Intelligence : A Logical Approach*, Oxford University Press, Oxford, 2004.

De Marchi: *Computational and Mathematical Modeling in the Social Sciences*, Cambridge University Press, Cambridge, 2001.

Dean, R. G., and R. A. Dalrymple: *Water Wave Mechanics for Engineers and Scientists*. World Scientific, 1991.

Di Lorenzo, E.: *Seasonal Dynamics of the Surface Circulation in the Southern*, California Current System, 2003.

Donald, E. Knuth: *Selected Papers on Computer Science*, CSLI Publications, Cambridge University Press, Cambridge, 1996.

G. Golub and C. Van Loan: *Matrix Computations*, Johns Hopkins University Press, Baltimore, 1996.

Gary, B. Fogel and David W. Corne: *Evolutionary Computation in Bioinformatics*, Elsevier India, 2003.

Gene, Golub and James M. Ortega: *Scientific Computing: An Introduction with Parallel Computing*, Academic Press, New York, 1993.

Gene, H Golub and Charles F Van Loan: *Matrix Computations*, Hindustan Book Agency, 2007.

Geoffrey, C. Fox, Parallel: *Computing and Education*, American Academy of Arts and Sciences, Winter, 1992.

Golub, G. H. and C. Van Loan: *Matrix Computations*, Johns Hopkins University Press, Baltimore, 1989.

H. D. Simon: *The Lanczos Algorithm with Partial Reorthogonalization*: *Mathematics of Computation*, 1984.

I. Dhillon, B. Parlett: *Multiple Representations to Compute Orthogonal Eigenvectors of Symmetric Tridiagonal Matrices*, Linear Algebra and its Applications, 2004..

J. L. Henessy, and D. A. Patterson: *Computer Architecture: A Quantitative Approach*, Morgan, Kaufmann, 1990.

J. H. Wilkinson: *Error Analysis of Floating-point Computation*: *Numerische Mathematik*, 1960.

J.P. Marc, G.Brassard, and C.H. Bennett: *How to Decrease Your Enemy's Information*, Springer, Berlin, 1986.

John, R. Berryhill: *C++ Scientific Programming - Computational Recipes at a Higher Level*, John Wiley, 2001.

John, R. Rice: *Academic Programs in Computational Science and Engineering,* Computational Science and Engineering, Spring, 1994.

Joseph, L. Zachary: *Introduction to Scientific Programming: Computational Problem Solving Using Maple and C,* Springer-Verlag, 1996.

———: *Introduction to Scientific Programming: Computational Problem Solving Using Mathematica and C,* Springer-Verlag, 1998.

Kai, Hwang: *Advanced Computer Architecture: Parallelism, Scalability, Programmability,* McGraw-Hill, London, 1993.

Keshab, Bhattarai: *Economic Theory and Models: Derivations, Computations and Applications for Policy Analyses,* Serials Pub, 2008.

Kevin, Dowd and Charles Severance: *High Performance Computing,* O'Reilly & Associates, 1998.

Lloyd, D. Fosdick, Elizabeth R. Jessup, Carolyn J. C. Schauble, and Gitta Domik: *Introduction to High-Performance Scientific Computing,* MIT Press, London, 1996.

M. Lotkin: *Mathematical Tables and Other Aids to Computation,* 1955.

Martha, Lee Ennis: *Update on the Status of Computational Science and Engineering in U.S. Graduate Programs,* The University of New Mexico, New Mexico, 1999.

Michael, T. Heath: *Scientific Computing,* McGraw-Hill, London, 2001.

Nielsen: *Quantum Computation & Quantum Information,* Cambridge University Press, 2003.

P. Ravinder Reddy, M.V.S. Murali Krishna and N.V. Srinivasulu: *Proceedings of International Conference on Computational Methods in Engineering and Sciences,* BS Publication, 2009.

P. J. Holmes: *Waves in Distributed Chemical Systems: Experiments and Computations,* SIAM, Philadelphia, 1980.

Pemmaraju: *Computational Discrete Mathematics: Combinatorics and Graph Theory with Mathematica,* Cambridge University Press, Cambridge, 2001.

Peter, J. Denning: *Great Principles in Computing Curricula,* Technical Symposium on Computer Science Education, 2004.

Peter, J. Denning: *Is Computer Science Sscience?,* Communications of the ACM, 2005.

R. B. Lehoucq: *Analysis and Implementation of an Implicitly Restarted Arnoldi Iteration,* Rice University, Houston, 1995.

Richard, C. Deonier, Simon Tavare, and Michael S. Waterman: *Computational Genome Analysis : An Introduction,* Springer, 2008.

Richard, E. Crandall: *Projects in Scientific Computation,* Springer-Verlag, 1994.

Robert, S. Wolff and Larry S. Yaeger: *Visualization of Natural Phenomena,* Springer-Verlag, 1993.

Ronald, F. Boisvert and Bruce R. Miller: *Enhancing the interactivity of software and data repositories with Java,* Berlin, 1997.

Rubin, H. Landau and M. J. Paez: *Computational Physics, Problem Solving With Computers,* John Wiley, London, 1997.

S.K. Dwivedy and Damodar Maity: *Recent Advances in Computational Mechanics and Simulations,* I.K. International, 2007.

Sauransu, Mukhopadhyay: *Optical Computation and Parallel Processing,* Classique Books, 2000.

Sivanandam and M. Janaki Meena: *Theory of Computation,* I.K. International, 2009.

Steven, E. Koonin and Dawn C. Meredith: *Computational Physics,* Addison-Wesley, 1990.

Susanne, Schmitt and Horst G. Zimmer: *Elliptic Curves : A Computational Approach,* Hindustan Book Agency, 2010.

T. A. Manteuffel: *An Incomplete Factorization Technique for Positive Definite Linear Systems: Mathematics of Computation,* 1980.

Tao, Pang: *An Introduction to Computational Physics,* Cambridge University Press, Cambridge, 1997.

V. Kumar, A. Grama, A. Gupta and G. Karypis: *Introduction to Parallel Computing: Design and Analysis of Algorithms,* Benjamin/Cummings, 1994.

Vibha Tandon, Shree: *Computational Chemistry,* 2005.

W. H. Press, B. P. Flannery, S. A. Teukolsky and W. T. Vetterling: *Numerical Recipes in Fortran: The Art of Scientific Computing,* Cambridge University Press, Cambridge, 1992.

W. Kahan: *Accurate Eigenvalues of a Symmetric Tridiagonal Matrix,* Computer Science Dept., Stanford University, 1966.

W. Kerner: *Large-scale Complex Eigenvalue Problems,* Journal of Computational Physics, 1989.

Wegner, P.: *Common myths and Preconceptions about Cambridge Computer Science"* Computer Science Department, Cambridge, 1976.

————: *Computer Science is the Study of all Aspects of Computer Systems, from the Theoretical Foundations to the very Practical Aspects of Managing Large Software Projects,* Massey University, 1976.

————: *Computer Science is the Study of Computation,* Saint John's University, Saint Benedict, 1978.

————: *Research Paradigms in Computer Science, Proceedings of the 2nd International Conference on Software Engineering,* San Francisco, California, 1976.

William, H. Press, Saul A. Teukolsky, Brian P. Flannery, and William T. Vetterling: *Numerical Recipes in C++: The Art of Scientific Computing,* Cambridge University Press, Cambridge, 2002.

Working group on CSE Education: *Graduate Education in Computational Science and Engineering,* SIAM Review, 2001.

Index

F

G

H

I

K

L

M

N

O

P

Q

R

S

T

U

V

W

□□□